Springer Proceedings in Physics 19

Springer Proceedings in Physics

Managing Editor: H.K.V. Lotsch

Springer Proceedings in Physics is a new series dedicated to the publication of conference proceedings. Each volume is produced on the basis of camera-ready manuscripts prepared by conference contributors. In this way, publication can be achieved very soon after the conference and costs are kept low; the quality of visual presentation is, nevertheless, very high. We believe that such a series is preferable to the method of publishing conference proceedings in journals, where the typesetting requires time and considerable expense, and results in a longer publication period. Springer Proceedings in Physics can be considered as a journal in every other way: it should be cited in publications of research papers as *Springer Proc. Phys.*, followed by the respective volume number, page number and year.

Lasers and Synergetics

A Colloquium on Coherence and Self-organization in Nature

Editors: R. Graham and A. Wunderlin

With 114 Figures

Springer-Verlag Berlin Heidelberg New York
London Paris Tokyo

Professor Dr. Robert Graham
Fachbereich 7, Universitäts-Gesamthochschule, Postfach 10 37 64,
D-4300 Essen, Fed. Rep. of Germany

Dr. Arne Wunderlin
Institut für Theor. Physik, Universität, Pfaffenwaldring 57/IV,
D-7000 Stuttgart 80, Fed. Rep. of Germany

ISBN-13:978-3-642-72760-3 e-ISBN-13:978-3-642-72758-0
DOI: 10.1007/978-3-642-72758-0

Library of Congress Cataloging-in-Publication Data. Lasers and synergetics. (Springer proceedings in physics; 19) "Extended versions of the contributions to a symposium in honour of the 60th birthday of Hermann Haken to be held in 1987 at the University of Stuttgart" – Pref. Includes index. 1. Lasers – Congresses. 2. System theory – Congresses. 3. Coherence (Optics) – Congresses. 4. Self-organizing systems – Congresses. 5. Haken, H. I. Graham, R. (Robert), 1942 –. II. Wunderlin, A. (Arne), 1947 –. III. Haken, H. IV. Series. QC685.L36 1987 535.5'8 87–12135

2153/3150-543210

Hermann Haken

Preface

The two terms "Laser" and "Synergetics" have been coined within the last thirty years, each in response to a distinct fundamental change of direction taken by contemporary physics. In optics, lasers mark the turn from incoherent optics, that is from a sophisticated use of optical noise (to borrow a phrase from Roy Glauber), to an equally sophisticated generation and use of coherent optical signals. In statistical physics, synergetics marks the turn from the statistical mechanics of equilibria to non-equilibrium physics encompassing instabilities, self-organization in space and time, and chaos. These two fundamental developments in physics are distinct, but not independent of each other. The invention and successful demonstration of lasers provided a powerful stimulus for statistical physics to enlarge its scope in order to deal with this new type of system, and thereby led to synergetics. Conversely, recent results on instabilities and the dynamical behaviour of non-equilibrium systems exerted their influence on laser physics and in turn transformed that field.

No one person has done more to shed light on these general connections and the underlying unity of the phenomena of coherence and self-organization, of lasers and synergetics, than Hermann Haken, to whom this volume is dedicated. The articles collected here present extended versions of the contributions to a symposium in honour of the 60th birthday of Hermann Haken to be held in 1987 at the University of Stuttgart. The contributions span a wide range of topics, from laser physics and coherent optics to general principles of self-organization in physical and biological systems. Written by authors in close contact with Hermann Haken at one time or another during their scientific careers, they mirror the wide range of problems to which Hermann Haken has contributed or which he has influenced. They also testify to the great integrative power of the concepts of synergetics.

It is our hope as editors that this juxtaposition of detailed accounts of recent advances in the field with the recurrent unifying "Leitmotiv" of synergetics will make this volume both a worthy tribute to a great scientist and a useful source of learning and insight.

Essen, Stuttgart
March 1987

R. Graham
A. Wunderlin

Contents

Part I

Introduction

Contributions of Hermann Haken to Our Understanding of Coherence and Selforganization in Nature

R. Graham

Fachbereich Physik, Universität-GHS Essen,
D-4300 Essen, Fed. Rep. of Germany

1. Introduction

We are about to begin our symposium on 'Lasers and Synergetics' which is held to honour Hermann Haken, on his 60th birthday, who developed a theory of the laser and founded synergetics as a discipline of science in its own right. It is with the greatest admiration for Hermann Haken as a scientist, teacher, and friend and for his seminal work that I take the opportunity to review on this occasion the important contributions which he made to our understanding of coherence and self-organization in nature. In view of his immense productivity this review can only be partial and sketchy, and is certainly biassed by my own background and limited expertise. A complete list of Haken's publications is given at the end of this book.

Hermann Haken started his scientific career in physics with his first published papers in the early fifties. It was the great time of the quantum theory of many particle systems. Superconductivity was being explored and finally explained. Quantum field theory proved itself by providing powerful tools to handle the new and often difficult questions which were asked and answered by a young generation of physicists. Hermann Haken was among them, then working in Erlangen on the quantum field theory of excitons and polarons. Already with those first papers he made a lasting impact: today the effective potential dynamical generated by the exciton-phonon coupling is well known as the Haken-potential. For reviews given by Haken see [1], [2].

Then, like a culture shock, came the advent of the laser, which like superconductivity, in the end was bound to change the life of every physicist. The first two physicists from Germany to experience and recognize that change were two young postdocs and friends, then working at the Bell Telephone Laboratories in the U.S., Wolfgang Kaiser and Hermann Haken. They immediately recognized that this was a discovery with far reaching consequences. Both have since then contributed immensely to laser physics, one experimentally, the other theoretically. The impressive sequence of ideas of Haken's work to be reviewed here in the following really started at that time.

The first task addressed by Hermann Haken was the development of a systematic theory of laser action which for the first time explained in a unified description the basic observed features of a laser-like thresholds, mode intensities, frequencies, mode competition, locking phenomena etc. [3]. This was the first systematic nonlinear laser theory, which went far beyond the rate equation approximation of all the earlier theories and which accounted correctly for the coherence of laser light and the radiating atomic dipole moment. The rate equation approach of others could be shown to follow from the new approach as a limiting case. Shortly after this theory had been created and published by Haken and his student Herwig Sauermann [4] it turned out that a similar theory had been developed independently at nearly the same time by Willis Lamb [5]. Despite differences in detail the two theories were later proven by Haken to be equivalent (see p. 196 of [6]).

These first theories neglected noise, which due to spontaneous emission is an inevitable byproduct of laser action required by quantum theory. In other words, the Haken-Lamb theory was a semi-classical theory of laser action. In particular, this theory could not yet account for the finite lower limit on the linewidth of the emitted laser light. The next big task addressed by Hermann Haken was therefore the creation of a nonlinear theory of laser noise [7], [8]. This was, indeed, a very big undertaking. From the success of the previous semi-classical theory it was clear that the new theory had to be nonlinear in an essential way, it had to be a quantum field theory in order to include the crucial effects of spontaneous emission, nevertheless it had to include the external pumping of the laser and dissipation in order to also account for the continuous give and take of energy in the laser system, and, above all, it had to be a statistical theory, since its very goal was to give a proper account of quantum fluctuations in a laser.

This was, indeed, a large order. Never before had a theory been tried which incorporated all these features at the same time. The existing theories of laser noise were all linear, and therefore at least incomplete, since the Haken-Lamb theory already demonstrated that nonlinearity was essential. Quantum statistics had so far only dealt with systems in thermodynamic equilibrium, e.g. in the theory of superconductivity, while here it was clear that the laser was a system driven very far away from thermodynamic equilibrium. The way to properly include pumping and dissipation in a quantum theory of the laser without violating fundamental principles like the uncertainty relation had to be invented and implemented before the theory could proceed.

The style of the very first fundamental papers by Hermann Haken on his new theory [7], [8], [9] still bring back the strong spirit of exploration and adventure he must have felt when doing the pioneering and sometimes daring steps necessary in order to overcome the immense difficulties before him. The final version of the theory [6], which emerged after years of combined efforts by Hermann Haken and the 'Stuttgart school' which rapidly grew around him, was much cleaner and more readily comprehensible than these very first papers, but the essential ideas are all there. Needless to say that this field became an extremely active area of theoretical physics all over the world after Haken's first pioneering papers had appeared, and there was no shortage of rather fierce competition from many quarters (cf. the many references given in [6]). Then experimental results came pouring in, in particular those by Armstrong and Smith and Arecchi and his group in Milano (cf. the reviews given in [10], [11]). They were found in beautiful agreement with Haken's results and with further results which Hans Risken, working in Haken's group, had been able to extract from the theory using the method of Fokker Planck equations [12]. These experimental and theoretical results taken together represent a cornerstone of our present day knowledge of the fundamental behavior of a laser when it passes the lasing threshold.

It was clear, then, that some of the most basic questions concerning lasers had been successfully answered - but only gradually did it dawn on us that actually much more had been achieved: For the first time a quantum statistical theory had been carried through successfully of a highly nontrivial system which made a transition from incoherent random emission to highly coherent, ordered emission when a parameter - the pumping parameter - was varied past a threshold value. We realized that in this theory we had before us a system, which exhibited all signs of a phase transition, at least when looked at from the point of view of the mean field theory of phase transitions developed by Landau, despite the fact that this system clearly was not in thermodynamic equilibrium but very far away from it. This new insight was gained in joint work of Hermann Haken and myself [13]. When we had published our paper we learned that the same conclusions had been drawn independently at the same time by Marlan Scully and Vittorio DeGiorgio [14].

The spontaneous ordering in a non-equilibrium system, of which laser action clearly was an example, was very exciting because clearly this is a phenomenon which is very widespread in nature and accounts for many seemingly mysterious processes of spontaneous self-organization, not only in physics but in other disciplines as well, most notably biology. We published this general idea in a paper [15] which on the insistence of Hermann Haken, was written in a popularizing non-technical way, because it had become clear to him that the main impact of these ideas might not even be in physics but rather in other fields like chemistry, biology or ecology, to name just a few. For this article Hermann Haken coined the notion 'synergetics' for the new program to explore the common mechanisms and properties of the phenomena of spontaneous self-organization in nature. This proved to be a challenging and extremely fruitful program which has since been strongly advocated and immensely developed by Hermann Haken and his coworkers, and the field has stimulated and attracted many scientists from different fields. An impressive account of much of the work which has been done is given by the Springer Series in Synergetics established and edited since 1977 by Hermann Haken [16]. An extremely important impetus for the whole development of the field was also the series of conferences on synergetics, most of them organized by Hermann Haken.

The most complete account of the theory he developed was given by Hermann Haken in the two books 'Synergetics, An Introduction' and 'Advanced Synergetics' [17], [18]. The staggering breadth which synergetics has achieved in roughly ten years can be inferred from a quotation from the preface of [18], where Haken writes:

"The central question in synergetics is whether there are general principles which govern the self-organized formation of structures and/or functions in both the animate and the inanimate world. When I answered this question in the affirmative for large classes of systems more than a decade ago and suggested that these problems be treated within the interdisciplinary research field of "synergetics", this might have seemed absurd to many scientists. Why should systems consisting of components as different as electrons, atoms, molecules, photons, cells, animals, or even humans be governed by the same principles when they organize themselves to form electrical oscillations, patterns in fluids, chemical waves, laser beams, organs, animal societies, or social groups? But the past decade has brought an abundance of evidence indicating that this is, indeed, the case, and many individual examples long known in the literature could be subsumed under the unifying concepts of synergetics. These examples range from biological morphogenesis and certain aspects of brain function to the flutter of airplane wings; from molecular physics to gigantic transformations of stars; from electronic devices to the formation of public opinion; and from muscle contraction to the buckling of solid structures. In addition, there appears to be a remarkable convergence of the basic concepts of various disciplines with regard to the formation of spatial, temporal, and functional structures."

Clearly, the following sections cannot have the aim to review the numerous contributions of Hermann Haken to the immense field of synergetics since the time it was created. Extensive reviews of these developments have been given by Hermann Haken in his books and further review articles. Instead I shall here try to review those of his ideas which occurred first in connection with the theory of the laser and which later led him to the foundation of synergetics.

2. Lasers at a Central Crossroads of Physics

The laser since its invention has proven itself as an extremely important tool in experimental physics, it has revolutionized spectroscopy, and has thereby contributed immensely to the general development of physics in the last decades. However, it is perhaps somewhat less appreciated that lasers are also extremely interesting physical systems in themselves - many particle systems of atoms or molecules, externally pumped to excited states and emitting coherent electromagnetic radiation by self-organization of their radiating dipole moments. These characteristic features ensure that lasers as physical systems occupy a very important place at one of the central crossroads in physics. I think it is one of Haken's greatest contributions to clearly recognize this fact and to do more than anybody else to bring this principal importance of lasers and the general lessons they can teach us to the attention of physicists and scientists of other disciplines.

To appreciate what gives lasers as physical systems such a unique importance beyond their technological relevance, let us consider some of the main roads along which physics has been expanded in this century. The first such road connects the microscopic world of quantum physics with the macroscopic world of classical physics. These two worlds are connected by Bohr's correspondence principle, but nevertheless they are very difficult to reconcile. The laws of linear superposition of probability amplitudes for quantum events and linear superposition of probabilities for classical events are not only different but even seem to exclude each other. The difficulty is also manifest in the quantum theory of measurement, where rules like the reduction of wavepackets and quantum jumps which seem to contradict the laws of quantum theory have to be justified by just these laws [19]. What happens, one may ask, if a microscopic system is made larger and larger? How can the quantum rules gradually change into the classical rules? (This question has a long history and was recently revived in connection with Josephson junctions by Leggett and coworkers [20].)

The laser is the first physical system where this question had to be analyzed and understood in detail. The process of emission from each atom in the laser is entirely determined by the rules of quantum theory. On the other hand, the resulting emission from the whole ensemble of laser atoms very nearly looks like a classical stochastic process, but the noise of this stochastic process has its origin in the quantum properties of each of the individual atoms. This shows that lasers occupy a very interesting place between the quantum world, in which the radiation of the laser is born, and the classical world where that radiation manifests itself in a macroscopic way. Haken's theory of the laser tells us how these worlds can be connected [21] - [24].

A second road along which physics is presently expanding at a tremendous rate leads us from the phenomena of thermodynamic equilibrium to non-equilibrium phenomena. The many-particle systems analyzed prior to the laser were systems in thermodynamic equilibrium. Magnets and superconductors are among the most important examples. With these systems lasers have in common that they are many-particle systems, too, but in addition lasers are strongly pumped from outside and they radiate and dissipate energy. Therefore they belong to the extremely broad and interesting class of many-particle systems far from thermodynamic equilibrium. The properties of systems of this kind in general are much richer and much more interesting than systems in thermodynamic equilibrium. Their potentialities are truly immense and far from being understood. They include all the phenomena known from systems in thermodynamic equilibrium, but also encompass new phenomena like oscillation, competition, selection, locking phenomena, chaos, pattern formation, self-organization, autocatalysis, self-reproduction, not to speak of the capability of life or even self-consciousness. The importance of the laser in this context is its minimal departure from a many-particle system in thermodynamic equilibrium. The only departure really is the presence of a source and a sink of energy, providing a flow of energy through the system. The strength of the external pump can be controlled and gives a convenient handle to vary the distance to thermodynamic equilibrium. Along the road from equilibrium thermodynamics to non-equilibrium physics, lasers therefore occupy a position which allows us to witness some of the new phenomena occurring as we move away from thermodynamic equilibrium.

The most important phenomenon one encounters along this way in the laser is the process of self-organization in the emission of light by the laser atoms. The laser threshold is the point, where this self-organization first occurs. Below threshold the laser atoms emit independently and incoherently, above threshold they emit in a collective, self-organized way and the emitted field is highly coherent. The process of self-organization is a result of nonlinearity, whose importance was clearly recognized and duly stressed in Haken's very first papers [7], [8], as we already mentioned. Today the process of self-organization occurring in a laser is probably the best analyzed and understood of all such processes, showing strong analogies to second order phase transitions in thermodynamic equilibrium [13], in particular in superconductors, which are quantitatively described by thermodynamic potentials of the Ginzburg Landau type. Haken was the first to write down the corresponding potential for the laser [7]. Non-equilibrium phase transitions are now commonplace in many other systems. On the road connecting equilibrium and non-equilibrium phase transitions the laser therefore again occupies a prominent place. E.g. the idea to introduce generalized thermodynamic potentials in order to describe non-equilibrium phenomena [25], [26] was suggested by the example of the laser. The general concept of the slaving of fast variables by slowly varying order parameters also first emerged there [27].

Another road of modern physics leads from regular predictable dynamics to chaotic unpredictable dynamics. The prototypical model connecting these two regimes of dynamical behavior is the Lorenz model [28] which was originally derived as an over-simplified physically unrealistic model of thermal convection. As Haken first pointed out, the Lorenz model is isomorphic to the equations of motions of a homogeneously broadened single mode laser [29]. What is more, recent experiments with far-infrared lasers have shown that in this new context the Lorenz model need no longer be considered as physically unrealistic even in the regime where chaotic behavior prevails [30].

The laser can therefore be seen at the crossroads between quantum and classical physics, between equilibrium and non-equilibrium phenomena, between phase-transitions and self-organization and between regular and chaotic dynamics. At the same time it is a system which we understand, on the basis of the theory initiated by Haken, both on a microscopic quantum mechanical and on a macroscopic classical level. It is a solid ground for discovering general concepts of non-equilibrium physics, and has therefore rightly been called by Haken a 'trail-blazer of synergetics' [31].

3. Quantum-Classical Correspondence

As the laser occupies such an important position at a central crossroads of physics, its theoretical analysis has led to theoretical methods and insights far more general than one might expect from such a concrete and specialized system. An example is the method of quantum-classical correspondence. Its foundations go back to Wigner's work on quasi-probability densities in quantum theory [32] and the work by Glauber [33] and Sudarshan [34], but the theory was considerably generalized [23], [24] and applied in an extremely successful way not only in the quantum theory of the laser but afterwards also in many other problems of quantum optics. Even today the method is still being extended further [35] and its potential does not seem to be exhausted.

Let us recall some of the salient points of the method and how it relates to lasers. The quantum nature of the emission of radiation from individual laser atoms makes it necessary to formulate the basic equations of motion of a laser quantum mechanically. As far as the conservative system is concerned, consisting of the electromagnetic field in the resonator and the radiating atoms, the equations of motion are provided by nonrelativistic quantum electrodynamics formulated either in the Heisenberg picture or the Schrödinger picture. A first difficulty arises from the necessity to include the effects of external pumping of the laser atoms and of dissipation due to losses of energy suffered by the atoms and the resonator modes. This difficulty is solved by accounting for these physical mechanisms explicitly in the Hamiltonian by coupling the modes and the atoms to external reservoirs. The elimination of these reservoirs in the Heisenberg picture ([7] - [9], [21], [22]) leads to a coupled set of operator Langevin equations for the field operators of the resonator modes and the operators of the laser atoms. This was the first method by which the theory of the laser proceeded. In fact, using linearization and quasi-linearization techniques, Haken was able to solve directly the operator Langevin equations in the regime below and above threshold [7] - [9], with the exception of a small neighbourhood of the laser threshold itself, thereby providing the first quantitative theory of laser noise and coherence which stood up to later experimental test [10].

The elimination of the reservoirs in the Schrödinger picture [36] leads to an equation of motion for the statistical operator ρ of the coupled field-atom system (master equation). This second formulation, which was favoured in the later developments of the theory, lends itself to further analysis by the method of quantum-classical correspondence [23]. The first step of this method consists in the choice of a complete basis of operators a_λ which form a closed algebra under commutation and in terms of which all observables $\Omega = \Omega(a)$ and $\rho = \rho(a)$ can be expressed. Here the set $\{a_\lambda\}$ is abbreviated by a. In the simplest case of a single-mode laser this basis is formed by the system of creation and annihilation operators of the modes b^+ and b, and the system of spin -1/2 operators

$$\tau_{\mu x} = \tau_\mu^+ + \tau_\mu^- \ , \ \tau_{\mu y} = \frac{1}{i}\,(\tau_\mu^+ - \tau_\mu^-) \ , \ \tau_{\mu z} \ ,$$

$\mu = 1,2,\ldots N$ associated with the two energy levels of each of the N laser atoms directly participating in the laser process [23]. The next step is to choose a definite ordering among the operators a_λ (e.g. all the b to the left of the b^+ and τ_μ^+ and τ_μ^- to the left and right of $\tau_{\mu z}$, respectively). Sticking to this order in all later calculations $\rho(a)$ may be replaced by a c-number function [23], [37]

$$\rho(a) \rightarrow f(\alpha) \tag{3.1}$$

with the quantum-classical correspondence of operators a to c-numbers

$$a_\lambda \rightarrow \alpha_\lambda \tag{3.2}$$

since no further commutation of operators can occur. However, in order to be able to maintain the chosen order of the operators a_λ in the solution $\rho(a,t)$ of the master equation, it is necessary to first introduce the chosen operator ordering in the master equation by employing the commutation relations among the a_λ. After this has been done the correspondence (3.1), (3.2) can also be introduced in the master equation, which is thereby converted to a c-number equation governing the function $f(\alpha)$. The interpretation of $f(\alpha)$ becomes clear when the final step of the quantum-classical correspondence is carried out, which establishes the rules for the calculation of expectation values. These rules are based on the fact that for each choice of order of the a_λ in $\rho(a)$ there exists a (generally different) dual choice of order of the a_λ in observables $\Omega = \Omega(a)$ such that expectation values $\langle\Omega\rangle$ can be calculated like classical averages

$$\langle\Omega(t)\rangle = \mathrm{Tr}\rho(t)\Omega = \int \prod_\lambda d\alpha_\lambda\, f(\alpha,t)\ \Omega(\alpha) \tag{3.3}$$

with the quantum-classical correspondence

$$\Omega \rightarrow \Omega(\alpha) \ . \tag{3.4}$$

Here $\Omega(\alpha)$ must be calculated by putting the observable Ω into the order dual to the one chosen for ρ using the commutation relations.

It is remarkable that this procedure can be carried through for very general master equations, as was shown by Haken [24]. For the laser this method leads also to a quantum mechanical understanding of the behavior of a laser near threshold [12], the region left out by the first method of Heisenberg Langevin equations. The expectation values and correlation functions of the electromagnetic field observed with photon counters in their ground states are all normally ordered [38], i.e. in Ω the b^+ are to the left of the b. This dual order in observables corresponds to anti-normal order in the density operator ρ, in which all b then have to be to the left of the b^+. In this case the function $f(\alpha)$ of (3.1) is the Glauber-Sudarshan P-representation of the density operator [33], [34]. The master-equation specialized to the region near threshold in quantum-classical correspondence yields a generalized Fokker Planck equation which is very well approximated by a classical stochastic process [12]. By this technique the theory is therefore able to go all the way from the fundamental quantum mechanical processes on which the laser is built to the observed macroscopic stochastic behavior of the output field.

4. From Equilibrium to Non-Equilibrium Statistical Mechanics

Lasers are non-equilibrium systems due to the simultaneous action of a pump mechanism and dissipation. The very remarkable new capabilities of such systems as compared to systems in thermodynamic equilibrium have already been mentioned, as have the analogies between the appearing instabilities and phase transitions. For the laser threshold this analogy is, in fact, strikingly complete. The reason for this completeness could be pinpointed in the rather amazing fact that the effective classical stochastic process which describes the laser near threshold within the method of quantum-classical correspondence (cf. section 3)satisfies a detailed balance symmetry [25], which is usually considered to be a hallmark of a state of thermodynamic equilibrium. In fact this symmetry is not obeyed by the laser equations far from the threshold region and there is no reason why it should. That this unexpected symmetry appears in the vicinity of the laser threshold can be seen as a consequence of Haken's slaving principle' [17], [18] according to which only the complex amplitude of the laser mode is dynamically relevant near threshold, while all the other variables describing the laser atoms follow that amplitude quasi-instantaneously. The global gauge symmetry of the stochastic equations of motion for the complex mode amplitude under arbitrary global phase changes is then sufficient to ensure also the detailed balance symmetry.

The equation of motion for a complex amplitude wave packet $\psi(x,t)$ formed from a continuum of laser modes near the central frequency of the laser transition and near a corresponding wavenumber then takes the form [13]

$$\frac{\partial\psi(x,t)}{\partial t} = \frac{-\delta\Phi}{\delta\psi^*(x,t)} + F(x,t) \qquad \text{with} \tag{4.1}$$

$$\Phi = \int dx\left[+\alpha|\psi(x,t)|^2+\beta|\psi(x,t)|^4+\xi_0^2\left|\frac{\partial\psi}{\partial x}\right|^2\right]. \tag{4.2}$$

Here $\emptyset$ acts as a generalized thermodynamic potential. α is the pump parameter, which is positive below and negative above threshold. β and ξ_0^2 are positive parameters. The term $F(x,t)$ is a complex Gaussian Langevin force with vanishing mean value. It has the only non-vanishing cumulant

$$\langle F(x,t)F^*(x',t')\rangle = 2Q\delta(x-x')\delta(t-t'), \tag{4.3}$$

where Q is a positive parameter. The form of Φ is identical to the form of the free energy functional in the Ginzburg Landau theory of superconductivity. In fact, the steady state probability density functional of the stochastic process (4.1) is given by

$$f([\psi]) = N_0 \exp\left(-\frac{\Phi}{Q}\right) \tag{4.4}$$

and looks entirely like the canonical ensemble of a 1-dimensional superconductor expressed in terms of its complex order parameter or macroscopic wave-function $\psi(x,t)$. Coherence in a laser above threshold is therefore seen to emerge just like coherence in a superconductor below its critical temperature.

The analogy between self-ordering in non-equilibrium systems and equilibrium phase transitions need not always be as closely knit as in the example of the laser. In particular the symmetry of detailed balance under the usual transformation of time-reversal is not present in the general case, which is already encountered in the laser when several modes are excited. Therefore, self-ordering phenomena in non-equilibrium systems are much richer in their dynamical behavior than corresponding phenomena in equilibrium systems. Another important difference between ordering phenomena in equilibrium and non-equilibrium systems, which also emerged from the laser example, is the fact that mean-field type theories are as a rule quantitatively correct far from equilibrium (near the transition point), while they are not quantitatively correct (with the important exception of superconductors) in thermodynamic equilibrium. The reason for this difference is a large difference in the basic correlation length ξ_0 which is microscopic in most equilibrium phase-transitions but usually macroscopic for ordering phenomena far from thermodynamic equilibrium.

The laser-phase transition analogy emerging from Haken's 1964 theory was a first detailed example whose great importance it was that it could be studied quite explicitly. Subsequently, it led many authors to establish similar analogies e.g. between first order phase transitions and dye lasers with saturable absorbers or with optically bistable systems, or in hydrodynamic and chemical systems. Haken continued to contribute immensely to further generalizations and applications of these ideas not only in hydrodynamics [39], [40] and in chemical systems [41], but even outside pyhsics [42]. The field of synergetics [40], [17], [18]was born.

5. From Competition and Evolution to Chaos

Beyond the process of self-organization occurring at threshold and leading to coherence and other signatures of order in their emitted light, lasers have other typical properties of non-equilibrium systems. Examples are competition and selection processes which are in evidence in multi-mode lasers [3] - [6]. The basic features of these processes are already contained in and well described by the original Haken-Lamb theory which treats the electromagnetic field classically and uses quantum theory only in the description of the atomic part of the laser. The rate equations following from that theory in the limit where phase-correlations are negligible are of the same type as equations used in mathematical models of Darwinian evolution by the mechanism of 'survival of the fittest.' The photons of different laser modes are found to reproduce by stimulated emission while competing for the gain provided by the external pump in much the same way as individuals of different species reproduce themselves while competing for external resources [15]. A mathematical model describing evolution on a molecular level was developed by Manfred Eigen [43], who arrived at rate equations very similar to those describing multi-mode lasers. In the context of the laser, the inclusion of noise in the quantized version of the theory even allowed to account for statistical aspects of competition and selection processes.

In my opinion, the full implications of this application of laser models in problems of evolution have not yet been exhausted. I believe this to be true for many of Haken's contributions, which will stimulate important research for a long time to come. In the present example the mathematical relationship between mode competition in lasers and the dynamics of evolution processes has so far been clarified only on the level of rate equations which concentrates on photon numbers of given modes or population numbers of a given species while neglecting all aspects of phase-coherence, which in lasers is a collective property only defined for many photons. Nevertheless, as shown by the Haken-Lamb theory, this collective property has a very important influence on competition and selection processes in real lasers, leading to completely new cycles and instabilities, which are not accounted for in rate equation models. It is natural to ask if, in view of the analogy established at the level of rate equations, there could not also exist analogous collective properties of large populations of a given species, analogous to coherence properties, which similarly influence their self-reproduction, competition and selection. Such generalized coherence properties of populations would have to be complementary to the population number just as frequency ω and time t, wavelength λ and position x, phase φ and photon number n are complementary concepts. One might argue that complementarity is a quantum mechanical concept which should have no relevance to large populations of macroscopic objects, but we note that the complementarity between the objects mentioned is actually completely independent of Planck's quantum of action,

$$\Delta\omega \cdot \Delta t \sim \frac{1}{\lambda^2} \Delta\lambda\Delta x \sim \Delta n \cdot \Delta\varphi \sim 1 \ .$$

Certainly one can think of many properties of populations which are complementary to the population number in the sense that they are well defined for large population numbers but cannot be defined at all in the case of a single individual. Most if not all of these properties are connected to some concept of information shared by individuals of the population, and indeed, it appears quite possible that such collective information has an important influence on its dynamics. If so, it would be important to generalize the models of population dynamics to include such influences. I think it is the beauty and power of synergetics that it induces and even compels us to make these further connections, and to ask new questions which would otherwise not occur to us, and to apply methods which have worked well in some examples to the solution of these new problems.

One of the new questions asked by Hermann Haken in the seventies was for instance: Can a laser also be turbulent? Turbulence was then a well-known phenomenon in hydrodynamics, the provocative paper by Ruelle and Takens [44] on chaos in low-dimensional systems had just appeared and had rekindled the interest in an otherwise forgotten paper by Lorenz [28] on chaos in a special model with three degrees of freedom. Recalling his early theoretical work on instabilities in single-mode bad-cavity lasers leading to 'random spiking' under conditions of extremely hard pumping [9], Haken was able to show that the Lorenz model and the model of a homogeneously broadened single-mode laser in complete resonance are, in fact, equivalent [29]. Lasers could in fact be turbulent. The long known phenomenon of random spiking in lasers could now be understood as signatures of just this phenomenon. This discovery once again opened the way to an entirely new field in quantum optics, the systematic study of instabilities and chaos in optical systems, which is at the moment extremely active [45]. Hermann Haken and his coworkers made also further significant contributions to this field [46] - [49]. The place of the laser at a central cross-roads of physics again makes this study particularly important. Quantum optical devices provide us with chaotic physical systems whose equations cannot only be derived from first principles i.e. from a microscopic theory, but can also be solved, and the results can be directly compared with experiment. A similarly close connection between a first principle theory and experiment is so far not possible in other fields where chaos is studied, such as hydrodynamics or the dynamics of chemical reactions, where theory and experiment therefore tend to emphasize universal aspects, which are independent of the special system considered. In addition, quantum optical devices raise the possibility that quantum effects on chaotic dynamics could be studied. This field is extremely active and far from being exhausted.

6. The Trail Blazer of Synergetics

The laser has once been called by Hermann Haken the trail blazer of synergetics [31], and I have tried above to explain why this is amply justified. Here, however, I would like to apply the same notion in a different way to Hermann Haken himself. Not only did he recognize and thereby create this new field of research when he saw the example of the laser, he also did more than anybody else to make inroads and open up trails in this new and uncharted territory. As fascinating lecturer and teacher for his students he introduced at the university of Stuttgart a whole generation of physicists into his theories and enabled them to contribute to the field. I was lucky enough to be one of them. We, his former students, are all deeply indebted to him for guiding us in our first steps on the various roads we took later on.

Beyond these activities at his home university and the vast research recorded in his papers, Hermann Haken acted as a trail blazer by organizing numerous conferences, bringing together scientists from all the different fields to which synergetics can contribute, writing monographs, text books, and popular books and editing the Springer series on synergetics. In short, the immense expansion and development of non-equilibrium physics which we have witnessed over the last decade was also stimulated by his work and owes much to his pioneering first steps and the trails he made which others could follow.

May he never stop discovering more excitements for all of us.

References

1. H. Haken, Forts. Phys. 6,271 (1958)
2. H. Haken, in 'Polarons and Excitons' ed. C.H. Kuper and G.D. Whitfield (Oliver and Boyd, Edinburgh p. 295 (1963)
3. H. Haken, Nonlinear Interactions of Laser Modes, talk given at the International Conference on Optical Pumping, Heidelberg 1962
4. H. Haken, H. Sauermann, Z. Physik 173,261 (1963); 176,47 (1963)
5. W.E. Lamb, Phys. Rev. 134,A1429 (1964)
6. H. Haken, Laser Theory, Encyclopedia of Physics, ed. S. Flügge, vol. XXV/2c ed. L. Genzel, Springer (Berlin 1970)
7. H. Haken, Phys. Rev. Lett. 13,329 (1964)
8. H. Haken, Z. Physik 181,96 (1964); 182,346 (1965)
9. H. Haken, Z. Physik 190,327 (1966)
10. J.A. Armstrong, A.W. Smith, in Progr. Optics vol. VI , ed. E. Wolf, North Holland (Amsterdam 1967)
11. F.T. Arecchi, in Quantum Optics - Proceedings of the 1967 Varenna School, Academic Press (New York, 1969), p. 57
12. H. Risken, in Progr. Optics vol. 8,239 (1970)
13. R. Graham, H. Haken, Z. Physik 237,31 (1970)
14. M.O. Scully, V. Degiorgio, Phys. Rev. A2,1170 (1970)
15. H. Haken, R. Graham, Umschau in Wiss. und Technik 6,191 (1971)
16. Springer Series in Synergetics, ed. H. Haken, vol. 1-32
17. H. Haken, 'Synergetics - An Introduction', Springer (Berlin 1977)
18. H. Haken, 'Advanced Synergetics', Springer (Berlin 1983)
19. 'Quantum Theory and Measurement', ed. J.A. Wheeler, W.Z. Zurek, Princeton University Press (Princeton 1983)
20. A.O. Caldeira, A.J. Leggett, Phys. Rev. Lett. 46,211 (1983)
21. J.R. Senitzky, Phys. Rev. 119,670 (1969); 124,642 (1961)
22. H. Haken, W. Weidlich, Z. Physik 189,1 (1966)
23. H. Haken, H. Risken, W. Weidlich, Z. Physik 206,355 (1967)
24. H. Haken, Z. Physik 219,411 (1969)
25. R. Graham, H. Haken, Z. Phys. 243,289 (1971); 245,141 (1971)
26. R. Graham, in Proceedings of the 1986 Varenna School, ed. G. Caglioti, H. Haken, L.A. Lugiato, Academic Press (New York, 1987)
27. H. Haken, A. Wunderlin, Z. Physik B47,179 (1982)
28. E.N. Lorenz, J. Atmos. Sci. 20,130 (1963)
29. H. Haken, Phys. Lett. 53A,77 (1975)
30. C.O. Weiss, J. Brock, Phys. Rev. Lett. 57,2804 (1986)
31. H. Haken, in 'Coherence and Quantum Optics IV', ed. L. Mandel, E. Wolf, Plenum (New York, 1978)
32. E.P. Wigner, Phys. Rev. 40,749 (1932)
33. R.J. Glauber, Phys. Rev. 130,2529 (1963); 131,2766 (1963)
34. E.G.G. Sudarshan, Phys. Rev. Lett. 10,277 (1963)
35. P.D. Drummond, C.W. Gardiner, J. Phys. A13,2353 (1980)
36. W. Weidlich, F. Haake, Z. Physik 185,30 (1965); 186,203 (1965)
37. M. Lax, W.H. Louisell, IEEE J. Quant. Electronics QE3,47 (1967)
38. P.L. Kelley, W.H. Kleiner, Phys. Rev. 136,A316 (1964)
39. H. Haken, Phys. Lett. 46A,193 (1973)
40. H. Haken, Rev. Mod. Phys. 47,67 (1975)
41. H. Haken, Z. Phys. B20,415 (1975)
42. H. Haken, H. Olbrich, J. Math. Biol. 6,317 (1978)
43. M. Eigen, Naturwissenschaften 58,465 (1971)

44. D. Ruelle, F. Takens, Comm. Math. Phys. 20,167 (1971)
45. 'Optical Instabilities', ed. R.W. Boyd, M.G. Raymer, L.M. Narducci, Cambridge University Press (Cambridge 1986)
46. H. Haken, H. Ohno, Opt. Commun. 26,117 (1978)
47. H. Haken, M. Renardy, Physica 8D,57 (1983)
48. Ji-yne Zhang, H. Haken, Z. Phys. B58,337 (1985)
49. Ji-yne Zhang, H. Haken, J. Opt. Soc. Am. B2,141 (1985)

Part II

Instabilities in Lasers

Laser Physics: Yesterday and Today *

F.T. Arecchi

Department of Physics, University of Florence and
Istituto Nazionale di Ottica, I-50125 Firenze, Italy

A survey is given of the theoretical contributions by H. Haken to laser physics, and their impact upon experimental investigation. The practical problems arising in the experimental implementation of laser chaos are discussed here, with particular reference to the recent observation of Shilnikov chaos.

1. Introduction

This is a personal tribute to a friend, Hermann Haken, whose intellectual achievements have strongly influenced my life as a physicist: first, through the indication of new frontiers in the theory of quantum optics, which appeared as challenges for the experimentalist; and second, through a long series of personal conversations and the reading of his books, which have revitalized some of the hot points of the scientific debate, along a tradition dating back to the "siècle de lumières" from the Bernoullis to Laplace, and then leading, via Helmoltz and Boltzmann, to our century, from Einstein and Planck up to the natural philosophy of life and artificial intelligence.

This contribution consists of two parts. The first (Sect. 2) is a personal recollection of those contributions of Haken which have been crucial for my line of research, plus a review of some experimental facts on laser chaos and hyperchaos (Sect. 3). The second (Sect. 4) is a homage of my experimental research group, showing our latest results, namely, a method of return time statistics which is the most efficient way of characterizing a Shilnikov instability. This instability, introduced as a theoretical object and observed quantitatively in chemical instabilities, is supported by fully quantitative and incontrovertible evidence from recent observations on a laser system to which our statistical method is applied here.

2. What we owe to Hermann Haken

Hermann Haken has devoted his efforts for three decades to the area of fluctuations, coherence and turbulence in nonequilibrium cooperative systems. His theoretical findings have been crucial in providing motivation for further theoretical work, or specific indications for experimentalists. More than emphasizing a specific single discovery, I wish to survey a long chain of contributions which have generated the area of cooperative phenomena in nonequilibrium systems (so-called synergetics), anticipating many effects later observed experimentally.
The keywords to classify his contributions should be: statistical theory of laser coherence; generalization of the theory to all cooperative nonequilibrium systems

---

(*) Symposium in honour of Prof. H. Haken.

(synergetics); extension of coherence from time to space patterns (morphogenesis); death of laser coherence (onset of deterministic chaos) beyond a second threshold. But let me be more specific, limiting the details to those areas where I have a personal working knowledge.

2.1 Microscopic laser theory

A preliminary outline of laser theory was given by Haken at a conference on optical pumping in Heidelberg (June 1962) and then a full account was published in two papers /1,2/.

This theory allowed the evaluation of the main parameters of a working laser. However, at the same time a more experimentally oriented approach had been developed by Lamb (first presentation: Int. Conf. on Quantum Electronics, Paris, 11-15 Feb. 1963) and published in /3/. This latter approach was more useful for dealing with the preliminary laser experiments (measurements of average values, no fluctuations).

2.2 Statistical laser theory, onset of order

It was soon realised that a laser description in terms of average parameters was insufficient. The crucial general questions on the statistics of the radiation field had already been addressed in 1956 /4/ and a general answer given by the introduction of a sensible theory /5/, which was soon tested experimentally /6/. However, a specific application to the laser required a detailed nonlinear statistical theory of such a physical system, which included the role of noise and explained the role of cooperation in inducing an ordered macroscopic state above a given threshold. This was provided in /7/ by Haken. Haken's theory was based on a technique of adiabatic elimination of variables with fast relaxations, already outlined in /2/. The procedure was already familiar in many-body physics as a qualitative approach, but Haken gave a rigorous treatment, taking fluctuations into account as well. Reference /7/ was the starting point for a fertile line of research. From Haken's school a more heuristic approach emerged /8/ which was the guideline for experimentalists /9/.

2.3 Phase transition analogies for nonequilibrium systems

The laser threshold showed phenomena such as critical fluctuations, slowing down, and insurgence of an order parameter, with the same critical exponents as a mean field theory. In fact Haken's theory was in end effect equivalent to a Landau approach, and a mean field approach was intrinsic not only to the laser threshold /10/ but generally to the onset of order in nonequilibrium many-body systems /11/.

Unlike Prigogine's approach to self-organization /12/based on entropy production arguments, Haken's approach consists in a statistical reduction starting from a microscopic theory, hence it appears conceptually more satisfactory.

2.4 Pattern formation (morphogenesis)

The single mode laser /7/ was the prototype of onset of order in a single-domain system. The crucial question was then addressed, whether a mean field (or Ginzburg-Landau) approach was also sufficient to account for space order or pattern formation in extended nonequilibrium systems. Preliminary reports on patterns of convection in fluids were already known (see /13/ for a review), however Haken provided a general approach /14,11/ including fluctuations.

2.5 Onset of deterministic chaos in lasers

The physics of chaos and related problems was covered in a recent Nobel Symposium /15/ in both its theoretical and experimental aspects.

The appearance of deterministic chaos in a single mode laser was first pointed out by Haken /16/, and that was the starting point for current interest in chaos in quantum optics.

3 Chaos in laser systems

A comprehensive review of this field is now available /17/. Here I present a short account of what has been done in the Florence group.

3.1 Various classes of chaos

The similarity of Maxwell-Bloch (MB) equations for a single laser mode resonant with a homogeneous gain line to Lorenz equations suggests the easy appearance of chaotic instabilities in single-mode, homogeneous-line lasers. However, time scale considerations rule out the full dynamics for most of the available lasers. Lorenz equations have damping rates within one order of magnitude. On the other hand, in most lasers the three damping rates are wildly different from one another.

Using as usual $\gamma_\perp$, $\gamma_\parallel$ and k for the three damping rates for polarization, population and field, the following classification has been introduced /18/

Class A (e.g. He-Ne, Ar, Kr, dye): $\gamma_\perp \simeq \gamma_\parallel >> k$

The two last MB equations can be solved at equilibrium (adiabatic elimination procedure) and one single nonlinear field equation describes the laser. N=1 means a fixed point attractor, and hence coherent emission.

Class B (e.g. ruby, Nd, CO_2): $\gamma_\perp >> k \gtrsim \gamma_\parallel$

Only polarization is adiabatically eliminated and the dynamics is ruled by two rate equations for field and population. N=2 allows also for periodic oscillations.

Class C (e.g. FIR lasers): $\gamma_\parallel \approx \gamma_\perp \approx k$

The complete set of MB equations has to be used, hence Lorenz-like chaos is feasible.

We have carried out a series of experiments on the birth of deterministic chaos in CO_2 lasers (Class B). In order to increase the number of degrees of freedom by at least 1, we have tested the following configurations:

(i) Introduction of a time-dependent parameter to make the system nonautonomous /19/. Precisely, an electro-optical modulator modulates the cavity losses at a frequency near the proper oscillation frequency Ω provided by a linear stability analysis, which for a CO_2 laser happens to lie in the 50-100kHz range, making an accurate set of measurements easy.

ii) Injection of the signal from an external laser detuned with respect to the main one, choosing the frequency difference near the above-mentioned Ω. With respect to the external reference the laser field has two quadrature components which represent two dynamical variables. Hence we reach N = 3 and observe chaos /18/.

(iii) Use of a bidirectional ring, rather than a Fabry-Perot cavity /20/. In the latter case the boundary conditions constrain forward and backward waves, by phase relations at the mirror, to act as a single standing wave. In the former case, forward and backward waves have to fill the total ring length with an integer number of wavelengths but there are no mutual phase constraints, hence they act as two separate variables. Furthermore, when the field frequency is detuned with respect to the center of the gain line, a complex population grating arises from interference of the two counter-propagating waves, and as a result the dynamics becomes rather complex, requiring N 3 dimensions.

(iv) Addition of an overall feedback, besides that provided by the cavity mirrors, by modulating the losses with a signal provided by the output intensity /21/. If the feedback has a time constant comparable with the population decay time, it provides a third equation sufficient to yield chaos.

Notice that while methods (i), (ii) and (iv) require an external device, (iii) provides intrinsic chaos. In any case, since feedback, injection or modulation are currently used in laser applications, the evidence of the existence of chaotic regions is a warning against optimistic trust in laser coherence.

Of course, the requirement of three coupled nonlinear equations does not necessarily restrict the attention to just Lorenz equations. In fact none of the explored cases i) to iv) corresponds to Lorenz chaos.

3.2 Role of noise in chaos: Hyperchaos

Nonlinear dissipative systems can have many simultaneously coexisting basins of attraction (generalized multistability /19/). This situation can be destabilized by changes of the control parameters, merging two independent attractors into a single one via an intermediate region which is only sporadically visited near the transition. The associated dynamics implies a low-frequency tail (deterministic diffusion) /22/. Conversely, when the above coexistence is stable, application of external noise may induce jumps between two otherwise disjoint regions of phase space.

The simultaneous presence of deterministic chaos and noise should not introduce new features within one attractor, since trajectories are already irregular. When however many attractors coexist for the same parameters, addition of noise makes it possible to leave a basin and go to another (which would be otherwise forbidden by the uniqueness theorem). This "hyperchaos" gives rise to low frequency spectra of 1/f type.

Clear evidence of generalized multistability was first shown in an electronic oscillator /23/ and then in the modulated laser /19/. In both cases, besides the qualitative appearance of different attractors in phase space, there was a low frequency spectral component due to noise-induced jumps between different attractors . Both measurements, however, might be considered as experimental artifacts. In fact, there is evidence of single attractors made of two sub-regions with infrequent passages from one to the other (see e.g. the Lorenz attractor). In such a case, the low-frequency tail corresponds to the sporadic passages, and does not require added noise (deterministic diffusion). As a matter of fact, power spectra do not permit discrimination between the two phenomena.

An analysis of the role of noise in GM was given for a cubic iteration map, allowing for two simultaneous attractors, and in the numerical studies of a forced Duffing oscillator with a double potential well, in a parameter region allowing for the simultaneous existence of more than one attractor /24/.

A recent solution of the 1/f spectral problem /25/ is based on the double randomness due to the irregular deterministic motion with long-lived transients, where the trajectory wanders near a fractal basin boundary, and in the presence of stochastic noise. In this case, 1/f behavior can be accurately traced over more than 3 decades in frequency.

4. Laser dynamics with competing instabilities /26/

By operating a CO_2 laser with feedback in a parameter range with three coexisting fixed points, we have found experimental evidence of competition between different instabilities, displaying the successive transition from a Hopf bifurcation to Shilnikov chaos and eventually to regular spiking, as a control parameter is monotonically increased. Each of these phenomena is due to the dominant attraction of one of the three fixed points. A global description could be achieved by composition of successive maps, each one describing the re-injection from and to a box around each of the fixed points /27/. However, because of the dominant character of one of the instabilities in each different range of the control parameter, we adopt a piecewise interpretation, in which we attribute most of the return time in a suitable Poincaré section to only one fixed point.

As stated in the previous section, a single-mode class B laser is governed by two coupled degrees of freedom (intensity x and population inversion y). Introduction of a third degree of freedom z by feedback leads to a three-dimensional system displaying oscillatory instabilities and chaos /21/.

The dynamical equations are given in /21/, where we also discuss the details of the feedback scheme. There are three control parameters, the bias B, and the gain r and damping rate β of the feedback loop, all the other parameters being fixed at the values reported in /21/.

The experimental information consists of phase space projections on the x-z plane, obtained by feeding an x-y oscilloscope with a photodetector signal proportional to the laser intensity and with the feedback signal applied to the modulator. For each phase space portrait the associated time plot x(t) is recorded on a digital oscilloscope.

In Figs. 1 and 2 we report experimental data for two different gain values, and in each case for three increasing B values. In the phase space portraits as

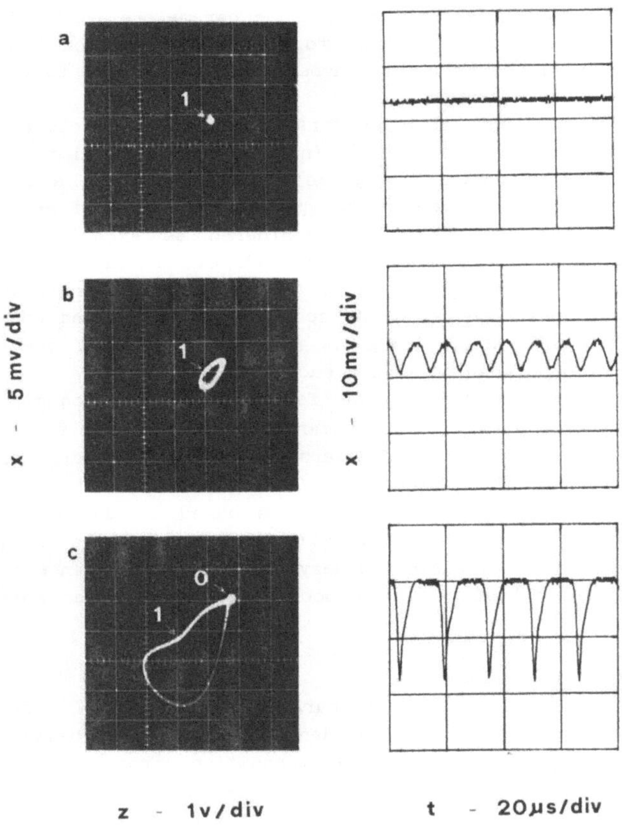

Fig. 1. Laser with feedback. Phase space projections z-x (feedback voltage - laser intensity) and time plots of the intensity x(t) for low feedback gain. Intensity increases downward. Normalization for B as in /21/; a) B=0.259; b) B=0.274; c) B=0.385. Approximate locations in the phase plane of points 1 and 0 are indicated.

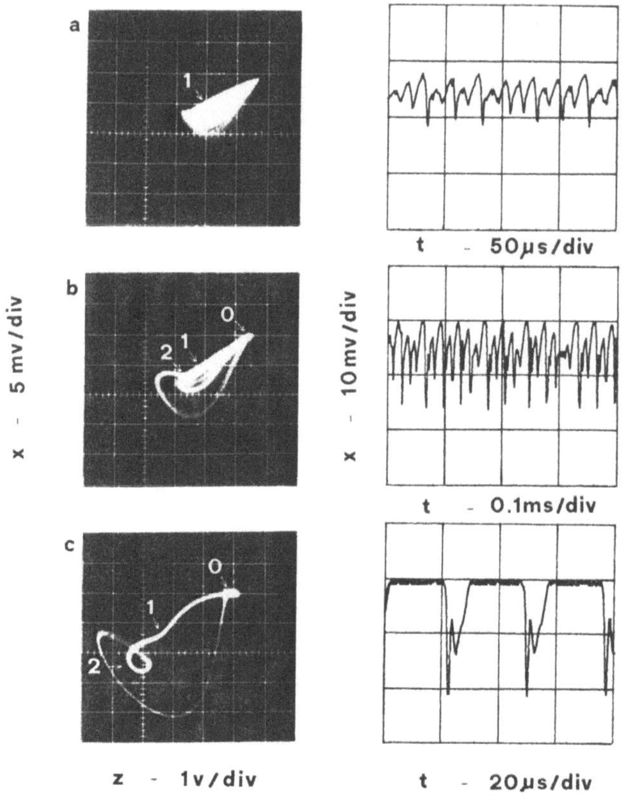

Fig. 2. Phase space projections z-x (feedback voltage - laser intensity) and time plots of the intensity x(t) for high feedback gain. a) B=0.296; b) B=0.311; c) B=0.411. Approximate locations of point 0, 1 and 2 in the phase plane are indicated.

well as in the time plots, 0, 1 and 2 denote the three fixed points. In Fig. 1 (low gain) as B increases, the first fixed point 1 becomes unstable through a Hopf bifurcation (Fig. 1b). For higher gain, the limit cycle becomes unstable and gives rise to chaotic trajectories, as we will see below in discussing Fig. 2b. As the loop gets wider, it eventually approaches 0 (the zero intensity solution). This implies a long pause, since the laser is nearly extinguished and must wait for a long build-up time before revisiting the region of phase space where the other fixed points are located. This pause has a stabilizing action since it washes out any memory of fluctuations associated with the previous cycle. Eventually the return time is dominated by this pause. The loop then corresponds to narrow spikes, slightly perturbed by the attraction of point 1, plus a long lethargy time in the vicinity of point 0 (Fig. 1c).

For high gain, two new important features appear. First (Fig. 2a) the Hopf bifurcation around 1 gives rise to a subharmonic route to a local chaos. As the loop widens, it is attracted by the fixed point 2 (Fig. 2b). The trajectories around 2 are subjected to large temporal fluctuations while spiralling (Shilnikov chaos). As the phase point approaches 2, the escape time gets longer, so that eventually the time spent around 2 is longer than that around 1 or around 0, thus characterizing the global behavior (Fig. 3). But, once 0 has been reached, its lethargy cancels the memory of the fluctuations around the Shilnikov instability, thus regularizing the return time and giving rise to narrow and equally spaced spikes (Fig. 2c).

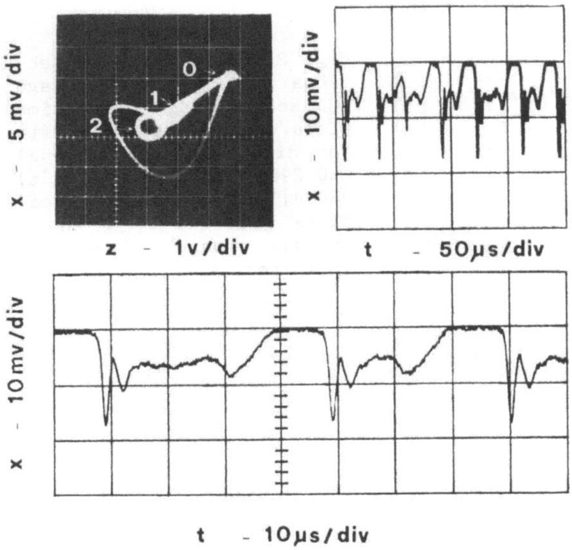

Fig. 3. Phase space projection and intensity vs. time x(t) for the same gain as Fig. 2 and for B=0.351. The time expanded plot x(t) shows clearly the role of the Shilnikov instability in yielding chaotic return times. Approximate locations of points 0, 1 and 2 in the phase plane are indicated.

If we now disregard the local features of the phase space motion and look for a global indicator, the most convenient one is the return time after one whole loop. This is measured experimentally by an average rate meter. Figure 4 shows the behavior of the average Poincaré frequency f_p (reciprocal of the return time) versus the bias B, for several gain values.

For low gain, f_p has a monotonic decrease versus B. Initially it corresponds to the Hopf frequency around 1, but, as the attraction of 0 prevails, f_p becomes the reciprocal of the occurrence time of the narrow spikes. For high gain, besides the irregular regime where f_ρ is undefined because of strong chaotic changes, a novel feature is the appearance of vertical step-like jumps, including hysteresis at each jump (see expanded inset in Fig. 4). This is equivalent to forbidden frequency values, just the opposite of what occurs in locking phenomena where there are horizontal steps in frequency.

In summary, the global features of the phase space motion with competing instabilities can be divided into three regimes, each corresponding to a dominant fixed point:

I) For low B the motion starts at 1 and for increasing B we have a Hopf bifurcation, followed by a subharmonic route to chaos.

II) For high B, the motion is a regular periodic spiking with no memory of other features (0 dominant).

III)At intermediate B (2 dominant), a Shilnikov instability provides chaotic fluctuations in the return time.

We can treat regime II) in the limit of large β , where the feedback voltage z can be abiabatically eliminated. In order to cope with the sharp spikes, it is more convenient to switch notation to a log representation of the intensity /28,29/. The two remaining equations are

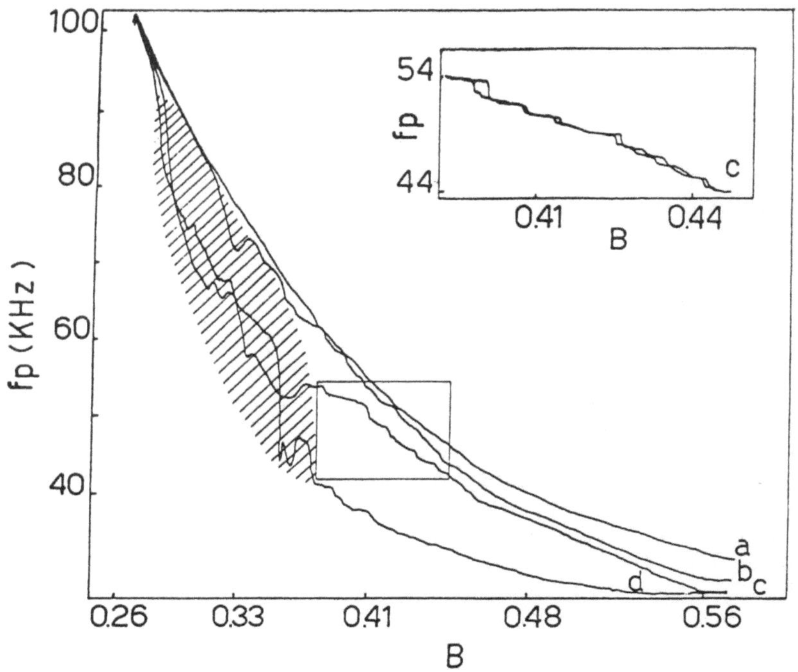

Fig. 4. Plot of the Poincaré frequency f_p versus bias B for four different gain values. Curves a and d refer to the gain values of Fig. 1 and Figs. 2,3, respectively. Curve c displays a step-like feature with hysteresis, as clearly shown in the expanded inset of the square region. The shaded region refers to a chaotic f_p, hence the traces reported within that region are just a single scan.

$$\frac{\dot{x}}{K} = - x\ (1 + \alpha \sin^2 z) + xy\ ,$$

$$\frac{\dot{y}}{\gamma_{\parallel}} = - x + A - xy\ , \qquad (1)$$

where the feedback voltage z is

$$z = B - rx\ . \qquad (2)$$

Now, with the transformation

$$s = \ln x\ , \qquad (3)$$

in the absence of feedback ($\alpha = 0$) Eqs. (1) reduce to

$$\frac{d^2 s}{d\tau^2} = - \varepsilon \frac{ds}{d\tau} (1 + e^s) + A - 1 - e^s\ , \qquad (4)$$

where $\tau = \sqrt{k\gamma}\ t$ and $\varepsilon = \sqrt{\gamma/k}$ are scaled parameters.

For a small ratio ε between population and photon decay rates (this ratio is 10^{-3} for our CO_2 laser) the equation reduces to that of a lossless Toda oscillator whose period is easily evaluated. It is in fact given by /29/

$$T = \int_0^{s_o} \frac{ds}{(E + (A-1)s - e^s)^{1/2}} , \qquad (5)$$

where the constants E and s_o are adjusted to experimental situations. By extending the above treatment to include the feedback we obtain an increasing Poincaré period T' which goes approximately as

$$\frac{T'}{T} \simeq \frac{1}{(1 - \frac{\alpha}{A} \sin^2 B)^{1/2}} , \qquad (6)$$

in qualitative agreement with the experimental data of Fig. 4, curve a.

In regime III we take the opposite limit, namely, that most of the return time is spent within a small distance of fixed point 2. The time $\bar{t}$ spent in the unit box around that point is shown in /27/ to vary as $\bar{t} \simeq \ln(1/z_o)$ where z_o is the offset at the box entrance along the expanding direction. As the system moves from 1 dominant to 2 dominant, it is reasonable to take $1/z_o$ proportional to the bias B, since for larger B the phase point comes closer to fixed point 2. Hence, the above solution yields a frequency $f_p = 1/\bar{t}$ monotonically decreasing with B, in qualitative agreement with the average trend of Fig. 4. However, to explain the steplike details, we must consider two contracting directions, that is, the three-dimensional character of the spiral at the Shilnikov instability. In this case the corresponding return map /27/ can be modeled in terms of a number of loops corresponding to the number of jumps in the expanded plot of Fig. 4, thus making it possible to evaluate the complex contraction rate and the real expansion rate of the Shilnikov instability.

Finally, let me stress the conceptual difference between Hopf chaos (Fig. 2a) and Shilnikov chaos (Fig. 3). The former case is the end point of a subharmonic sequence. Many amplitudes appear, corresponding to different loops around the unstable point 1.

The latter case yields practically equal-height pulses but with different time separations. For Shilnikov chaos, the interplay between the interaction along the stable manifold and the expansion along the unstable direction manifests itself in a spread of return times at the same Poincaré section. Such a situation is the natural evolution of the staircase region of Fig. 4c,d, for higher gain values. In such a case, the statistical properties of the return time allow reconstruction of the theoretical map of /27/.

A detailed correspondence between the model of /27/ and our experimental time statistics is given in a forthcoming paper /30/, together with a global theoretical picture of the phenomena reported here.

REFERENCES

1. H. Haken and H. Sauermann, Z. Phys. 173, 261 (1963).
2. H. Haken and H. Sauermann, Z. Phys. 176, 47 (1965).
3. W.E. Lamb Jr., Phys. Rev. 134, A1429 (1964).
4. R. Hanbury Brown and R.Q. Twiss, Nature 177, 27 (1958); Proc. Roy. Soc. A242, 300 (1957) and A243, 291 (1957).
5. R.J. Glauber, Phys. Rev. Lett. 10, 84 (1963); Phys. Rev. 130, 2529 (1963) and 131, 2766 (1963).
6. F.T. Arecchi, Phys. Rev. Lett. 15, 912 (1965); F.T. Arecchi, A. Bernè , and P. Burlamacchi, Phys. Rev. Lett. 16, 32 (1966).

7. H. Haken, Z. Phys. 181, 96 (1964).
8. H. Risken, Z. Phys. 186, 85 (1965).
9. F.T. Arecchi, G.P. Rodari and A. Sona, Phys. Lett. 25A, 59 (1967);
F.T. Arecchi, M. Giglio and A. Sona, Phys. Lett. 25A, 341 (1967).
F.T. Arecchi, V. Degiorgio and B. Querzola, Phys. Rev. Lett.19, 1168 (1987)
10. R. Graham and H. Haken, Z. Phys. 237, 71 (1970).
11. H. Haken, Rev. Mod. Phys. 47, 67 (1975).
12. I. Prigogine, G. Nicolis, J. Chem. Phys. 46, 3542 (1967);
I. Prigogine, P. Lefever, J. Chem. Phys. 48, 1695 (1968).
13. F.H. Busse, Rep. Prog. Phys. 41, 1925 (1978).
14. H. Haken, Z. Phys. B21, 105 (1975).
15. Stig Lundqvist (ed.) - The physics of chaos and related problems, Physica Scripta T9, (1985).
16. H. Haken, Phys. Lett. 53A, 77 (1975).
17. F.T. Arecchi and R.G. Harrison (Eds.), Instabilities and Chaos in Quantum Optics, Springer Ser. Syn., Vol. 34 (Springer, Berlin, Heidelberg 1987).
18. F.T. Arecchi, G.L. Lippi, G.P. Puccioni and J.R. Tredicce, Optics Comm. 51, 308 (1984).
19. F.T. Arecchi, R. Meucci, G. P. Puccioni and J.R. Tredicce, Phys. Rev. Lett. 49, 1217 (1982).
20. G.L. Lippi, J.R. Tredicce, N.B. Abraham and F.T. Arecchi, Optics Comm. 53, 129 (1985).
21. F.T. Arecchi, W. Gadomski and R. Meucci, Phys. Rev. A34, 1617 (1986).
22. T. Geisel and J. Nierwetberger, Phys. Rev. Lett. 48, 7 (1982).
23. F.T. Arecchi and F. Lisi, Phys. Rev. Lett. 49, 94 (1982) and 50, 1328 (1983).
24. F.T. Arecchi, R. Badii, and A. Politi, Phys. Lett. A29, 1006 (1984); F.T. Arecchi, R. Badii and A. politi, Phys. Rev. A32, 402 (1985).
25. F.T. Arecchi, and A. Califano, Europhysics Lett. 3, 5 (1987).
26. F.T. Arecchi, R. Meucci and W. Gadomski (to be published).
27. A. Arneodo, P.H. Coullet, E.A. Spiegel, and C. Tresser, Physica 14D, 327 (1985).
28. A.M. Ratner, Spectral, spatial and temporal properties of lasers, (Plenum, New York 1972).
29. G.L. Oppo, and A. Politi, Z. Phys. B59, 11 (1985).
30. F.T. Arecchi, A. Lapucci, R. Meucci, J.A. Roversi and P. Coullet (to be published).

Recent Advances in Laser Instabilities

L.A. Lugiato[1], *L.M. Narducci*[2], *D.K. Bandy*[2], *J.R. Tredicce*[2], *and P. Ru*[2]

[1]Dipartimento di Fisica del Politecnico di Torino, Italy
[2]Physics Department, Drexel University, Philadelphia, PA 19104, USA

1. Introduction

One of the unexpected features of early laser systems was the emergence of output pulsations even under steady (or nearly steady) pumping conditions [1]. Actually, spiking was observed in some masers [2] even before the discovery of their optical counterparts, but it became a virtually universal feature of solid state optical devices [3]. It did not take long before this surprising result was recognized as an important aspect of laser phenomenology. In fact, proposals linking the output pulsations to the emergence of dynamical instabilities appeared even in the very early literature on this subject [4]. To be sure, significant discrepancies existed between the experimental and calculated thresholds for unstable behavior, but this did not appear to be a point of serious concern until much later when, guided by a wealth of new insights about dynamical systems and especially because of the discovery [5] and analytic description [6] of optical bistability, the field of laser instabilities acquired a physiognomy of its own.

The current efforts in lasers and, more generally, optical instabilities [7] seek the answer, in a limited context, to a deep and provocative query: what is the origin and the functioning of evolving structures, the true vis viva of inanimate and, perhaps also, living systems? Are there general principles, universal laws, if you wish, that allow or, in fact, demand the growth of specific structures, that confer stability, and that differentiate among the many possible macroscopic behaviors of a large collection of elementary units?

Behind these efforts is the vision of Hermann Haken, one of the distinguished scientists of our time. Stimulated by the successful interpretation of the laser as a prototype system capable of displaying the emergence of an ordered state and by the identification of the spontaneous growth of spatial and temporal patterns in innumerable other systems, Haken created an impressive body of evidence for the existence of unifying links among seemingly different phenomena [8]. There is a strong esthetic value in his view of nature's behavior and a forceful call to reverse the trend for fragmentation in modern research.

During the last ten years Synergetics has provided both the motivation and the technical framework for the development of an organized assault on the problem of laser instabilities. It is a pleasure to take this festive opportunity to review our present understanding of this problem and to offer suggestions on the most promising directions for future research.

Some of the earliest theoretical models of laser action were based on the description of the energy exchanges between a collection of inverted two-level atoms and the cavity field. In their simplest version, the rate equations were designed to couple the population difference, $D=N_2-N_1$, between the two active levels and the photon number, n, according to the nonlinear system of equations

$$\frac{dD}{dt} = -2WnD - \frac{1}{T_1}(D-D_o) \tag{1.1a}$$

$$\frac{dn}{dt} = WnD - \frac{1}{T_c}n \tag{1.1b}$$

where W is proportional to the absorption and stimulated emission rate; T_1 is the relaxation time that characterizes the exponential approach of the population difference to its equilibrium value, D_o, under the action of both spontaneous emission and the external pump; T_c is the escape time of the electromagnetic energy out of the optical cavity. With regard to the temporal behavior of the emitted laser intensity, the main prediction of Eqs. (1.1) is that, above a certain pump level (threshold inversion), the output intensity becomes independent of time after a transient which can either involve a monotonic or damped oscillatory approach to steady state. Thus, the rate equations are inadequate to provide a useful description of the observed instabilities [9].

Our current theoretical view of this problem connects the failure of the rate equation approach to its lack of consideration for the coherent coupling between radiation and matter. Actually, this important point did not go unnoticed even in some of the earliest contributions. For example, in their lucid 1964 contribution Grazyuk and Oraevskii [4f] announced that the single-mode coherent laser equations, a predecessor of the more modern Maxwell-Bloch theory, predicted persistent oscillations under well defined conditions involving the magnitude of the unsaturated gain of the active medium and the relative size of the cavity and the atomic relaxation rates. This point will be discussed in greater detail in Section 2; here we only stress that the numerical solutions of the single-mode equations yielded random looking trains of pulses reminiscent of the observed spiking effect, at least in a superficial way.

More recent investigations of the laser equations [10] revealed the existence of pulsation phenomena in the form of periodic oscillations even under multimode operating conditions. These investigations were re-opened by Lugiato, Narducci, Eschenazi, Bandy and Abraham [11] who reported not only periodic oscillations but also much more complicated patterns.

Perhaps one of the most significant advances in this field was recorded in 1975 by Haken [12], who proved the existence of an isomorphism between the single-mode laser model and the Lorenz equations [13]. The Lorenz equations, originally derived to simulate the onset of convective instabilities, were already well known in the mathematical literature of that time as the source of a most peculiar behavior: the so-called deterministic chaos [14]. Thus the isomorphism between the laser and the Lorenz equations determined in one stroke that the laser itself could also be the source of chaotic behavior, at least as long as the single-mode approximation was sufficiently accurate. In fact, the chaotic instability is by far the most common type of behavior for the resonant single-mode model in the unstable domain. Over a restricted range of parameters, however, periodic oscillations can also develop, as shown very recently by Narducci, Sadiky, Lugiato and Abraham [15].

An important new type of instability, the so-called phase instability, was discovered by Hendow and Sargent [16] and independently by Narducci, Tredicce, Lugiato, Abraham and Bandy [17]. This is also a multimode instability, in a way a relative of the one discovered by Graham and Haken [10a] and by Risken and Nummedal [10b,c], but its dynamical origin lies in the loss of stability of the field phase rather than its amplitude. The experimental confirmation of this phenomenon showed good qualitative agreement between the theoretical predictions and the observed phenomenology [18].

While our emphasis, so far, has been with homogeneously broadened models, i.e. lasers whose active atoms have all the same transition frequency, experiments have shown the frequent appearance of regular and chaotic pulsations also in inhomogeneously broadened lasers. This area of investigation was pioneered by Casperson [19] and further advanced in more recent times by Abraham and collaborators [20].

Even on the basis of the few results touched upon in the preceeding paragraphs, the evidence strongly supports the essential significance of the coherent interaction between the field and the active medium. It is also very clear from a survey of the available experiments that significant areas of disagreement still exist between the theoretical predictions of the Maxwell-Bloch model and the laboratory data. The discrepancies observed with solid state lasers may be ascribed to the fact that nobody really knows how to model the active medium with sufficient accuracy. There are gas lasers, on the other hand, that can operate with a homogeneously broadened gain line and, yet, show behaviors that are simply not within the reach of the ordinary theoretical descriptions. A case in point is the CO_2 laser whose unstable behavior near the threshold for the ordinary laser action stands in striking contrast to the theoretical creeds [21]. In fact even the recent exact solution [22] of the linearized Maxwell-Bloch equations failed to record significant departures from the predictions of approximate treatments: the plane-wave Maxwell-Bloch approach for both single- and multimode lasers requires that the unsaturated gain of the active medium be several times larger than the threshold value in order for instabilities to develop. As a result of this statement, instabilities should be the exception and not the rule.

The plane-wave approximation, long viewed as adequate for capturing the essential aspects of laser dynamics, may be a much stronger theoretical constraint than anticipated, and could be even unsuitable for a close quantitative match between theory and experiments. Important warning signals in this direction came from the lack of quantitative agreement between the predictions of the plane-wave stationary theory of optical bistability [6,23] and the careful steady state experiments performed by Kimble and collaborators [24] and by the failure of the time-dependent plane-wave calculations [25] to match the observed pulsing patterns [26]. In addition, significant qualitative improvements resulted from the inclusion of a Gaussian profile for the cavity field [27].

The combined message of the small but growing number of experimental [28] and theoretical [29] reports supports the belief that transverse effects, i.e. effects related to the departure from the plane-wave configuration, play a non-negligible role in problems of this type and, in fact, may be even more influential when the optical resonator contains an active rather than a passive medium. If one accepts this premise, the inclusion of transverse degrees of freedom in the existing plane-wave theory is not just a cosmetic improvement but a necessary ingredient for a more accurate description of laser dynamics.

For the purpose of a balanced presentation of existing results and proposed improvements, we review first the traditional Maxwell-Bloch theory of the laser for both the multimode and single mode operation in Section 2; then we give a brief assessment of the experimental situation in Section 3, and conclude, in Section 4 with a sketch of some very recent calculations of ours, which attempt to describe the steady state and the linear stability of a ring cavity laser with spherical mirrors.

2. The Maxwell-Bloch Theory of a Ring Laser

The basic working model of a laser [30] rests on a self-consistent approach. An incident electromagnetic field interacts with a collection of microscopic dipoles and creates a macroscopic polarization. In turn, this acts as the source of a radiated field, which interacts again with the microscopic dipoles, etc. Energy is fed into the medium by an external source, the pump, and is extracted in the form of electromagnetic energy radiated by the macroscopic polarization.

The mathematical basis for this model is provided by the classical wave equation for the slowly varying complex amplitude of the electromagnetic field, and the Schroedinger equation for the collection of two-level systems. This mix of classical and quantum equations (the Maxwell-Bloch theory) forms the essence of the semiclassical theory of the laser. Its self-consistency derives from the identification of Maxwell's macroscopic polarization with the ensemble average of the quantum mechanical dipole operator. The presence of a resonant cavity is simulated by appropriate boundary conditions. The traditional theory, at this point, adds the assumption that the cavity field can be represented by a plane-wave to a good approximation.

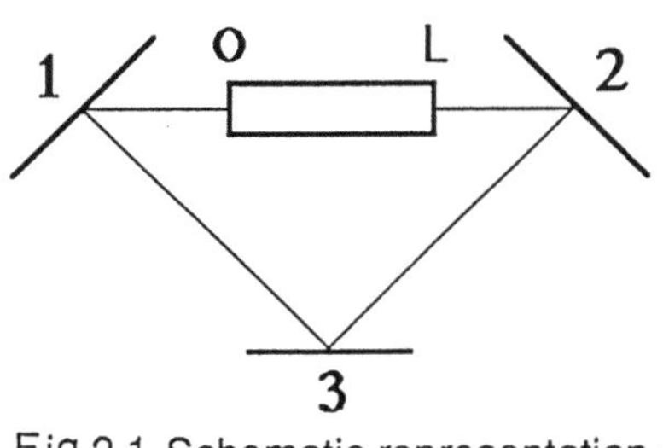

Fig.2.1 Schematic representation of a ring cavity of length Λ. The medium is confined within $0 \leq z \leq L$; mirrors 1 and 2 have a power reflectivity R, while mirror 3 is an ideal reflector.

In this paper we confine our attention to a ring cavity of the type shown in Fig. 2.1, with a total round-trip length Λ. The optical cavity is characterized by an infinite number of equispaced resonances at frequencies $\alpha_n = n2\pi c/\Lambda$ $(n=0,\pm1,\pm2,...)$. The active medium has a transition frequency ω_A and a homogeneously broadened lineshape with a width $\gamma_\perp$. For convenience of terminology, the empty cavity mode whose frequency lies closest to the atomic transition frequency is called the resonant mode; all the other modes are labelled off-resonant modes.

The Maxwell-Bloch equations for a ring laser are

$$\frac{\partial F}{\partial z} + \frac{1}{c}\frac{\partial F}{\partial t} = -\alpha P \tag{2.1a}$$

$$\frac{\partial P}{\partial t} = -\gamma_\perp \{ FD + (1+i\tilde{\delta}_{AC})P \} \tag{2.1b}$$

$$\frac{\partial D}{\partial t} = -\gamma_\parallel \{ -\frac{1}{2}(F^*P + FP^*) + D - 1 \} \tag{2.1c}$$

where F is the slowly varying complex amplitude of the cavity field, P is the amplitude of the macroscopic polarization, and D is the population difference between the upper and lower levels; α denotes the small signal gain constant per unit length, $\tilde{\delta}_{AC}$ is the frequency detuning between the atomic transition frequency and the resonant mode in units of atomic linewidth $\gamma_\perp$, and $\gamma_\parallel$ is the decay rate of the atomic population. Equations (2.1) are supplemented by the boundary conditions

$$F(0,t) = R\,F(L,t-\Delta t) \tag{2.2}$$

where R is the reflection coefficient of mirrors 1 and 2 (mirror 3, for simplicity, is assumed to be an ideal reflector), and $\Delta t = (\Lambda - L)/c$ is the propagation time of light through the empty segment of the cavity.

2a. Multimode operation

The possible stationary states of Eqs. (2.1) have the form [17]

$$F(z,t) = F_{st}(z)\exp(-i\delta\Omega t) \tag{2.3a}$$

$$P(z,t) = P_{st}(z)\exp(-i\delta\Omega t) \tag{2.3b}$$

$$D(z,t) = D_{st}(z) \tag{2.3c}$$

where F_{st}, P_{st}, and D_{st} are solutions of the equations

$$\frac{dF_{st}}{dz} - i\frac{\delta\Omega}{c}F_{st} = -\alpha P_{st} \tag{2.4a}$$

$$0 = F_{st}D_{st} + (1+i\Delta)P_{st} \tag{2.4b}$$

$$0 = -\frac{1}{2}(F_{st}^* P_{st} + F_{st} P_{st}^*) + D_{st} - 1 \tag{2.4c}$$

and where the detuning parameter Δ is defined as

$$\Delta = \tilde{\delta}_{AC} - \frac{\delta\Omega}{\gamma_\perp} \tag{2.5}$$

The atomic variables can be calculated at once in terms of the stationary field profile with the result

$$P_{sr}(z) = -F_{st}(z)\frac{1-i\Delta}{1+\Delta^2+|F_{st}(z)|^2} \tag{2.6a}$$

$$D_{st}(z) = \frac{1+\Delta^2}{1+\Delta^2+|F_{st}(z)|^2} \tag{2.6b}$$

where $F_{st}(z)$ is the solution of Eq. (2.4a) subject to the boundary condition

$$F_{st}(0) = R\, F_{st}(L)\, \exp(i\delta\Omega(\Lambda-L)/c) \tag{2.7}$$

A simple calculation shows that, in general, multiple steady state solutions do exist, each with its own value of the output intensity

$$|F_j(z)|^2 = \frac{2}{1-R^2}\left\{\alpha L - (1+\Delta_j^2)\,|ln\ R|\right\} \tag{2.8a}$$

and operating frequency

$$\delta\Omega_j = \frac{\kappa\,\delta_{AC} + j\,\alpha_1\gamma_\perp}{\gamma_\perp+\kappa} \tag{2.8b}$$

where the symbol κ, defined by

$$\kappa = \frac{c|ln\ R|}{\Lambda} \tag{2.9}$$

represents the field damping rate and α_1 is the cavity free spectral range. The quantity $\delta\Omega_j$ represents the frequency offset between the j-th steady state laser field and the resonant cavity mode which has been selected as a reference marker. Thus the possible stationary solutions oscillate with carrier frequencies given by the mode-pulling formula

$$\omega_c + \delta\Omega_j = \frac{(\omega_c + j\alpha_1)\gamma_\perp + \omega_A\kappa}{\gamma_\perp + \kappa} \tag{2.10}$$

The range of j is constrained by the threshold requirement $\alpha L>(1+\Delta_j^2)|ln\ R|$, [Eq. (2.8a)]

Depending on the values of the unsaturated gain and of the intermode spacing, we can identify two rather different situations, as shown in Figs. 2.2a,b. In the first case the intermode spacing is large enough that only one nontrivial steady state solution exists for any value of the detuning parameter. In the second case, more than one steady state solution (two, in this figure) can satisfy the lasing condition for selected values of $\tilde{\delta}_{AC}$.

The meaning of Fig. 2.2a is unambiguous: if one performs a detuning scan, beginning with the resonant configuration $\tilde{\delta}_{AC}=0$, the output intensity is maximum at first and then decreases until, eventually, the laser is driven below threshold. A further increase in the detuning causes the next steady state to satisfy its own threshold condition; the corresponding output intensity then increases monotonically until $\tilde{\delta}_{AC}$ equals a full spectral range of the ring cavity (α_1) and the process repeats itself. Note that the carrier frequency of the operating stationary state changes as the laser switches from the steady state j=0 to j=1. This is a consequence of the fact that, during the detuning scan, the laser is driven below threshold for a certain value of the detuning and then begins to operate again in correspondence to a different cavity resonance.

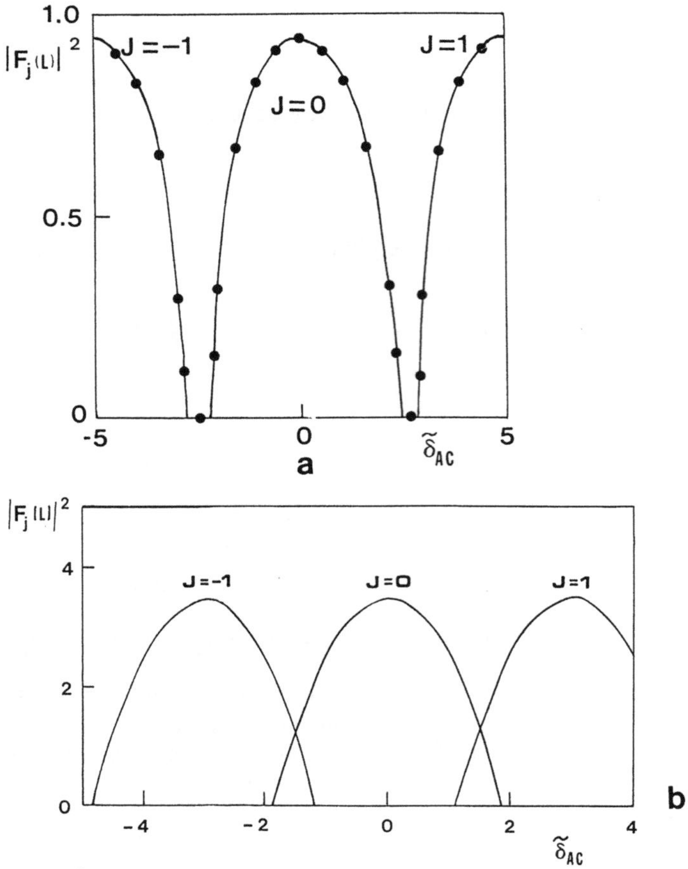

Fig.2.2(a) Steady state output intensity $|F_j(L)|^2$ for three stationary states labelled j=0, ±1 as a function of the detuning parameter $\tilde{\delta}_{AC}$. The value $\tilde{\delta}_{AC}= 0$ corresponds to resonance between the atomic transition frequency and the reference cavity mode. The dots correspond to steady state values obtained by direct integration of the Maxwell-Bloch equations. The parameters used in this calculation are $\alpha L=2.0$, R=0.5, $\tilde{\gamma}=\gamma_{\parallel}/\gamma_{\perp}=2.0$, $\tilde{\alpha}_1=5.0$. (The parameter $\tilde{\gamma}=2.0$ is irrelevant for the calculation of the steady state values, but is necessary for solving the Maxwell-Bloch equations). (b) Steady state output intensity $|F_j(L)|^2$ for three stationary states labelled j=0, ±1, as a function of the detuning parameter $\tilde{\delta}_{AC}$. The parameters used in this calculation are $\alpha L=2.0$, R=0.5 and $\tilde{\alpha}_1=3.0$.

The actual realizability of the various steady states is controlled by their stability properties. In our case the Maxwell-Bloch equations (2.1) involve five dynamical variables (two each for the complex field and polarization, and one for the population difference). Each variable can be associated with a discrete infinity of degrees of freedom (one degree of freedom for each cavity resonance) so that, at first sight, the solution of the linearized equations of motion appears to be hopelessly complicated.

Recently we have solved this problem exactly [22], and found only small quantitative differences with respect to earlier approximate treatments. Thus, for the sake of simplicity, we limit the following discussion to the uniform field limit [31]. As already discussed in the companion paper on passive optical systems [32] this limit corresponds to a situation where

$$\alpha L \to 0, \quad T \to 0 \quad \text{with} \tag{2.11a}$$

$$\frac{\alpha L}{T} \equiv 2C = \text{const} \tag{2.11b}$$

Here, as usual, T is the transmittivity of the mirrors and C is the gain parameter. In this case the problem can be reduced to an infinite number of decoupled sets of five equations each, that can be solved rather easily by numerical techniques [17]. Figure 2.3a shows, for each modal frequency $\tilde{\alpha}_n$ the two eigenvalues with the largest real parts (the other eigenvalues have large and negative real parts and are irrelevant for our purposes. For clarity we have drawn Re λ as a continuous function of the frequency variable $\tilde{\alpha}_n$, although the only physical values of relevance correspond to the positive or negative multiples of $\tilde{\alpha}_1$ [Note that Fig. 2.3a displays only the positive range of the frequency axis because Re $\lambda(\tilde{\alpha}_n) = \text{Re}\ \lambda(-\tilde{\alpha}_n)$]. Clearly for the case shown in this figure

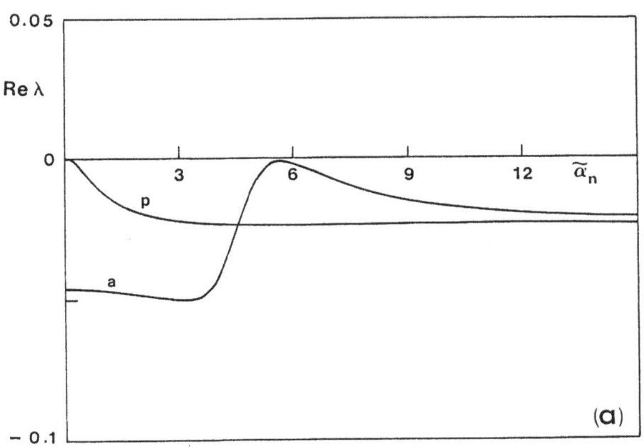

Fig.2.3 (a) The two largest real parts of the eigenvalues of the linearized equations are plotted as functions of $\tilde{\alpha}_n$ viewed as a continuous variable for αL=0.8, R=0.95, $\tilde{\alpha}_1$=3.0, $\tilde{\gamma}$=1.5 and $\tilde{\delta}_{AC}$=0. These eigenvalues describe the linear response of the system around the stationary state j=0 which is stable because Re $\lambda < 0$ for all values of $\tilde{\alpha}_n$. The other two possible steady state j=±1, which are above threshold for the chosen parameters, are both unstable. Under the conditions of this simulation the j=0 steady state would be stable for long times. The line marked (a) denotes the amplitude eigenvalue; the line marked (p) denotes the phase eigenvalue. (b) Same as (a) with αL=2.0. For sufficiently high values of the gain, the real part of the amplitude eigenvalue becomes positive and the j=0 steady state becomes unstable by developing sidebands at $\tilde{\alpha}_{\pm 3} = \pm 3\tilde{\alpha}_1$. Because the stationary states j=±1 are also unstable under these conditions, the system is expected to develop self-pulsing for sufficiently long times.

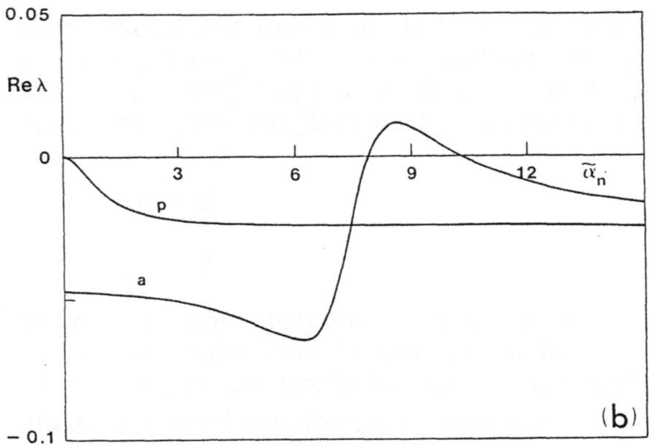

Figure 2.3b corresponds to a larger value of the unsaturated gain. In this case we see that the cavity mode at frequency $\tilde{\alpha}_3 = 3\tilde{\alpha}_1$ is unstable. This implies that an arbitrary perturbation at this frequency can grow exponentially during the linear regime, and force the cavity field to become a superposition of the steady state field, whose carrier frequency is ω_c, and of a growing pair of sidebands at frequencies $\omega_c \pm \alpha_3$. The total output intensity will then exhibit the characteristic beat pattern due to the interference between these frequency components, which appears as spontaneous self-pulsing. An example of this situation, obtained by solving the full set of Maxwell-Bloch equations, is shown in Fig. 2.4.

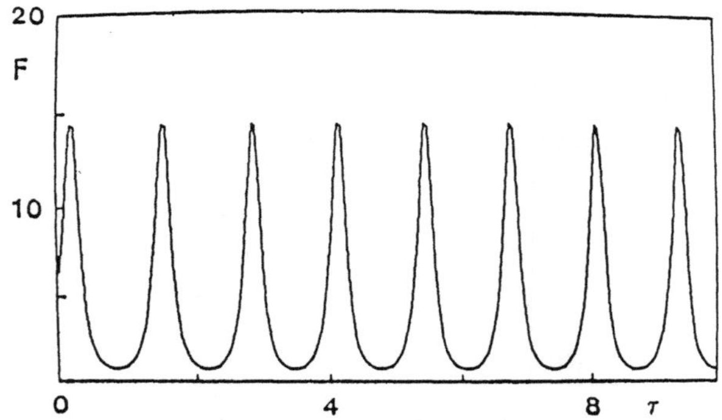

Fig.2.4 Time dependence of the output field modulus using the parameters αL=2.0, R=0.95, $\tilde{\gamma}$=0.5, $\tilde{\delta}_{AC}$=0, and $\tilde{\alpha}_1$=5.0. The solution is displayed after all the transients have died off. The time is measured in units of $\gamma_\perp^{-1}$.

all the complex rate constants have negative real parts, so that the steady state is actually stable.

Usually, more than one cavity sideband can become unstable at higher values of the gain parameter αL. In this case the nonlinear dynamics can be very complicated, as shown in Ref. 33 and further emphasized by the sequence of time dependent snapshots in Figs. 2.5a,b,c, where 3 cavity modes are unstable according to the linear stability

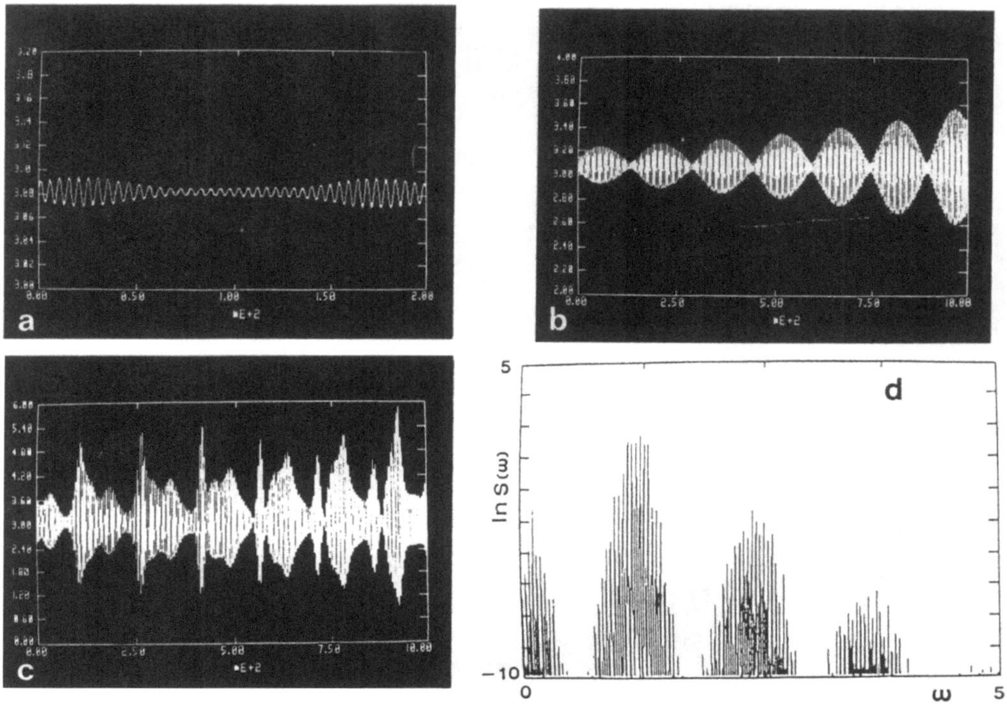

Fig.2.5 Time evolution of the output field modulus corresponding to the parameters $\alpha L=0.54$, $R=0.95$, $\tilde{\gamma}=0.1$, $\tilde{\delta}_{AC}=0$, and $\tilde{\alpha}_1=0.05$. The time is measured in units of $\gamma_\perp^{-1}$. Figures a,b,c represent snapshots of 1000 units of times, each, during the transient evolution and into steady state; the time frames are not consecutive. (d) Homodyne power spectrum of the time dependent solution. The individual spectral component are spaced by a full free spectral range which in this simulation is $\tilde{\alpha}_1=0.05$.

analysis [11]. Yet, for all the complicated beat patterns of the output intensity, the power spectrum of the signal is relatively simple-looking (See Fig. 2.5d), as it contains only a band of oscillating frequencies and harmonic components, in addition to a low-frequency part due to the interference among neighboring modes. Higher values of the gain can change the spectral picture dramatically, as shown in Figs. 2.6a,b Here the temporal oscillations are not noticeably more complicated than in the previous case, but the power spectrum of the output intensity (Fig. 2.6c) acquires the characteristic broad structure that is indicative of dynamical chaos.

We now turn our attention back to Fig. 2.2b. Here the situation is less obvious than in the corresponding case of Fig. 2.2a. However, it is reasonable to expect that a detuning scan should lead to two possible scenarios:

(i) The lasing mode retains control of the laser emission well into the coexistence region, or until, perhaps, its losses overcome the available gain, or

(ii) The two coexisting steady states compete with one another.

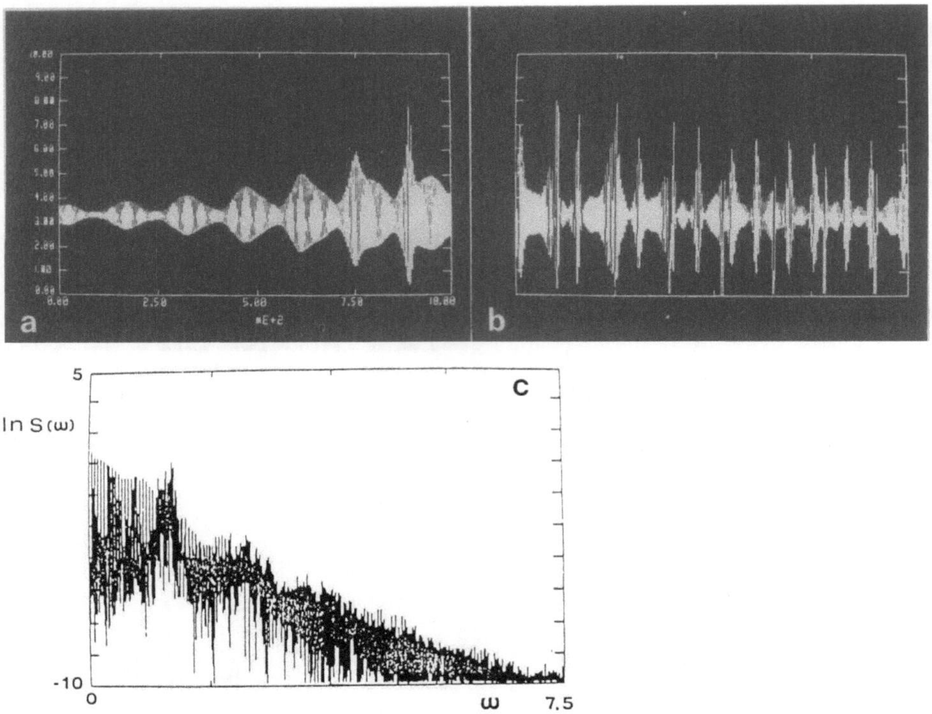

Fig.2.6 Time dependence of the modulus of the output field corresponding to the parameters $\alpha L=0.62$, R=0.5, $\tilde{\gamma}=0.1$, $\tilde{\delta}_{AC}=0$, and $\tilde{\alpha}_1=0.05$. Figure (b) is actually chaotic as shown by the broadband nature of the power spectrum in (c).

In case (i) we expect discontinuous jumps of intensity and frequency, and hysteresis upon reversing the direction of the scan; in case (ii) the competing steady states have different frequencies and, thus, can be expected to cause undamped beats at a frequency approximately equal to the intermode spacing (we say "approximately" because of the smallness of the mode pulling effect in the good cavity limit).

The results of the linear stability analysis [17] confirm this expectation and show that case (i) is favored when $\tilde{\gamma}$ is sufficiently smaller than unity, while case (ii) is more typical of situations where $\tilde{\gamma}$ is closer to one. In both cases the mechanism for the emergence of an instability is quite different from the previous situation. In Figs. 2.7a,b we show again the two largest real parts of the linearized eigenvalues plotted as functions of the mode frequency $\tilde{\alpha}_n$ viewed as a continuous variable. For small values of the detuning, the lasing mode is stable (Fig. 2.7a); for larger values of $\tilde{\delta}_{AC}$ one of the real parts of the eigenvalues (the one corresponding to the cavity eigenfrequency $\tilde{\alpha}_1$ in Fig. 2.7b) becomes positive and the lasing mode develops an instability. At this point the subsequent evolution of the system is determined by the stability properties of the coexisting stationary solutions (In the case of Fig. 2.2b at most two steady states can

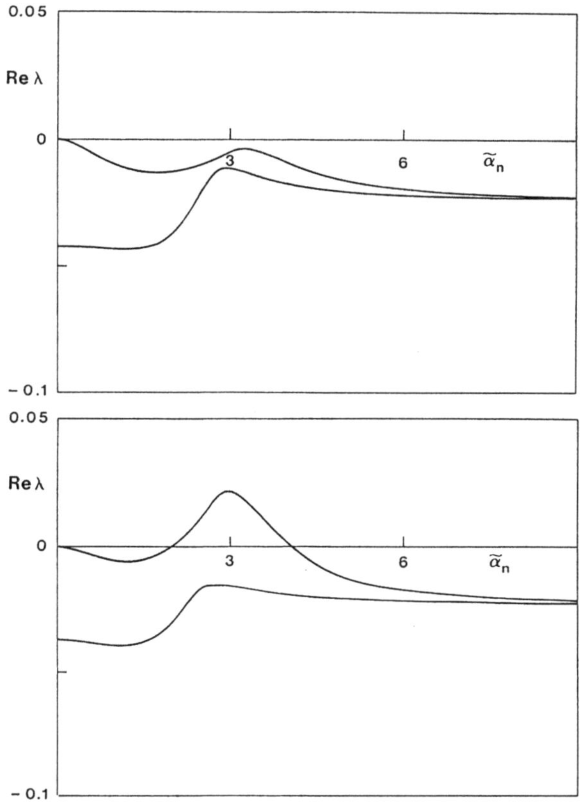

Fig.2.7 (a) For increasing values of the detuning parameter $\tilde{\delta}_{AC}$ the phase eigenvalue eventually develops a positive real part. In this case, the detuning is not sufficiently large to create an instability. The chosen parameters are αL=0.5, R=0.95, $\tilde{\alpha}_1$=3.0, $\tilde{\gamma}$=0.8 and $\tilde{\delta}_{AC}$=0.7. (b) Same as (a) with $\tilde{\delta}_{AC}$=1.2.

coexist). If the coexisting solution is stable, the laser operation is transfered from the unstable to the stable solution with a change of output intensity and operating frequency. If, on the other hand, the coexisting solution is itself unstable, undamped self-pulsing will develop. A schematic illustration of the behavior of the intensity and operating frequency as functions of the detuning parameter is shown in Figs. 2.8a,b for the case of a stable sideband.

Direct inspection of the behavior of the linearized eigenvalues in Figs. 2.3b and 2.7b shows immediately that, while both kinds of unstable behavior are of the multimode type, in the sense that they involve the running laser mode and at least one of the sidebands, the type of eigenvalue whose real part becomes positive is different in these two cases. In Fig. 2.3b the unstable eigenvalue is associated with the linear response of the field amplitude; in the case of Fig. 2.7b the unstable eigenvalue is connected with the field phase. For this reason, we label the first type of behavior "amplitude instability", and the second "phase instability" [34].

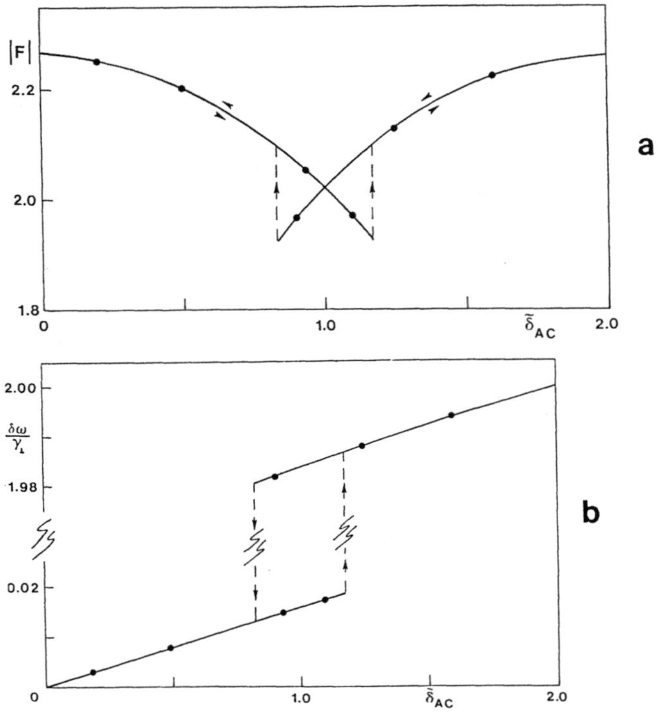

Fig. 2.8 (a) Simulated detuning scan corresponding to the parameter $\alpha L=0.3$, $R=0.95$, $\tilde{\gamma}=2.0$, $\tilde{\alpha}_1=2.0$. The dots represent the steady state values calculated numerically. The solid line is a plot of the modulus of the output field. (b) The variation of the frequency $\widetilde{\delta\Omega}$ of the output field. The dots represents the operating frequency, measured from the selected reference frequency ω_c, and obtained from the behavior of the real part of the output field.

2b. Single-mode operation

The single-mode laser is historically the oldest to have displayed unstable behaviors [4d-h]. Two questions come to mind immediately:

(i) How can one talk about modes in a lossy cavity, and
(ii) how does the single-mode model follow from the Maxwell-Bloch equations?

The answer to the first question was provided in Ref. 35 with a reformulation of the Maxwell-Bloch equations. This version of the theory, in fact, allows us to introduce the concept of mode even in the presence of cavity losses. As a first step we introduce two new variables z' and t' according to the definitions

$$z' = z \tag{2.12a}$$

$$t' = t + \frac{\Lambda - L}{c}\frac{z}{L} \tag{2.12b}$$

Next we define the new dependent variables

$$F'(z',t') = F(z',t') \exp(\frac{z'}{L} \ln R) \tag{2.13a}$$

$$P'(z',t') = P(z',t') \exp(\frac{z'}{L} \ln R) \tag{2.13b}$$

$$D'(z',t') = D(z',t') \tag{2.13c}$$

We note that the original boundary conditions (2-2) become

$$F'(0,t') = F'(L,t') \tag{2.14}$$

when expressed in terms of the new space-time coordinates and the new field variable. This type of boundary conditions appears frequently in one-dimensional linear vibration problems. The Fourier decomposition

$$F'(z',t') = e^{-i\,\delta\Omega t'} \sum_{n=-\infty}^{+\infty} e^{ik_n z'} e^{-i\alpha_n t'} f_n(t') \tag{2.15a}$$

$$P'(z',t') = e^{-i\delta\Omega\, t'} \sum_{n=-\infty}^{+\infty} e^{ik_n z'} e^{-i\alpha_n t'} p_n(t') \tag{2.15b}$$

$$D'(z',t') = \sum_{n=-\infty}^{+\infty} e^{ik_n z'} e^{-i\alpha_n t'} d_n(t') \tag{2.15c}$$

is automatically consistent with the new boundary conditions provided that

$$k_n = 2\pi n/L \quad (n=0,\pm1,\pm2\ldots) \tag{2.16}$$

The frequency offset $\delta\Omega$ must be calculated from the steady state equation, and it represents the frequency difference between the operating laser carrier and the reference cavity mode. Because Eq. (2-15a) is consistent with the boundary conditions for all values of the mirror reflectivity, it is clear that the space-dependent functions $\exp(ik_n z')$ play the role of the appropriate cavity modes in the new reference system, while the set of time-dependent coefficients $f_n(t')$ are the associated modal amplitudes.

The single-mode laser equations can be obtained through the following steps. First we impose the uniform field limit

$$\alpha L \to 0, \quad T \to 0, \quad \text{with} \quad 2C \equiv \alpha L/T < \infty \tag{2.17}$$

whence the modal equations for the field and atomic amplitudes become

$$\dot{f}_n = i\delta\Omega\, f_n - \kappa\,(f_n + 2C\, p_n) \tag{2.18a}$$

$$\dot{f}_n^* = -i\delta\Omega\, f_n^* - \kappa\,(f_n^* + 2C\, p_n^*) \tag{2.18b}$$

$$\dot{p}_n = -\gamma_\perp \{ \sum_{n'} f_{n'} d_{n-n'} + [1 + i(\tilde{\delta}_{AC} - \widetilde{\delta\Omega} - \tilde{\alpha}_n)] p_n \} \tag{2.18c}$$

$$\dot{p}^*_n = -\gamma_\perp \{ \sum_{n'} f^*_{n'} d^*_{n-n'} + [1 - i(\tilde{\delta}_{AC} - \widetilde{\delta\Omega} - \tilde{\alpha}_n)] p^*_n \} \tag{2.18d}$$

$$\dot{d}_n = i\alpha_n d_n - \gamma_\parallel \{ -\frac{1}{2} \sum_{n'} (f^*_{n'} p_{n+n'} + f_{n'} p^*_{n'-n}) + d_n - \delta_{n,o} \} \tag{2.18e}$$

Next, we require that all the modal amplitudes f_n, p_n, and d_n, with $n \neq 0$, vanish identically at $t'=0$, and that all the sidemodes be stable under infinitesimal perturbations. Under these conditions, the Maxwell-Bloch equations reduce to the single-mode model

$$\dot{f} = -\kappa (f + 2Cp) \tag{2.19a}$$

$$\dot{p} = -\gamma_\perp (fd + p) \tag{2.19b}$$

$$\dot{d} = -\gamma_\parallel (-fp + d - 1) \tag{2.19c}$$

where, for simplicity, we have made the further assumption of resonance between the center of the atomic line and the only remaining cavity mode. The single-mode laser equations, also known as the Haken-Lorenz model, are unstable under the following two conditions [4]

i) $\kappa > \gamma_\perp + \gamma_\parallel$ (bad cavity condition) (2.20a)

ii) $2C > 1 + (\kappa + \gamma_\perp + \gamma_\parallel)(\kappa + \gamma_\perp)/(\kappa - \gamma_\perp - \gamma_\parallel)$
(high gain condition) (2.20b)

The bad cavity condition is a constraint for the cavity linewidth relative to the sum of the atomic decay rates, and should not be interpreted literally as an indication that the cavity design must be of poor optical quality. It is true, in fact, that for most optical lasers, the atomic linewidth $\gamma_\perp$ is so large, that Eq. (2.20a) can only be satisfied by a cavity with a very large damping rate. However, certain lasers, mainly those operating in the far-infrared, are characterized by very narrow atomic linewidths so that the actual quality of the optical cavity, in this case, does not have to be poor in absolute terms. A useful graphical display of Eqs. (2.20) is shown in Fig. 2.9 which displays the domain of instability of a single-mode laser in the space of the parameters 2C and $\tilde{\kappa} = \kappa/\gamma_\perp$. For each value of $\tilde{\gamma}$, the solid curves represent the instability boundary of the laser. It is clear that, even under the most favorable conditions, the ratio $\kappa/\gamma_\perp$ must be sufficiently larger than unity, and the gain must exceed the ordinary laser threshold by about a factor of 10. Note that the limiting boundary corresponding to $\tilde{\gamma} \to 0$ is very close to curve (b) so that the selection of small values of $\tilde{\gamma}$ does not help in lowering the instability threshold in any significant way.

It is interesting to summarize some of the dynamical aspects of the Haken-Lorenz model. For low values of the unsaturated gain, the laser approaches a stable steady state monotonically, while, for higher gain values, the transient is characterized by damped relaxation oscillations. Eventually, above the instability threshold (often called

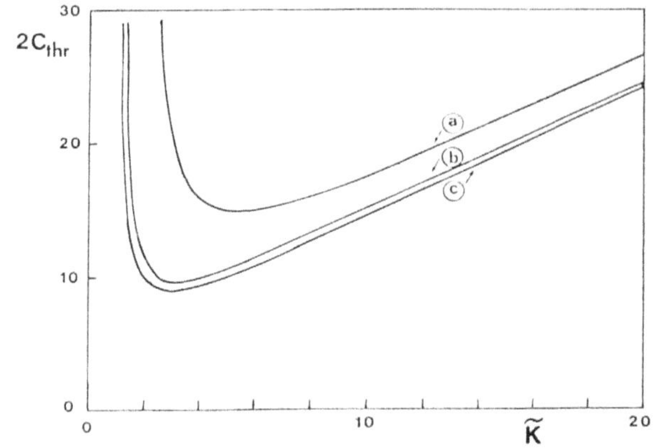

Fig.2.9 The instability boundary of the single-mode laser model is plotted in the plane of the parameters $\tilde{\kappa}$ and 2C for (a) $\tilde{\gamma}$=1.0, (b) $\tilde{\gamma}$=0.1, (c) $\tilde{\gamma}$=0.01. The optimum configuration (i.e., the unstable configuration with the lowest threshold) corresponds to $\tilde{\kappa} \cong 3$ and $2C_{thr} \cong 9$. Note that the threshold for ordinary laser action is $2C_{thr} = 1$.

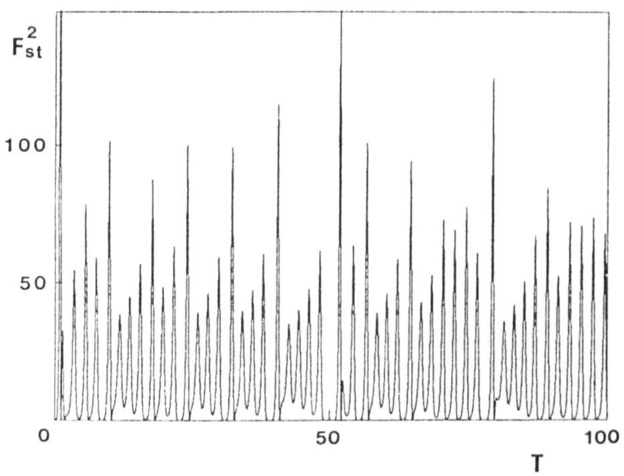

Fig.2.10Time evolution of the output intensity according to Eqs. (2.19). When 2C exceeds the second threshold value, the laser output often develops undamped chaotic pulsations. The parameters chosen in this simulation are 2C=15.0, $\tilde{\gamma}$=0.5, $\tilde{\kappa}$=4.0.

the second laser threshold), undamped pulsations set in. Usually they take the form of erratic oscillations [12] as shown in Fig. 2.10, but for selected values of the parameters, periodic behaviors can also develop [15]. These are common when $\tilde{\gamma}$ is smaller than about 0.2. In Fig. 2.11a [15] we show a periodic solution of the Haken-Lorenz equations whose electric field amplitude executes symmetric oscillations along the positive and negative direction of the vertical axis. We call this type of solution symmetric for reasons that are especially clear if one inspects the corresponding phase-space portrait [Fig. 2.11b]. A small change in $\tilde{\gamma}$ brings about a symmetry-breaking transition, that leads to unequal positive and negative excursions of the electric field, as shown in Fig. 2.12.

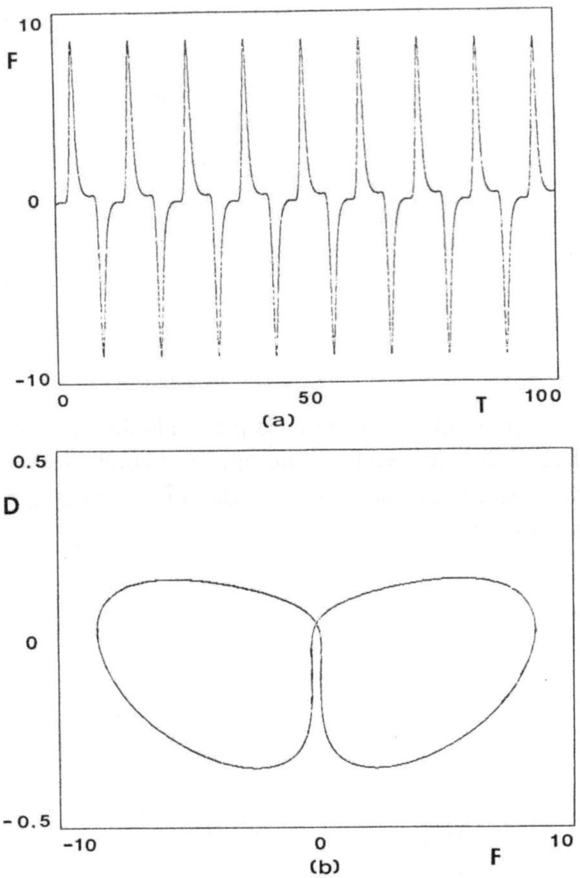

Fig.2.11(a) Time dependence of the electric field amplitude f(t) for $\tilde{\kappa}=4$, $\tilde{\gamma}=0.14$ and 2C=12. This is an example of a symmetric solution. (b) Projection of the phase-space trajectory in the (f,d) plane.

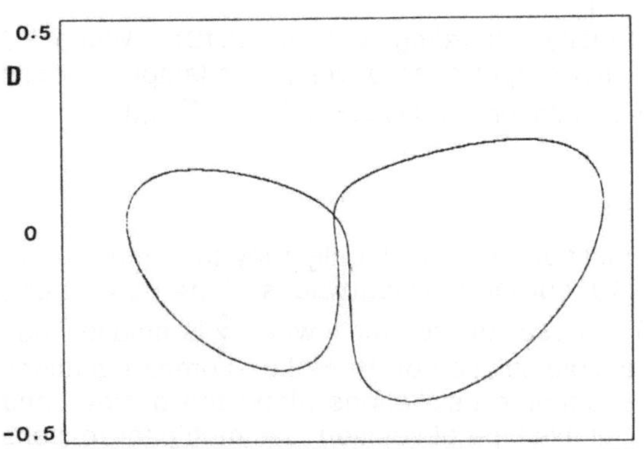

Fig.2.12Projection of the phase space trajectory in the (f,d) plane for $\tilde{\kappa}=4$, $\tilde{\gamma}= 0.17$ and 2C=12. This is an example of an asymmetric solution with period one.

It is interesting to note that the intensity patterns corresponding to Figs. 2.11 and 2.12 would lead one to believe that a period doubling bifurcation has taken place. In fact, this is not the case, as one sees easily from the numerical calculations. Because, however, the electric field amplitude is not directly observable experimentally, one is led to ask how a symmetry-breaking transformation can be identified in a traditional measurement. The answer lies in the application of heterodyne mixing techniques using a stable reference source. Consider, in fact, the superposition of the signal of interest with a coherent reference wave. The total electric field at the detector takes the form

$$E_{tot}(t) = f(t)\exp(-i\omega t) + A\exp(-i\omega_0 t) + cc. \tag{2.21}$$

where A and ω_0 are the amplitude and the carrier frequency of the reference signal, respectively. The photocurrent produced by a standard detector is given by

$$i(t) = \text{const} \times \left\{ |A|^2 + |f(t)|^2 + f(t)A^* \exp[-i(\omega-\omega_0)t] + f(t)^* A \exp[i(\omega-\omega_0)t] \right\} \tag{2.22}$$

The power spectrum of i(t) contains a delta-function contribution at zero frequency, due to the constant background intensity $|A|^2$, the homodyne power spectrum of $|f(t)|^2$ in the low frequency range, and finally the spectrum of the electric field f(t) centered at the frequency difference $\omega-\omega_0$ between the two optical carriers. The main difference between a symmetric and an asymmetric solution is that the former has a zero average electric field, while the latter does not. Hence, the heterodyne spectrum of a symmetric solution displays symmetric frequency components around $|\omega-\omega_0|$, but no spectral power at $|\omega-\omega_0|$, while the distinguishing signature of an asymmetric solution is the appearance of a line at the frequency $|\omega-\omega_0|$, in addition to symmetric sidebands. Further changes of $\tilde{\gamma}$ produce real bifurcations and eventually chaos. An example of a period-doubled asymmetric solution is shown in Fig. 2.13.

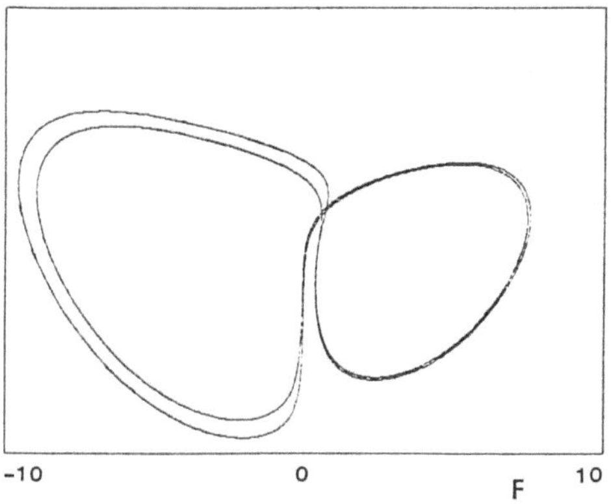

Fig.2.13Projection of the phase space trajectory in the (f,d) plane for $\tilde{\kappa}=4$, $\tilde{\gamma}= 0.19$ and 2C=12. This is an example of an asymmetric solution with period two.

3. A brief assessment of the experimental situation for homogeneously broadened lasers

3a. Multimode instabilities

There is still no definitive experimental confirmation of the multimode instability of the amplitude type [10] for a homogeneously broadened laser. However, a very interesting type of multimode instability was reported by Hillman, Krasinski, Boyd and Stroud [36] in 1984, and was analyzed in greater details by Stroud [37] and Lawandy [38] together with their respective collaborators. The most intriguing aspects of this phenomenon are reviewed in Ref. 37c: an Argon laser-pumped dye laser designed as a ring cavity with very high reflectivity mirrors begins lasing at about 2 Watts of Argon pump power. At a pump level of about 5.5 Watts the operating laser line disappears and is replaced by two symmetric sidebands in an apparently discontinuous way; these sidebands then move further apart from one another at higher pump power.

A preliminary analysis [11] of this experiment vis-a-vis the theoretical predictions of the Maxwell-Bloch theory failed to show convincing evidence that the observations by Hillman, Krasinski, Boyd and Stroud could be interpreted as a standard multimode amplitude instability. The experiments show the onset of unstable behavior at pump levels about three times as large as required for ordinary laser action. It is true that the gain of the active medium is not necessarily a linear function of the pump power, but it is still reasonable to infer from the data that the unsaturated gain at the instability threshold is smaller than expected on theoretical grounds. In addition, the instability is characterized by a relatively sharp, almost discontinuous onset, and by the disappearance of the operating laser component. Stroud and collaborators speculate that their observations may be the result of mode-locking. On the other hand, questions have been raised [38] about the possible role of bidirectional propagation in this effect, so that more work may be needed before the theorists can begin to sort out the available evidence. Still, on the face of what we now have, it would appear that the traditional Maxwell-Bloch equations may need some modifications to produce the observed phenomenology.

The phase instability, instead, seems to be understood and confirmed at a much higher level of confidence by the experiments of Tredicce and collaborators [18]. This effect can develop even at low pump powers and can be viewed as the result of mode-mode competition in homogeneously broadened lasers.

3b. Single-mode instabilities

The Haken-Lorenz model is predicted to undergo unstable oscillations under stringent experimental conditions. On the one hand, the cavity linewidth must be sufficiently larger than the sum of the atomic relaxation rates, and on the other, the unsaturated gain must exceed the laser threshold by at least a factor of 10. If we consider that for most lasers it is difficult to exceed the threshold gain by more than a factor of two or three, it is easy to see why gain values of the order of 10 or 20 times above threshold represent an enormous challenge.

Much to their credit, Weiss [39] and Lawandy [40] and their respective collaborators realized that perhaps the only way of testing the single-mode theory was to use far infrared (FIR) transitions as active laser lines. FIR lasers are characterized by very high gain, and linewidths that are much narrower than their optical counterparts. As a result, the bad cavity condition can be satisfied without unusually large cavity losses, and a very high level of inversion can be produced. One of the drawbacks of these types of lasers is that they must be pumped by another laser whose coherence may couple more than the two levels considered in the Haken-Lorenz model [41]. Interesting experimental results [39,40] have revealed significant points of contact with some of the theoretical predictions; still, the feeling persists that additional work may be needed before the matter of comparison between the single-mode model and real laser systems can be regarded as being satisfactory. In addition, other homogeneously broadened lasers, such as the CO_2 laser, have exhibited behaviors that are strongly at variance with the predictions of the theory [21]; low threshold oscillations, self Q-switching, quenching of pulsations at higher pump levels, etc. are all incompatible with the predictions of the simple Haken-Lorenz model.

In view of these comments, we must conclude that the origin of the pulsations in lasers is still not entirely understood. It is no longer in question that these phenomena are the result of internal instabilities, at least for the most part, but the quantitative and often qualitative disagreement with the Maxwell-Bloch theory suggests the existence of important missing pieces.

4. The inclusion of transverse effects in the uniform field limit

Several important physical effects and design features of laser resonators are automatically excluded from consideration in the plane-wave approximation:

i) the diffraction caused by the finite cross section of the cavity field and by its transverse variations of amplitude and phase;
ii) the wavefront distortion introduced by curved optical surfaces;
iii) the lack of transverse uniformity in the population inversion;
iv) the confinement of the active medium to a column of finite radius.

A generalization of the Maxwell-Bloch equations that allows arbitrary variations of the field and atomic variables along the x and y transverse directions would be very desirable. However, the analytic and numerical complexities of such an approach are so great that, as a compromise, it is more reasonable to limit ourselves to an axially symmetric configuration, whose only significant transverse variations are along the radial direction. In general, this is not a very realistic setting for a practical laser system where even small departures from complete axial symmetry tend to be amplified and become dominant features under steady state conditions [42]. Yet, as a first step, and in an attempt to gain some useful insight into the role of transverse effects we shall investigate the implications of this simpler model.

In the paraxial approximation we can introduce the various features listed above under (i)-(iv) with a few modifications of the plane-wave equations and boundary conditions. The new Maxwell-Bloch equations take the form

$$\frac{\partial F}{\partial \eta} + \frac{1}{v}\frac{\partial F}{\partial \tau} = \frac{i}{4}\left(\frac{\partial^2 F}{\partial \rho^2} + \frac{1}{\rho}\frac{\partial F}{\partial \rho}\right) - \alpha L\, P \qquad (4.1a)$$

$$\frac{\partial P}{\partial \tau} = -\left(FD + (1+i\tilde{\delta}_{AC})\, P \right) \qquad (4.1b)$$

$$\frac{\partial D}{\partial \tau} = -\gamma \left\{ -\frac{1}{2}\left(F^*P + FP^* \right) + D - \chi(\rho) \right\} \qquad (4.1c)$$

where the scaled independent variables are defined by

$$\eta = \frac{z}{L}, \quad \rho = \left(\frac{\pi}{L\lambda}\right)^{1/2} r, \quad \tau = \gamma_{\perp} t \qquad (4.2)$$

L is the length of the active medium and $v=c/L\gamma_{\perp}$. The function $\chi(\rho)$ gives the radial profile of the equilibrium population difference; all the other symbols are defined as in the plane wave theory.

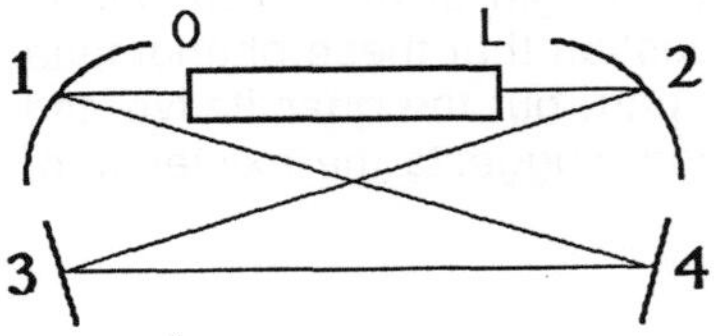

Fig.4.1 Schematic representation of a ring cavity with spherical mirrors. The curved mirrors have power reflectivity R; the plane mirrors are ideal reflectors. The medium is confined to the region $0 \le z \le L$. The figure-eight shape of the cavity minimizes aberration errors upon reflection.

We consider a ring resonator of the type sketched in Fig. 4.1 with a total length Λ and spherical mirrors having the same radius of curvature $\rho_0=R_0/L$ and reflectivity coefficient R; the plane mirrors, for simplicity are assumed to be ideal reflectors. The modes of the empty cavity $A_p(\rho,\eta)$ with p=0,1,2... are Gauss-Laguerre functions. They are solutions of the differential equation

$$\frac{\partial}{\partial \eta} A_p(\rho,\eta) = \frac{i}{4}\left(\frac{\partial^2}{\partial \rho^2} + \frac{1}{\rho}\frac{\partial}{\partial \rho}\right) A_p(\rho,\eta) \qquad (4.3)$$

and obey the orthonormality relation

$$\int_0^\infty d\rho\, \rho\, A_p^*(\rho,\eta)\, A_{p'}(\rho,\eta) = \delta_{p,p'} \qquad (4.4)$$

Their explicit form over the region occupied by the active medium is

$$A_p(\rho,\eta) = \frac{2}{\gamma(\eta)} \exp\left(-\frac{\rho^2}{\gamma^2(\eta)}\right) L_p\left(\frac{2\rho^2}{\gamma^2(\eta)}\right) \exp\left(i\frac{\rho^2}{u(\eta)}\right) x$$

$$x \exp\left(-i\,(2p+1)\tan^{-1}\left(\frac{\eta}{\eta_0}\right)\right) \qquad (4.5)$$

where

$$v(\eta) = \sqrt{\eta_0}\,\sqrt{1+\left(\frac{\eta}{\eta_0}\right)^2} \qquad (4.6a)$$

$$u(\eta) = \frac{1}{\eta}\left(\eta^2 + \eta_0^2\right) \qquad (4.6b)$$

$$\eta_0 = \frac{1}{2}\sqrt{\frac{(\rho_0 - 1)(\rho_0 - 1 + f)}{f(1+\rho_0) - 1}} \qquad (4.6c)$$

and $f = L/\Lambda$ is the filling factor of the cavity. The empty cavity eigenfrequencies are given by

$$\omega_{n,p}\frac{\Lambda}{c} = 2\pi n + 2(2p+1)\left\{\tan^{-1}\left(\frac{1}{2\eta_0}\right) + \tan^{-1}\left(\frac{1/f-1}{2\eta_1}\right)\right\} \qquad (4.7)$$

where

$$\eta_1 = \frac{1}{\rho_0^2\eta_0}\left(\left[\eta_0\left(\frac{1}{f} - 1 + \rho_0\right)\right]^2 + \left[\frac{1}{2}\left(\frac{\rho_0 - 1}{f} + 1\right)\right]^2\right) \qquad (4.8)$$

and $n = 0, \pm 1, \pm 2, \ldots$ is the axial index of the resonator.

If we expand the slowly varying field amplitude according to the relation

$$F(\rho,\eta,\tau) = \sum_p f_p(\eta,\tau)\, A_p(\rho,\eta) \qquad (4.9)$$

and take advantage of the orthonormality property (4.4) of the Gauss-Laguerre modes, the wave equation expressed in terms of the modal amplitudes $f_p(\eta,\tau)$ takes the form

$$\frac{\partial f_p}{\partial \eta} + \frac{1}{v}\frac{\partial f_p}{\partial \tau} = -\alpha L \int_0^\infty d\rho\, \rho\, A_p^* P \qquad (4.10)$$

The atomic equations maintain the same structure as in Eqs. (4.1b,c) with $F(\rho,\eta,\tau)$ given by Eq. (4.9). The appropriate boundary conditions for the ring resonator are

$$f_p(-1/2, \tau) = R\, e^{-i\delta_p} f_p\left(1/2, \tau - \gamma_\perp \frac{\Lambda - L}{c}\right) \qquad (4.10b)$$

where $\delta_p = (\omega_p - \omega_{p=0})\,\Lambda/c$. From now on we shall assume that the laser operates in a single axial mode so that the index n can be omitted from our formulas

4a. Steady state

In steady state we seek solutions of the type

$$F(\rho,\eta,\tau) = e^{-i\delta\Omega\tau}\,\overline{F_{st}} = e^{-i\delta\Omega\tau}\sum_p A_p \overline{f_p} \qquad (4.11)$$

$$P(\rho,\eta,\tau) = e^{-i\delta\Omega\tau}\,\overline{P}(\rho,\eta,\tau)$$

where $\delta\Omega$ is the unknown frequency difference between the laser carrier and the reference cavity mode (n,p=0) in units of $\gamma_\perp$. From Eq. (4.10) and (4.1b,c) we obtain

$$\frac{\partial}{\partial\eta}\overline{f}_p = i\frac{\delta\Omega}{V}\overline{f}_p + \alpha L\int_0^\infty d\rho\,\rho\,A_p^*\,\chi(\rho)\,\overline{F}_{st}\frac{(1-i\Delta)}{1+\Delta^2+|\overline{F}_{st}|^2} \tag{4.12}$$

where $\Delta = \delta_{AC} - \delta\Omega$ and

$$\overline{F}_{st} = \sum_p A_p\overline{f}_p$$

The boundary conditions for the modal amplitudes f_p are

$$\overline{f}_p(-1/2) = R\,e^{-i\delta_p}\,\overline{f}_p(1/2)\,e^{i\delta\Omega\gamma_\perp\frac{\Lambda-L}{c}} \tag{4.13}$$

In the uniform field limit, with the additional constraint $\delta_{p\neq 0} = O(1)$, and with the help of Eq. (4.13), it is easy to see that the magnitude of the modal amplitudes $\overline{f}_p$ for $p\neq 0$ are of the order of αL, while the magnitude of $\overline{f}_0$ is of order unity. Hence the steady state in the uniform field limit is a single radial mode solution whose amplitude satisfies the state equation

$$1 = 2C\int_{-1/2}^{1/2} d\eta\int_0^\infty d\rho\,\rho\,\chi(\rho)\frac{|A_0(\rho,\eta)|^2}{1+\Delta^2+|A_0(\rho,\eta)\,\overline{f}_0|^2} \tag{4.14}$$

and whose oscillation frequency is given by the usual mode-pulling formula

$$\delta\Omega = \tilde{\kappa}\,\frac{\delta_{AC}}{1+\tilde{\kappa}} \tag{4.15}$$

A more detailed discussion of these results will be presented elsewhere. Here we mention only that the various geometry factors of the ring resonator play a much more significant role than they do in the plane-wave case.

4b. Linear stability analysis

In this section we analyze the following problem. Given that the ring laser is in a steady state configuration characterized by a TEM_{00} mode, an amplitude given by the steady state Eq. (4.14), and an operating frequency assigned by Eq. (4.15), is it possible for a fluctuation to grow in correspondence with one of the radial modes p=0,1,2,... etc? The problem is similar, conceptually, to the one solved in Refs. 10, but it also quite different in practical details; in addition, our preliminary reports indicate instability thresholds that are considerably closer to experimental levels than are available from a plane-wave theory.

We begin by introducing the fluctuation variables δf_p, δP, and δD defined by

$$F(\rho,\eta,\tau) = e^{-i\delta\Omega\tau}\left(\overline{F}_{st}(\rho,\eta) + \delta F\right) = e^{-i\delta\Omega\tau}\sum_p A_p(\overline{f}_p+\delta f_p) \tag{4.16a}$$

$$P(\rho,\eta,\tau) = e^{-\delta\Omega\tau}\left(\overline{P_{st}} + \delta P\right) \tag{4.16b}$$

$$D(\rho,\eta,\tau) = \overline{D_{st}} + \delta D \tag{4.16c}$$

In the limit of infinitesimal fluctuations, the Maxwell-Bloch equations reduce to the linearized form

$$\frac{\partial}{\partial\eta}\delta f_p + \frac{1}{V}\frac{\partial}{\partial\tau}\delta f_p = i\frac{\delta\Omega}{V}\delta f_p - \alpha L\int_0^\infty d\rho\, \rho\, A_p^*\, \delta P \tag{4.17a}$$

$$\frac{\partial}{\partial\tau}\delta P = -\left(\delta F\, D_{st} + \overline{F_{st}}\,\delta D + (1+i\Delta)\,\delta P\right) \tag{4.17b}$$

$$\frac{\partial}{\partial\tau}\delta D = -\gamma\left\{-\frac{1}{2}\left(\delta F^*\,\overline{P_{st}} + \overline{F_{st}^*}\,\delta P + c.c\right) + \delta D\right\} \tag{4.17c}$$

plus similar equations for $\delta f_p{}^*$ and δP^* obtained by complex conjugation of Eqs. (4.17a,b). The field fluctuation variables are subject to the boundary conditions

$$\delta f_p(-1/2,\tau) = R\, e^{-i\delta_p}\, e^{i\delta\Omega\gamma_\perp\frac{\Lambda-L}{c}}\, \delta f_p(1/2, \tau-\gamma_\perp\frac{\Lambda-L}{c}) \tag{4.18}$$

Next, we introduce the new independent variables

$$\eta' = \eta \tag{4.19a}$$

$$\tau' = \tau + \gamma_\perp\frac{\Lambda-L}{c}(\eta+1/2) \tag{4.19b}$$

whose objective is to make the boundary conditions isochronous and we seek solutions of the type

$$\begin{pmatrix} \delta f_p(\eta',\tau') \\ \delta f_p^*(\eta',\tau') \\ \delta P(\rho,\eta',\tau') \\ \delta P^*(\rho,\eta',\tau') \\ \delta D(\rho,\eta',\tau') \end{pmatrix} = e^{\lambda\tau'} \begin{pmatrix} \delta\varphi_p(\eta') \\ \delta\varphi_p^*(\eta') \\ \delta p(\rho,\eta') \\ \delta p^*(\rho,\eta') \\ \delta d(\rho,\eta') \end{pmatrix} \tag{4.20}$$

The atomic equations reduce to a set of three linear algebraic equations for δp, δp^*, and δd that can be solved easily with the result

$$\delta p = T_1\sum_{p'} A_{p'}\,\delta\varphi_{p'} + T_2\sum_{p'} A_{p'}^*\,\delta\varphi_{p'}^* \tag{4.21a}$$

$$\delta p^* = T_2^*\sum_{p'} A_{p'}\,\delta\varphi_{p'} + T_1^*\sum_{p'} A_{p'}^*\,\delta\varphi_{p'}^* \tag{4.21b}$$

where T_1 and T_2 depend on the steady state field and atomic variables as well as the eigenvalues λ. At this point the linearized field equations become

$$\frac{\partial}{\partial\eta'}\delta\varphi_p + \gamma_\perp \frac{\Lambda}{c}\lambda\,\delta\varphi_p = i\,\frac{\delta\Omega}{\gamma}\delta\varphi_p - \alpha L \int_0^\infty d\rho\,\rho\,A_p^*$$

$$\left\{ T_1 \sum_{p'} A_{p'}\delta\varphi_{p'} + T_2 \sum_{p'} A_{p'}^*\delta\varphi_{p'}^* \right\} \tag{4.22a}$$

$$\frac{\partial}{\partial\eta'}\delta\varphi_p^* + \gamma_\perp \frac{\Lambda}{c}\lambda\,\delta\varphi_p^* = -i\,\frac{\delta\Omega}{\gamma}\delta\varphi_p^* - \alpha L \int_0^\infty d\rho\,\rho\,A_p$$

$$\left\{ T_2^* \sum_{p'} A_{p'}\delta\varphi_{p'} + T_1^* \sum_{p'} A_{p'}^*\delta\varphi_{p'}^* \right\} \tag{4.22b}$$

In the uniform field limit and through a fairly lengthy algebraic process, we can show that Eqs. (4.22) reduce to the infinite set of linear equations

$$L\,\delta\vec{v} = \lambda\,\delta\vec{v} \tag{4.23}$$

where L is a matrix of the form $L = L^{(0)} + \kappa L^{(1)}$ and κ is of order T. In the limit as $T \to 0$ the required eigenvalues can be expressed as

$$\lambda = \lambda^{(0)} + \kappa\lambda^{(1)} \tag{4.24}$$

The result of this procedure for $p \neq 0$ is

$$\lambda^{(0)}_{p\neq 0} = -i\,\frac{c}{\Lambda\gamma_\perp}\delta_p \tag{4.25}$$

whence it follows that the real part of the first order correction $\lambda_p^{(1)}$ determines the stability of the p radial sideband. The explicit expression for Re $\lambda_p^{(1)}$ is fairly complicated but it can be obtained explicitly through simple calculations.

Some interesting observations can be made at once as a result of the steady state and linear stability calculation:

(i) the gain thresholds for the radial modes are not evenly spaced, so that under normal conditions (especially for resonant, or near resonant operation) only few radial modes (e.g. p=0,1, and 2) appear to play an important role in the dynamics of the system;

(ii) the instability threshold for the radial sidebands can be only slightly higher than the threshold for ordinary laser action (say 20-30% higher);

(iii) the instability is enhanced in the presence of a flat (or flatter) inversion profile; conversely, a more radially concentrated gain region appears to enhance the stability of the laser.

These preliminary results, taken together, give encouraging indication that the low instability thresholds observed experimentally may be related to the nonuniform transverse profile of the field and of the atomic variables and that control of these instabilities may reside in some appropriate tailoring of the geometrical and pump parameters of the system.

Acknowledgements

It is a special pleasure to dedicate this work to Hermann Haken, teacher and friend with gratitude for his encouragement and inspiration.

We are grateful to Neal B. Abraham, of Bryn Mawr College and Jon Davis and Michael Rankin of the Naval Air Development Center, Warminster, Pa., for allowing us to air our often incomplete and tentative results, for listening patiently and providing timely and useful suggestions. This work was partially supported by the US Army Research Office (Durham, North Carolina) and by the EEC twinning project on Dynamics of Nonlinear Optical Systems.

References

1. Several recent reviews have dealt with the subject of laser instabilities. They include :
   (a) A.N. Oraevskii, Kvant. Elekt. 8, 130 (1981);
   (b) N.B. Abraham, Laser Focus (May) 19, 73 (1983);
   (c) L.W. Casperson, in Laser Physics, edited by J.D. Harvey and D.F. Walls, Springer Lecture Notes in Physics, Vol. 182 (Springer-Verlag, Berlin, 1983);
   (d) J.C. Englund, R.R. Snapp, and W.C. Schieve, in Progress in Optics, vol XXI, edited by E. Wolf (North-Holland, 1984) p. 355;
   (e) N.B. Abraham, L.A. Lugiato, and L.M. Narducci, J. Opt. Soc. Am. B2, 7 (1985);
   (f) J.R. Ackerhalt, P.W. Milonni, and M.L. Shih, Phys. Rept. 128, 205 (1985);
   (g) L.W. Casperson, in Optical Instabilities, edited by R.W. Boyd, M.G. Raymer, and L.M. Narducci (Cambridge University Press, Cambridge, 1986) p. 58;
   (h) N.B. Abraham, in Optical Instabilities, edited by R.W. Boyd, M.G. Raymer, and L.M. Narducci (Cambridge University Press, Cambridge, 1986) p. 46;
   (i) F.T. Arecchi, in Optical Instabilities, edited by R.W. Boyd, M.G. Raymer, and L.M. Narducci (Cambridge University Press, Cambridge, 1986) p. 183;
   (j) L.A. Lugiato, and L.M. Narducci, in Optical Instabilities, edited by R.W. Boyd, M.G. Raymer, and L.M. Narducci (Cambridge University Press, Cambridge, 1986) p. 34;
   (k) R.G. Harrison and D.J. Biswas, Prog. Quantum Electron. 11, 127 (1986);
   (l) F.T. Arecchi, in Optical Chaos, edited by J. Chrostowski and N.B. Abraham, SPIE vol. 667 (SPIE, Bellingham, 1986) p. 139;
   (m) L.A. Lugiato, and L.M. Narducci, in Optical Chaos, edited by J. Chrostowski and N.B. Abraham, SPIE vol. 667 (SPIE, Bellingham, 1986) p. 132;
   (o) N.B. Abraham, P. Mandel, and L.M. Narducci, in Progress in Optics, edited by E. Wolf (to be published).
2. See, for example:
   (a) G. Makhov, C. Kikuchi, J. Lambe, and R.W. Terhune, Phys. Rev. 109, 1399 (1958);
   (b) C. Kikuchi, J. Lambe, G. Makhov, and R.W. Terhune, J. Appl. Phys. 30, 1061 (1959);
   (c) G. Makhov, L.G. Cross, R.W. Terhune, and J. Lambe, J. Appl. Phys. 31, 936 (1960).
3. (a) R.J. Collins, D.F. Nelson, A.L. Schawlow, W. Bond, C.G.B. Garrett, and W. Kaiser, Phys. Rev. Lett. 5, 303 (1960);
   (b) D.F. Nelson, and W.S. Boyle, Appl. Opt. 1, 181 (1962);
   (c) T.N. Zubarev, and A.K. Sokolov, Sov. Phys. Doklady, 9, 1006 (1965).

4. Some of the early maser studies dealing with unstable behaviors include:
(a) K.Y. Khaldre and R.V. Khokhlov, Izv. Vyss. Uchebn. Zaved. Radiofiz. 1, 60 (1958);
(b) A.G. Gurtovnick, Izv. Vyss. Uchebn. Zaved. Radiofiz. 1, 83 (1958);
(c) A.N. Oraevskii, Radio Eng. Electron. Phys. (USSR) 4, 718 (1959);
The first instability studies for homogeneously broadened laser systems appear to have been carried out by:
(d) A.V. Uspenskii, Radio Eng. Electron. Phys. (USSR) 8, 1145 (1963); ibid 9, 605 (1964);
(e) V.V. Korobkin and A.V. Uspenskii, Sov. Phys. JETP 18, 693 (1964);
(f) A.Z. Grazyuk and A.N. Oraevskii, in Quantum Electronics and Coherent Light, edited by P.A. Miles (Academic Press, New York, 1964); Radio Eng. Electron. Phys. (USSR) 9, 424 (1964);
(g) H. Haken, Z. Phys. 190, 327 (1966);
(h) H. Risken, C. Schmidt and W. Weidlich, Z. Phys. 194, 337 (1966).
5. H.M. Gibbs, S.L. McCall and T.N.C. Venkatesan, Phys. Rev. Lett. 36, 1135 (1976).
6. (a) R. Bonifacio and L.A. Lugiato, Opt. Comm. 19, 172 (1976);
(b) R. Bonifacio and L.A. Lugiato, Phys. Rev. A18P, 1129 (1978).
7. For an update of this rapidly expanding field see:
(a) Optical Instabilities, edited by R.W. Boyd, M.G. Raymer and L.M. Narducci (Cambridge University Press, Cambridge, 1986);
(b) Optical Chaos, edited by J. Chrostowski and N.B. Abraham, SPIE Vol. 667 (SPIE, Bellingham, 1986).
8. (a) H. Haken, Synergetics-An Introduction, (Springer-Verlag, Berlin, 1977);
(b) H. Haken, Advanced Synergetics, (Springer-Verlag, 1983).
9. See, for example, E. Hofelich-Abate and F. Hofelich, Z. Phys. 211, 142 (1968) and references therein.
10. (a) R. Graham, and H. Haken, Z. Phys. 213, 420 (1968);
(b) H. Risken, and K. Nummedal, J. Appl. Phys. 39, 4662 (1968);
(c) H. Risken and K. Nummedal, Phys. Lett. 26A, 275 (1968).
11. L.A. Lugiato, L.M. Narducci, E.V. Eschenazi, D.K. Bandy, and N.B. Abraham, Phys. Rev. A32, 1563 (1985).
12. H. Haken, Phys. Lett. 53A, 77 (1975).
13. E. Lorenz, J. Atm. Sci. 20, 130 (1963).
14. For an extensive discussion of the Lorenz chaos see C.T. Sparrow, The Lorenz Equations: Bifurcations, Chaos, and Strange Attractors (Springer-Verlag, Berlin, 1982).
15. L.M. Narducci, H. Sadiky, L.A. Lugiato, and N.B. Abraham, Opt. Comm. 55, 370 (1985).
16. (a) S.T. Hendow, and M. Sargent, III, Opt. Comm. 43, 59 (1982);
(b) S.T. Hendow, and M. Sargent, III, J. Opt. Soc. Am. B2, 84 (1985).
17. L.M. Narducci, J.R. Tredicce, L.A. Lugiato, N.B. Abraham, and D.K. Bandy, Phys. Rev. A33, 1842 (1986).
18. J.R. Tredicce, L.M. Narducci, D.K. Bandy, L.A. Lugiato, and N.B. Abraham, Opt. Comm. 56, 435 (1986).
19. (a) L.W. Casperson, IEEE J. Quantum Electron. QE-14, 756 (1978);
(b) L.W. Casperson, Phys. Rev. A21, 911 (1980);
(c) L.W. Casperson, Phys. Rev. A23, 248 (1981);
(d) L.W. Casperson, J. Opt. Soc. Am. B2, 62 (1985);
(e) L.W. Casperson, J. Opt. Soc. Am. B2, 73 (1985);
(f) L.W. Casperson, J. Opt. Soc. Am. B2, 993 (1985).
20. (a) J. Bentley, and N.B. Abraham, Opt. Comm. 41, 52 (1982);

(b) M. Maeda, and N.B. Abraham, Phys. Rev. A26, 3395 (1982);
(c) N.B. Abraham, T. Chyba, M. Coleman, R.S. Gioggia, N. Halas, L.M. Hoffer, S.N. Liu, M. Maeda, and J.C. Wesson, in Laser Physics, edited by J.D. Harvey and D.F. Walls, Springer Lecture Notes in Physics, Vol. 182 (Springer-Verlag, Berlin, 1983) p. 107.
(d) R.S. Gioggia, and N.B. Abraham, Phys. Rev. Lett. 51, 650 (1983);
(e) R.S Gioggia, and N.B. Abraham, Opt. Comm. 47, 278 (1983);
(f) R.S. Gioggia, and N.B. Abraham, Phys. Rev. A29, 1304 (1984).
21. G.L. Lippi, N.B. Abraham, J.R. Tredicce, L.M. Narducci, G.P. Puccioni, and F.T. Arecchi, in Optical Chaos, edited by J. Chrostowski and N.B. Abraham, SPIE vol. 667 (SPIE, Bellingham, 1986) p. 41;
22. L.A. Lugiato, L.M. Narducci and M.F. Squicciarini, Phys. Rev. A34, 3101 (1986).
23. L.A. Lugiato, Theory of Optical Bistability, in Progress in Optics, edited by E. Wolf, (North-Holland, Amsterdam, 1984).
24. (a) D.E. Grant, and H.J. Kimble, Opt. Lett. 7, 353 (1982);
(b) D.E. Grant, and H.J. Kimble, Opt. Comm. 44, 415 (1983).
25. L.A. Lugiato, L.M. Narducci, D.K. Bandy, and C.A. Pennise, Opt. Comm. 43, 281 (1982).
26. (a) A.T. Rosenberger, L.A. Orozco and H.J. Kimble, Phys. Rev. A28, 2569 (1963);
(b) A.T. Rosenberger, L.A. Orozco, and H.J. Kimble, in Fluctuations and Sensitivity in Nonequilibrium Systems, edited by W. Horsthemke and D.K. Kondepudi (Springer, Berlin, 1984);
(c) L.A. Orozco, A.T. Rosenberger and H.J. Kimble, Phys. Rev. Lett. 53, 2547 (1984);
(d) L.A. Orozco, M.G. Raizen, A.T. Rosenberger, and H.J. Kimble, in Optical Bistability III, edited by H.M. Gibbs, P. Mandel, N. Peyghambarian, and S.D. Smith (Springer, Berlin, 1986), p. 327.
27. L.A. Lugiato, R.J. Horowicz, G. Strini, and L.M. Narducci, Phys. Rev. A30, 1366 (1984), and L.A. Lugiato, M.L. Asquini (unpublished).
28. (a) M. LeBerre, E. Ressayre, A. Tallet, K. Tai, H.M. Gibbs, M.C. Rushford, N. Peyghambarian, J. Opt. Soc. Am. B1, 591 (1984);
(b) J.F. Valley, H.M. Gibbs, M.W. Derstine, R. Pon, K. Tay, M. LeBerre, E. Ressayre, A. Tallet, in Optical Bistability III, edited by H.M. Gibbs, P. Mandel, N. Peyghambarian, and S.D. Smith (Springer-Verlag, 1986) p. 327;
(c) M.W. Derstine, H.M. Gibbs, F.A. Hops, J.F. Valley, in Optical Bistability III, edited by H.M. Gibbs, P. Mandel, N. Peyghambarian, and S.D. Smith (Springer-Verlag, 1986) p. 342.
29. (a) D.W. McLaughlin, J.V. Moloney and A.C. Newell, Phys. Rev. Lett. 54, 681 (1985);
(b) J.V. Moloney, M.R. Belic, H.M. Gibbs, Opt. Comm. 41, 379 (1982);
(c) J.V. Moloney, Phys. Rev. Lett. 53, 556 (1984).
30. (a) H. Haken and H. Sauermann, Z. Phys. 173, 261 (1963);
(b) H. Haken and H. Sauermann, Z. Phys. 176, 47 (1963);
(c) W.E. Lamb, Jr., Phys. Rev. 134, 1429 (1964);
(d) M. Lax, in Statistical Physics, Phase Transitions and Superfluidity, Vol II, Brandeis University Summer Institute in Theoretical Physics, edited by M. Chretien, E.P. Gross, and S. Deser (Gordon and Breach, New York, 1967) p. 269.
(e) H. Haken, Laser Physics, in Handbuch der Physik, edited by L. Genzel, Vol XXV/2c (Springer-Verlag, Berlin, 1970).
31. R. Bonifacio and L.A. Lugiato, Lett. Nuovo Cimento 21, 505 (1978).
32. L.A. Lugiato, L.M. Narducci and R. Lefever, this volume.
33. (a) H. Haken and H. Ohno, Opt. Comm. 16, 205 (1976);
(b) H. Haken and H. Ohno, Opt. Comm. 26, 117 (1978);
(c) H. Ohno and H. Haken, Phys. Lett. 59A, 261 (1976).

34. (a) L.A. Lugiato and L.M. Narducci, Phys. Rev. A32, 1576 (1985);
(b) L.A. Lugiato, M.L. Asquini and L.M. Narducci, in Optical Chaos, edited by J. Chrostowski and N.B. Abraham, SPIE Vol. 667 (SPIE, Bellingham, 1986), p. 132;
(c) L.A. Lugiato, D.K. Bandy, J.R. Tredicce and L.M. Narducci, in Advances in Laser Science I, edited by W.C. Stwalley and M. Lapp, AIP Conference Proceedings 146 (1986).
35. (a) L.A. Lugiato, Opt. Comm. 33, 108 (1980);
(b) L.A. Lugiato, Lett. Nuovo Cimento 29, 375 (1980).
36. L.W. Hillman, J. Krasinski, R.W. Boyd, and C.R. Stroud, Jr., Phys. Rev. Lett. 52, 1605 (1984).
37. (a) L.W. Hillman, J. Krasinski, K. Koch, and C.R. Stroud, Jr., J. Opt. Soc. Am. B2, 211 (1985);
(b) C.R. Stroud, Jr., K. Koch, and S. Chakmakjian, in Optical Instabilities, edited by R.W. Boyd, M.G. Raymer, and L.M. Narducci (Cambridge University Press, Cambridge, 1986) p. 274;
(c) C.R. Stroud, Jr., K. Koch, S. Chakmakjian and L.W. Hillman, in Optical Chaos, edited by J. Chrostowski and N.B. Abraham, SPIE vol. 667 (SPIE, Bellingham) p. 47.
38. N.M. Lawandy, R.S. Afzal, and W.S. Rabinovich (private communication).
39. (a) C.O. Weiss, and W. Klische, Opt. Comm. 50, 413 (1984);
(b) C.O. Weiss, and W. Klische, Opt. Comm. 51, 47 (1984);
(c) C.O. Weiss, J. Opt. Soc. Am. B2, 137 (1985);
(d) C.O. Weiss, W. Klische, P.S. Ering, and M. Cooper, Opt. Comm. 52, 405 (1985);
(e) W. Klische, and C.O. Weiss, Phys. Rev. A31, 4049 (1985);
(f) E. Hogenboom, W. Klische, C.O. Weiss, A. Godone, Phys. Rev. Lett. 55, 2571 (1985)
(g) C.O. Weiss, in Optical Chaos, edited by J. Chrostowski and N.B. Abraham, SPIE vol. 667 (SPIE, Bellingham, 1986) p. 26.
(h) C.O. Weiss, F. Spiezeck, H.R. Telle, and H. Li, Opt. Comm. 58, 193 (1986).
40. (a) N.M. Lawandy, and G.A. Knoepf, IEEE J. Quantum Electron. QE-16, 701 (1980);
(b) N.M. Lawandy, IEEE J. Quantum Electron. QE-18, 1992 (1982);
(c) N.M. Lawandy, J. Opt. Soc. Am. B2, 108 (1985).
41. M.A. Dupertuis, R.R.E. Salomaa and M.R. Siegrist, Opt. Comm. 57, 410 (1986). Chrostowski and N.B. Abraham, SPIE vol. 667 (SPIE, Bellingham, 1986) p. 41.
42. In discussing the various modal configurations that are accessible to an ordinary Fabry-Perot laser many textbooks emphasize that it is much more difficult to sustain oscillations in the axially-symmetric Gauss-Laguerre modes than in their non-symmetric Gauss-Hermite counterparts. See, for example A.E. Siegman, An Introduction to Lasers and Masers, (McGraw-Hill, New York, 1971).

Instabilities, Spatial and Temporal Patterns in Passive Optical Systems

L.A. Lugiato[1], *L.M. Narducci*[2], *and R. Lefever*[3]

[1]Dipartimento di Fisica del Politecnico di Torino, Italy
[2]Physics Department, Drexel University, Philadelphia, PA 19104, USA
[3]Service de Chimie-Physique II, Université Libre Bruxelles, Belgium

1. Introduction

Optical Bistability (OB), one of the fastest growing areas of quantum optics, began its explosive rise about ten years ago when the first experimental observation with sodium vapor [1] made this effect a strong potential candidate for the practical realization of optical transistors, logic units and, perhaps, even optical computers at some future time. It did not take long for the theoretical community to realize that optically bistable devices are marvellous examples of nonlinear systems, capable of a wide variety of static and dynamic effects. In fact, this aspect of optical bistability was one of the main motivating factors behind the extensive theoretical activity of the last ten years. At this point it is reasonable to say that, after the laser, optical bistability is the most investigated topic in quantum optics.

The development of a theory of bistability was aided by the pioneering efforts and outstanding advances in laser physics produced by Hermann Haken and his Stuttgart School [2,3]. Haken's influence on this new area of research was not accidental because lasers and bistable systems share a number of theoretical links, in addition to their common requirement for a suitable nonlinear coupling between radiation and matter inside an optical cavity. Of course there are also some important differences, as well; in the case of the laser, the medium is pumped into a state of inversion; a bistable system, instead, is not inverted and acts as a regular absorber. This essential difference can be emphasized by classifying lasers as active systems and optically bistable devices are their passive counterparts.

The foundation of Synergetics by Haken [4,5] provided both the conceptual and technical background for the formulation and development of the theory of optical bistability. The concept of cooperation, for example, plays a central role in a wide spectrum of optical phenomena emerging from lasers and bistable systems, and of course it is one of the cornerstones of Synergetics. In this article we wish to review and emphasize those aspects of lasers and bistable systems which are most directly manifestations of atomic cooperation.

The emission of laser light is the consequence of an instability. When the pump parameter exceeds a suitable threshold value, the nonlasing state becomes unstable and laser emission begins. This instability is of the soft-mode type, as it arises when a real eigenvalue of the linear stability analysis becomes equal to zero. The behavior of the laser in the threshold region is analogous to a second order phase transition in equilibrium systems [6,7].

Optical bistability is related similarly to a soft-mode instability. When the bistability parameter is raised beyond a certain critical value, the steady state curve that governs the behavior of the transmitted intensity as a function of the incident intensity acquires an S-shaped form where the segment with a negative slope is unstable. The behavior of a bistable system under these conditions is analogous qualitatively to a first-order phase transition in equilibrium systems [8,9].

In the laser, when the pump parameter is raised several times above threshold, the stationary lasing state becomes unstable (this is often called the second laser threshold) and the output breaks into spontaneous oscillations [10-13]. The behavior of the output intensity is typically chaotic, and the reason for this fact is rooted in the equivalence between the single-mode laser and the Lorenz model of hydrodynamics [15]. This simple and profound result, established by Haken for the first time in 1975 [14], endowed the laser at once with the same chaotic features which had been known for some time in hydrodynamics systems. The instability that is responsible for the second laser threshold is of the hard-mode type, i.e. it originates from the vanishing of the real parts of a pair of complex conjugate eigenvalues of the linear stability analysis.

Also in the case of Optical Bistability, a suitable increase in the bistability parameter and appropriate variations of the detuning and mistuning parameters can cause a hard-mode instability and the emergence of spontaneous oscillations [16-20]. The single-mode version of this instability [19,20] has been the subject of accurate experimental observations [21] that will be reviewed in our discussion.

The hard-mode instabilities mentioned in the preceding paragraphs correspond to the emergence of <u>temporal</u> structures because the destabilized stationary state undergoes an oscillatory behavior with the subsequent break of time translational invariance. For the case of passive systems we have discovered [22] a soft-mode instability that is capable of generating a stationary <u>spatial</u> structure. Phenomena of this type are known to occur, for example, in appropriate nonlinear chemical reactions and developmental biology [23,4]; our prediction is, possibly, the first of its kind in optics. We must point out, however, that the spatial patterns discussed in References [23,4] arise from the interplay between the nonlinearities inherent in the chemical reactions and spatial diffusion; in our case this phenomenon is induced by nonlinear dispersion effects and diffraction.

2. The single-mode model - The stationary state picture

We consider a Fabry-Perot cavity (Fig. 1) containing a homogeneously broadened system of two-level atoms. In the case of the laser the medium is pumped to a state of population inversion by an external mechanism. In the case of optical bistability no pump action is required; instead, a coherent field E_I is injected into the cavity. In our discussion we assume that both the input and the internal cavity fields can be represented by plane-waves, so that their slowly varying envelopes are only functions of the longitudinal coordinate z.

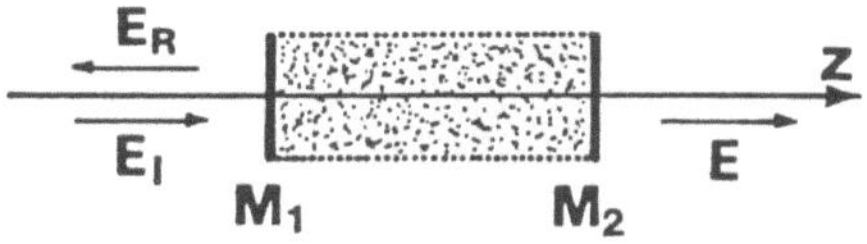

Fig.1. Schematic representation of a Fabry-Perot cavity filled with a nonlinear medium. The mirrors M_1 and M_2 have transmittivity coefficient T. E_I, E and E_R are the input, output and reflected fields, respectively. In the case of the laser the input field E_I vanishes.

We describe the dynamics of this system by means of the so-called single-mode model, formulated for the laser in the early 1960's by the Stuttgart school [2,3] and later generalized to the case of optical bistability [8,9]. This model holds in the uniform field limit [24] (also called mean field limit in the literature)

$$\alpha L \ll 1, \quad T \ll 1, \quad \text{with } \frac{\alpha L}{T} \text{ arbitrary} \tag{1}$$

where α is the absorption coefficient of the material, L is the length of the atomic sample and T is the transmission coefficient of the mirrors. The internal electric field has the single-mode standing-wave structure

$$e(z,t) \propto E(t)\ \cos(k_z z)\ \exp(-i\omega_0 t) + \text{c.c.}$$
$$k_z = \pi n_z / L \tag{2}$$

where n_z is the integer that labels the cavity mode, E(t) is the slowly varying envelope and ω_0 is a reference frequency. In the case of optical bistability it is convenient to select the reference frequency as the carrier of the injected signal, while in dealing with the laser ω_0 is the oscillation frequency in the stationary regime. Hence in this case, the parameter ω_0 is not fixed a priori but must be calculated from the model.

Strictly speaking, the model considered here is designed to hold for a ring cavity, but under appropriate conditions it can also be used for a Fabry-Perot configuration [25]. The equations of motion are

$$\frac{dE}{dt} = -\kappa\left[E(1+i\theta) - E_I + 2CP\right] \tag{3.1}$$

$$\frac{dP}{dt} = -\gamma_\perp\left[-ED + P(1+i\Delta)\right] \tag{3.2}$$

$$\frac{dD}{dt} = -\gamma_\parallel\left[\frac{1}{2}(E^*P + EP^*) + D - 1\right] \tag{3.3}$$

where E is the slowly varying envelope of the electric field normalized to the square root of the saturation intensity, P is the normalized atomic polarization, and D the normalized difference between the populations of the lower and the upper levels. The variables E^* and P^* obey the complex conjugate equations of (3.1) and (3.2), respectively. The scaled amplitude E_I of the input field is set equal to zero in the case of the laser. The parameter C is defined as

$$C = -\frac{\alpha L\sigma}{2T} \tag{4}$$

where σ is the population inversion per atom. Hence we select $\sigma = -1$ for the case of OB and refer to C as the bistability parameter; for the laser we select $\sigma = 1$ and call $|C|$ the pump parameter. The cavity damping constant κ, or cavity linewidth, is defined as

$$\kappa = \frac{cT}{2L} \tag{5}$$

$\gamma_\perp = T_2^{-1}$ and $\gamma_\parallel = T_1^{-1}$ are the atomic relaxation rates. The cavity and the atomic detuning parameters are defined as

$$\theta = \frac{\omega_C - \omega_0}{\kappa}, \qquad \Delta = \frac{\omega_A - \omega_0}{\gamma_\perp} \tag{6}$$

respectively, where ω_C is the frequency of the cavity mode and ω_A is the atomic transition frequency.

The last term in Eq. (3.1) and the nonlinear terms in Eqs. (3.2), (3.2) describe the coherent interaction between the electric field and the atoms and are the sources of the atomic cooperation. Equation (3.1) originates from Maxwell's wave equation and describes how the cavity mode is generated by the input field, how it is absorbed or amplified by the atomic system (term containing C), and damped by the finite reflectivity of the mirrors (term containing κ).

Equations (3.2), (3.3) are the optical analogue of the Bloch equations for nuclear magnetic resonance. The nonlinear terms -ED and $1/2(E^*P+EP^*)$ describe how the atomic system is driven by the electric field. The other terms have a dissipative character and take into account the effects of spontaneous emission and of collisions between the atoms. In particular, the relaxation rate $\gamma_\perp$ coincides with the atomic linewidth.

The stationary states follow at once by setting all the time derivatives equal to zero. In the case of the laser ($E_I=0$, $\sigma=1$) one obtains two different kinds of solutions [2,3]. The first is

$$E = P = 0, \quad D = 1$$

which corresponds to the nonlasing state. The second stationary solution is characterized by the well-known mode-pulling formula for the operating frequency ω_0

$$\omega_0 = \frac{\kappa\omega_A + \gamma_\perp\omega_C}{\kappa+\gamma_\perp} \qquad (7)$$

which implies $\theta = -\Delta$. Furthermore, the steady state intensity $|E|^2$ is given by

$$|E|^2 = 2|C| - 1 - \Delta^2 \qquad (8)$$

while the phase of E remains arbitrary. The trivial stationary solution is stable for all values of the pump parameter |C| smaller than the threshold value $(1+\Delta^2)/2$. At threshold the lasing state, characterized by Eqs. (7) and (8), bifurcates from the trivial solution and laser emission starts (See Fig. 2).

In the case of optical bistability ($E_I \neq 0$, $\sigma = -1$) we take the input field E_I real and positive for definiteness. At steady state, the frequency of the output field coincides with that of the input field, and the stationary solutions are characterized by the following relation between the input and output intensities E_I^2 and $|E|^2$ [26-28]

$$E_I^2 = |E|^2 \left\{ \left(1+\frac{2C}{1+\Delta^2+|E|^2}\right)^2 + \left(\theta - \frac{2C\Delta}{1+\Delta^2+|E|^2}\right)^2 \right\} \qquad (9)$$

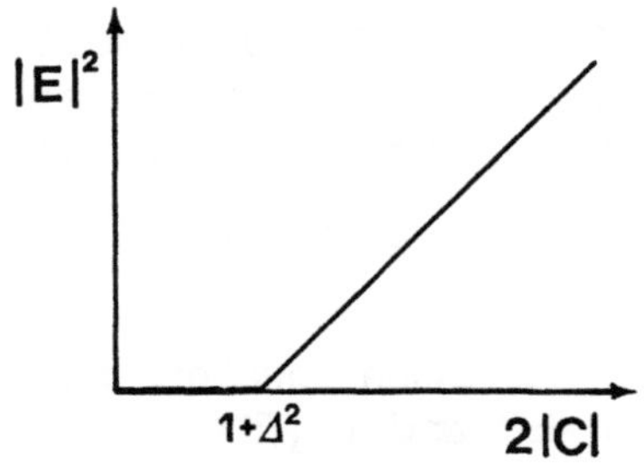

Fig. 2. The steady state output intensity of the laser is plotted as a function of the pump parameter $|C|$.

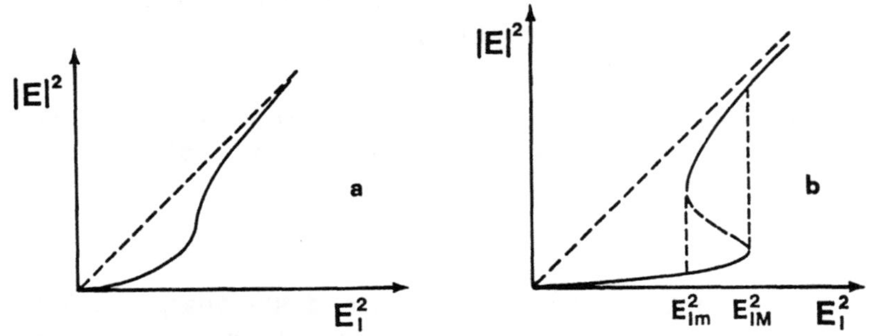

Fig. 3. Optical bistability. The steady state output intensity is plotted as a function of the input intensity for (a) $C < C_{min}$, and (b) $C > C_{min}$.

For each value of E_I^2 Eq.(9) yields a cubic equation for $|E|^2$. When the bistability parameter is smaller than a certain threshold value C_{min} (see, for example, Ref. [27]) that depends on the detuning parameters Δ and θ, the steady state dependence of the output intensity $|E|^2$ on the input intensity is single-valued (Fig. 3a). When, instead, $C > C_{min}$ the steady state curve becomes S-shaped and the negative-slope portion is unstable. It follows that the system is bistable over the interval $E_{Im}^2 < E_I^2 < E_{IM}^2$ (Fig. 3b). If we increase the incident intensity slowly from zero up to and beyond E_{IM}^2, and then retrace our path backward, the system scans a hysteresis loop with two distinct states of transmission. This phenomenon is called optical bistability and has become the focus of attention in view of its possible applications for logic systems [25,29].

The two boundaries, E_{Im}^2 and E_{IM}^2, of the hysteresis cycle are points of marginal stability. Here, the instability is of the saddle-point type and it corresponds to the coalescence of a stable (positive slope) and an unstable (negative slope) branch of the steady state curve.

3. Purely absorptive and purely dispersive bistability

Equation (9) contains two distinct contributions: the first $2C/(1+\Delta^2+|E|^2)$ arises from the absorptive component of the atomic susceptibility, the second, $2C\Delta/(1+\Delta^2+|E|^2)$, from the dispersive part. This is connected with the fact that bistability can be induced by the two very different mechanisms of nonlinear absorption and nonlinear dispersion [25,27,28].

In fact, under resonance conditions ($\Delta=\theta=0$), Eq. (9) yields purely absorptive bistability which is governed by the state equation [8]

$$E_I^2 = E^2\left(1+\frac{2C}{1+E^2}\right)^2 \qquad (10)$$

with a value of C_{min} equal to 4. If we consider, instead, values of the parameters such that (a) the nonlinear absorption term can be neglected, and (b) that the approximation

$$\frac{1}{1+\Delta^2+|E|^2} = \frac{1}{1+\Delta^2} - \frac{|E|^2}{(1+\Delta^2)^2}$$

holds, the state equation (9) reduces to the approximate form

$$E_I^2 = |E|^2\left\{1+\left(\theta' + \frac{2C\Delta}{(1+\Delta^2)^2}|E|^2\right)^2\right\} \qquad (11)$$

where

$$\theta' = \theta - \frac{2C\Delta}{1+\Delta^2} \qquad (12)$$

Equation (11) describes purely dispersive (or refractive) bistability and was proposed for the first time in Reference [1].

As a preliminary to our discussion of Sections 5 and 6, we derive from Eqs. (3) a model that holds under conditions of purely dispersive bistability and which is able to describe not only the steady-state behavior but also the dynamical evolution. This model requires the validity of the so-called "good cavity" condition

$$\kappa \ll \gamma_\perp, \gamma_\parallel \qquad (13)$$

which can be imposed with the help of Haken's adiabatic elimination principle [4]. Technically, this amounts to setting the time derivatives of the atomic variables in Eqs. (3.2) and (3.2) equal to zero, and results in the following approximate relation between the polarization, the population difference, and the cavity field

$$P = \frac{ED}{1+i\Delta} \qquad (14)$$

After substitution of Eq. (14) into Eqs. (3.1) and (3.3) we obtain

$$\kappa^{-1}\frac{dE}{dt} = -E\,(1+i\theta) + E_I - \frac{2C}{1+i\Delta}ED \tag{15.1}$$

$$0 = -\frac{|E|^2 D}{1+\Delta^2} - D + 1 \tag{15.2}$$

In the case of purely dispersive bistability almost all the atoms are in the lower level [27], so that, to a good approximation, we can set D=1-n, where the quantity n<<1 is proportional to the intensity-dependent part of the refractive index of the medium. After neglecting the absorption terms in Eq. (15.1) and setting D=1 in the first term of Eq. (15.2), our set of equations takes the form

$$\kappa^{-1}\frac{dE}{dt} = -E\,(1+i\theta') + E_I - i\frac{2C\Delta}{1+\Delta^2}En \tag{16.1}$$

$$0 = \frac{-|E|^2}{1+\Delta^2} + n \tag{16.2}$$

where θ' is defined by Eq. (12). After elimination of the variable n, we obtain the self-contained equation for the field variable E

$$\kappa^{-1}\frac{dE}{dt} = -E + E_I - \left(\theta' + \frac{2C\Delta}{1+\Delta^2}|E|^2\right)E \tag{17}$$

Clearly, at steady state, Eq. (17) reduces to Eq. (11). In order to minimize the number of parameters involved in the calculation, it is convenient to rescale the variables as follows

$$\bar{E} = \left(\frac{2C|\Delta|}{1+\Delta^2}\right)^{1/2} E\,, \qquad \bar{E}_I = \left(\frac{2C|\Delta|}{1+\Delta^2}\right)^{1/2} E_I$$

$$\bar{\theta} = \eta\,\theta' \tag{18}$$

where $\eta = +1$ (-1) for $\Delta < 0$ ($\Delta > 0$). In terms of the new variables the equation for the electric field becomes

$$\kappa^{-1}\frac{d\bar{E}}{dt} = -\bar{E} + \bar{E}_I + i\eta\,(|\bar{E}|^2 - \bar{\theta})\,\bar{E} \tag{19}$$

In the following development we are going to omit the overbar for notational simplicity; however, E, E_I, and θ from now on will denote the scaled variables. The steady state equation has the form

$$E_I^2 = |E|^2\left\{1 + (|E|^2 - \theta)^2\right\} \tag{20}$$

regardless of η. For $\theta < \sqrt{3}$ the steady state curve for $|E|^2$ as a function E_I^2 is single-valued, while for $\theta > \sqrt{3}$ the curve becomes multivalued leading to bistable behavior (Fig. 4). The model based on Eq. (19) holds also for a Kerr medium [27].

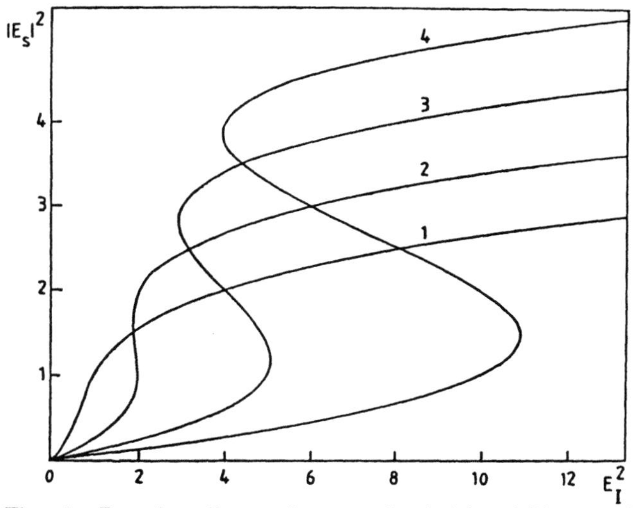

Fig. 4. Purely dispersive optical bistability. Steady state curves of the transmitted versus incident intensity for $\theta = 1,2,3,4$.

4. Self-oscillatory behavior

Equations (3.1), (3.2), (3.3), plus the complex conjugates of (3.1) and (3.2), form a set of five nonlinear dynamical equations. If we linearize this system around a stationary state, and seek elementary exponential solutions for the linearized equations, the result is a fifth-order algebraic eigenvalue equation that governs the stability of the system in the neighborhood of the selected steady state. To be more precise, the stationary state is unstable when at least one of the roots of the eigenvalue equation has a positive real part.

Focusing our attention on the laser case ($E_I > 0$, $\sigma = 1$, $\Delta = -\theta$), we consider the resonant configuration $\Delta = 0$ in which the eigenvalue equation splits into two factors, a cubic and a quadratic polynomial. On examining the cubic equation, one finds an instability if the following two conditions are met [10,11]

$$\kappa > \gamma_\perp + \gamma_{\parallel} \tag{21}$$

$$2|C| \geq 1 + \frac{(\kappa+\gamma_\perp)(\kappa+\gamma_\perp+\gamma_{\parallel})}{\gamma_\perp(\kappa-\gamma_\perp-\gamma_{\parallel})} \tag{22}$$

The inequality (21) is usually called the "bad cavity" condition, just the opposite of Eq. (13); the condition expressed by Eq. (22) prescribes that the pump parameter must be several times larger than its threshold value in order to obtain unstable behavior. This equation identifies the second threshold of the laser where the output changes from a cw to an oscillatory character.

In the case of optical bistability by appropriate variation of the external control parameters, one can force a segment of the positive slope part of the stationary curve to become unstable; in correspondence to the unstable portion, the transmitted intensity exhibits oscillatory behavior. This possibility was first advanced by Bonifacio and Lugiato [16] and by McCall [30] in 1978. One year later, Ikeda [18] showed that under appropriate conditions the spontaneous oscillations emerging from the instability are not periodic, but chaotic.

In optical bistability one recognizes several types of instabilities that lead to spontaneous oscillations; most of them are associated with the multimode regime of operation. Here we discuss the instabilities that are connected to the single-mode model (3) [19,20]. When the incident field, the atoms and the cavity are exactly in resonance ($\Delta = \theta = 0$), no positive-slope instability exists. For increasing values of the mismatch between the carrier frequency of the incident field and the cavity frequency, the system may no longer be able to follow the driving action of the external signal, and the stationary state becomes destabilized.

This instability can arise both in the presence (S-shaped stationary curve) or in the absence of bistability (single-valued stationary curve). A characteristic signature of this instability is that it requires atomic and cavity detunings with opposite signs, i.e. that $\Delta\theta < 0$. When $C < 150$, the spontaneous oscillations that result from this instability are regular with a frequency of the order of the cavity damping constant, κ, and a nearly sinusoidal shape (Fig. 5). Even if these oscillations are very simple in structure, they are still quite interesting because they suggest that a bistable system may function as a transformer of cw into pulsed radiation, while acting, at the same time, as an optical clock.

The experimental observation of this self-oscillatory behavior has been achieved by Orozco, Rosenberger and Kimble [21]. In order to fulfil the condition of homogeneous broadening and reach simultaneously the values of the bistability parameter that are required to observe the instability ($C > 50$) these researchers used ten parallel atomic beams of sodium at right angle with the axis of the cavity. Figure 6 shows an example of this experimental observation. The incident power P_I is slowly increased; when one arrives to the right boundary of the hysteresis cycle the system jumps, as expected, to the upper branch. Here, however, the transmitted intensity is no longer stationary, but it oscillates. In fact, the fuzzy segment is a

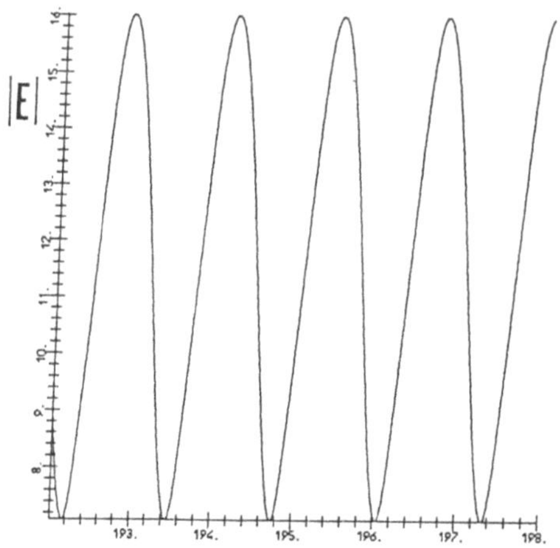

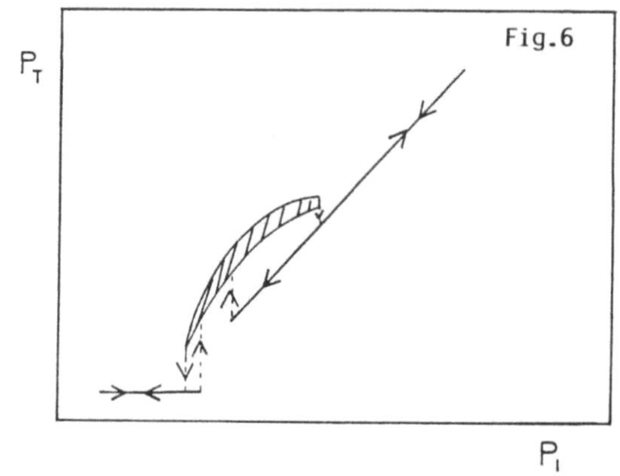

Fig. 5. Self-oscillations in optical bistability. The modulus |E| of the normalized electric field is plotted as a function of $\gamma_{\perp}t$ ($C = 75$, $\Delta = 1$, $\theta = -9$, $\kappa/\gamma_{\perp} = 0.5$, $\gamma_{\|}/\gamma_{\perp} = 2$).

Fig. 6. Behavior of the self-pulsing instability observed with atomic sodium beams by Kimble and collaborators [21]. The drawing of the hysteresis cycle shows the pulsing instability as an arch separated from the otherwise straight-line segment of the upper branch ($C= 155$, $\Delta = -0.7$, $\theta = 13$, $\kappa/\gamma_{\perp} = 0.4$, $\gamma_{\|}/\gamma_{\perp} = 1.76$).

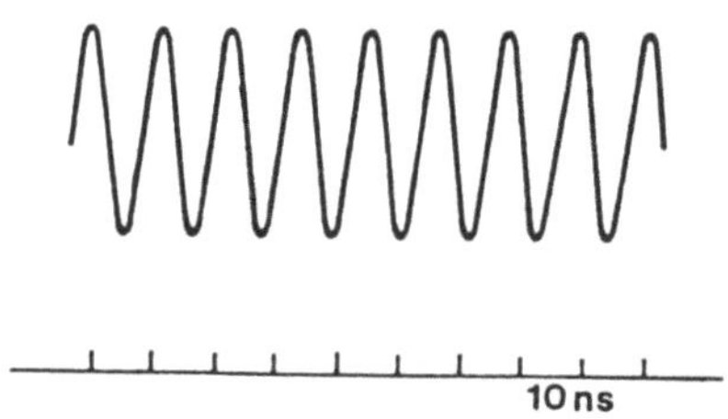

Fig. 7. Self-oscillations observed in the experiment of reference [21].

signature of oscillations that are only partially resolved because of the insufficient bandwidth of the detecting system. The fully resolved oscillations are shown in Fig. 7.

These experimental observations are in good qualitative and semi-quantitative agreement with the theory, and, as expected, the observed pulsations are nearly sinusoidal. We remark however that, even if the oscillations have such a simple structure, this experiment is still very significant because it provides the first observation of a Lorenz-type instability in a homogeneously broadened system. In fact, this instability emerges from equations [Eqs. (3)] that generalize the Lorenz-Haken model to the case of optical bistability.

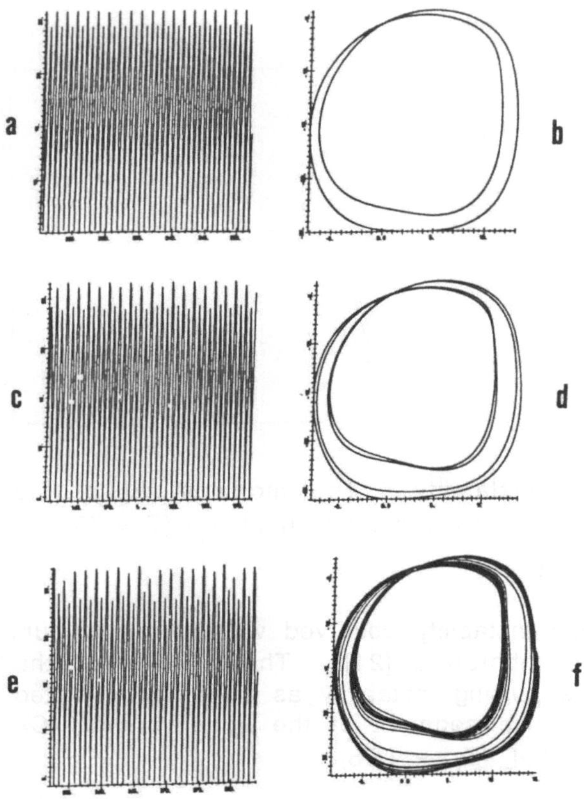

Fig. 8. Self-oscillations in optical bistability from the numerical solution of Eqs. (3) for C = 400, $\Delta = -1$, $\theta = 20$, $\kappa/\gamma_\perp = 0.4$, $\gamma_\parallel/\gamma_\perp = 1.76$. Figures (a), (c), (e) display |E| as a function of $\gamma_\perp t$; figures (b), (d), (f) show the projection of the phase-space trajectory in the plane (Re E, Im E); (a) period 2, $E_I = 309$, (b) period 4, $E_I = 303$, (c) chaos, $E_I = 297.3$.

The theory governed by Eqs. (3) predicts that, if the bistability parameter is increased above a value of about 300, the system can exhibit period-doubling and chaos (Fig. 8). The experiments, so far, have not revealed temporal structures of this type even if the value of C is well within the required range. The apparent discrepancy can be traced to the plane-wave approximation on which the model (3) is based. A real cavity field, on the other hand, has significant radial variations of intensity and phase which are typically of the Gaussian type.

To account for this deviation from the plane-wave approximation we have included a transverse Gaussian variation of the electric field into the model [20]. The numerical solutions of the new equations show that the

Gaussian average washes out higher order temporal structures, such as period-doubling and chaos, and favors only the simpler solutions of period-1 type. This analysis is in much better agreement with the experiments and shows that,in order to observe period doubling and chaos, it is necessary to assemble a setting that fulfils the uniform-field condition in the transverse direction.

5. The effect of diffraction

The models discussed in the previous sections are based on the plane-wave approximation, and exclude the possibility that diffraction may play any role from the very beginning. We now consider again Eq. (19), obtained in the purely dispersive limit, and include the effects of diffraction by completing the equation as follows

$$\frac{\partial}{\partial t} E(x,y,t) = -\kappa\{ E(x,y,t) - E_I - i\eta E(x,y,t)\left(|E(x,y,t)|^2 - \theta\right)\}$$
$$+ i\frac{c}{2k_z}\left(\frac{\partial^2}{\partial x^2} + \frac{\partial^2}{\partial y^2}\right)E(x,y,t) \qquad (23)$$

where x and y are the transverse coordinates and where we allow, in general for a dependence of E on x and y. We note that for $\eta = 1$ (i.e. $\Delta < 0$) the nonlinearity is of the self-focusing type, while for $\eta = -1$ (i.e. $\Delta > 0$) it has a self-defocusing character.

Because we assume that the input field E_I is independent of the transverse coordinates, Eq. (23) has stationary solutions that are homogeneous along the x and y directions and which coincide, in fact, with those calculated in Sect. 3 using Eq. (19) as a starting point [See Eq. (20)]. In general, when E_I is a plane wave, it is common to assume that also the internal field also is independent of x and y, so that Eq. (23) reduces to Eq. (19). This assumption, however, is not always correct, as one can verify from a study of the stability properties of the homogeneous stationary solution. The trouble with Eq. (19) is that it allows us to study the stability properties of this system only in the presence of fluctuations which are uniform along the transverse directions. Random perturbations, on the other hand, need not be uniform at all, so that a more general linear stability analysis is needed using Eq. (23) as the starting point. The importance of diffraction in connection with the stability properties of plane-wave stationary solutions was already emphasized in Ref. [31].

The presence of second space derivatives in Eq. (23) requires that we specify additional boundary conditions for our problem. We consider a cavity defined by four mirrors: two are orthogonal to the longitudinal axis, are separated by a distance L from one another, and have a transmission

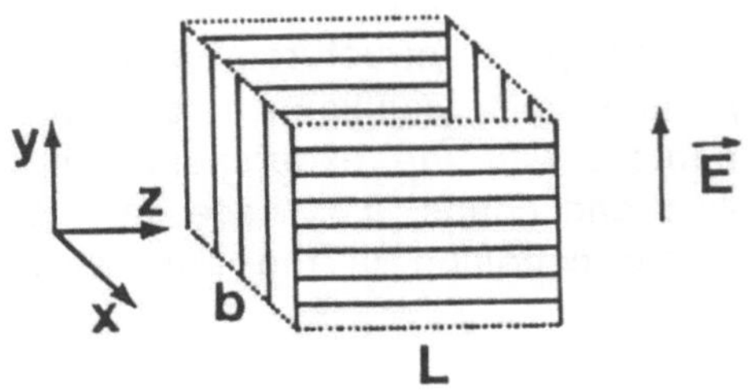

Fig. 9. Schematic representation of the cavity configuration discussed in Sects. 5 and 6. The two mirrors orthogonal to the z-axis have transmission coefficient $T << 1$; the two mirrors orthogonal to the x-axis have 100% reflectivity. The cavity is open in the y direction, which is also the direction of polarization of the electric field.

coefficient $T << 1$; the other two are orthogonal to the x-axis, at a distance b, and have 100% reflectivity (see Fig. 9). The cavity is open along the y direction. We assume that both the input and the cavity field are linearly polarized along the y direction. From this requirement and from the transversality condition, it follows that the internal field is independent of y. In this case, Eq. (23) reduces to

$$\frac{\partial}{\partial t} E(x,t) = -\kappa\left\{ E(x,t) - E_I - i\eta\, E(x,t)\left(|E(x,t)|^2 - \theta\right)\right\} + i\frac{c}{2k_z}\frac{\partial^2}{\partial x^2} E(x,t) \tag{24}$$

The field obeys reflecting boundary conditions on the two mirrors that are orthogonal to the x-axis; these prescribe that $\partial E/\partial x$ will vanish on reflection.

From these considerations it follows that the cavity supports modes of the type

$$\cos(k_x x), \qquad k_x = \pi n/b \tag{25.1}$$

and

$$\cos(k_z z), \qquad k_z = \pi n_z/L \tag{25.2}$$

where n and n_z are nonnegative integer numbers. Note that Eq. (2) fixes the value of n_z, while the integer n remains free to vary over the range 0,1,2,... Therefore, Eq. (24) is a single-mode model from the longitudinal point of view, but it is capable of accounting for an infinite number of transverse modes. The stationary solution governed by Eq. (20) corresponds to selecting n=0; however, as we shall show in the next section, solutions of this type can become unstable under appropriate conditions and allow the growth of transverse modes with $n \neq 0$.

In order to reduce the number of parameters in Eq. (24) it is convenient to scale the independent variables. Thus, we define

$$\bar{t} = \kappa t, \quad \bar{x} = \frac{x}{b} \tag{26}$$

With the help of Eq. (5) and the equality $k_z=2\pi/\lambda$, where λ is the wavelength of the radiation, Eq. (24) takes the form

$$\frac{\partial}{\partial \bar{t}} E(\bar{x},\bar{t}) = - E(\bar{x},\bar{t}) + E_I + i\eta\, E(\bar{x},\bar{t})\left(|E(\bar{x},\bar{t})|^2 - \theta\right)$$

$$+ i a \frac{\partial^2}{\partial \bar{x}^2} E(\bar{x},\bar{t}) \tag{27}$$

The parameter a is defined as

$$a = (2\pi TF)^{-1} \tag{28}$$

where F, the Fresnel number is given by $F = b^2/\lambda L$. Because this model requires that T << 1, while we want the parameter a to remain arbitrary, we complete the definition (1) of the uniform field limit by requiring also that

$$F >> 1, \qquad FT = \text{arbitrary} \tag{29}$$

6. Spatial diffractive patterns

We study the stability of the homogeneous stationary solution governed by Eq. (20) by linearizing Eq. (27) around the steady state. We label with δE the deviation of the field E from its uniform stationary configuration. The ansatz

$$\begin{pmatrix} \delta E(\bar{x},\bar{t}) \\ \delta E^*(\bar{x},\bar{t}) \end{pmatrix} = \exp(\lambda \bar{t}) \cos(\pi n \bar{x}) \begin{pmatrix} \delta E_n \\ \delta E_n^* \end{pmatrix} \tag{30}$$

leads to a homogeneous algebraic set of equations for δE_n and δE_n^* and to a quadratic eigenvalue equation of the form

$$\lambda^2 + 2\lambda + \zeta(|E|^2, a(n)) = 0, \tag{31.1}$$

with

$$a(n) = a\pi^2 n^2, \tag{31.2}$$

The stationary solution is unstable when one of the roots of equation (31.1) has a positive real part for at least one choice of n. This is equivalent to requiring that the constant term ζ be negative. Explicitly, this condition reads

$$1+(|E|^2-\theta)(3|E|^2-\theta)+a(n)\left[a(n)-2\eta(2|E|^2-\theta)\right]<0 \qquad (32)$$

One notes:

(i) The first two terms in Eq. (32) are equal to the slope $dE_I^2/d|E|^2$ of the homogeneous steady state curves as given by Eq. (20). Accordingly, an instability is possible with respect to homogeneous perturbations, i.e. corresponding to n=0, only if the steady state curves have a segment with negative slope; this implies that θ must exceed $\theta_c=\sqrt{3}$.

(ii) In the positive-slope regions of the steady state curve, the system is unstable with respect to inhomogeneous perturbation modes lying within the interval

$$a^{(-)}(|E|^2)<a(n)<a^{(+)}(|E|^2) \qquad (33)$$

where

$$a^{(\pm)}(|E|^2)=\eta(2|E|^2-\theta)\pm(|E|^4-1)^{1/2} \qquad (34)$$

and

$$|E|^2\geq 1\,,\quad \eta(2|E|^2-\theta)>0 \qquad (35)$$

The unstable domain is shown in Fig. 10b for $\eta=1$, $\theta=1$ (See also Fig. 10a). This robust instability arises for the following reasons. If we consider only the longitudinal modes of the cavity, for T << 1 the modes are well separated and therefore the input field selects the nearest mode (resonant mode) for its operating state. On the other hand, when the parameter a is of the order of unity, transverse modes develop whose frequency separation from the resonant homogeneous mode is of the order of the cavity linewidth κ. As a result, these modes compete with the resonant mode and through the instability they give rise to a spatial coexistence, quite unlike the temporal coexistence that arises in multimode oscillatory behavior [12,13,16,17,32].

In the neighborhood of the bifurcation point at which the first unstable mode n_c develops, i.e. for

$$|E|^2=1,\quad E_I=E_{Ic}=\left[1+(1-\theta)^2\right]^{1/2},\quad a(n_c)=\eta(2-\theta) \qquad (36)$$

the small-amplitude inhomogeneous stationary solutions of Eq. (27) can be calculated analytically using the methods of bifurcation theory. Here we only quote the final result. We let, first, $E_1=\mathrm{Re}(E)$, and $E_2=\mathrm{Im}(E)$, and set $E_1=E_{1,s}+u(\bar{x})$, $E_2=E_{2,s}+v(\bar{x})$, where $E_{1,s}$ and $E_{2,s}$ refer to the homogeneous stationary solution and u and v represent the inhomogeneous part of the new stationary state; the explicit form of the functions $u(\bar{x})$ and $v(\bar{x})$ is

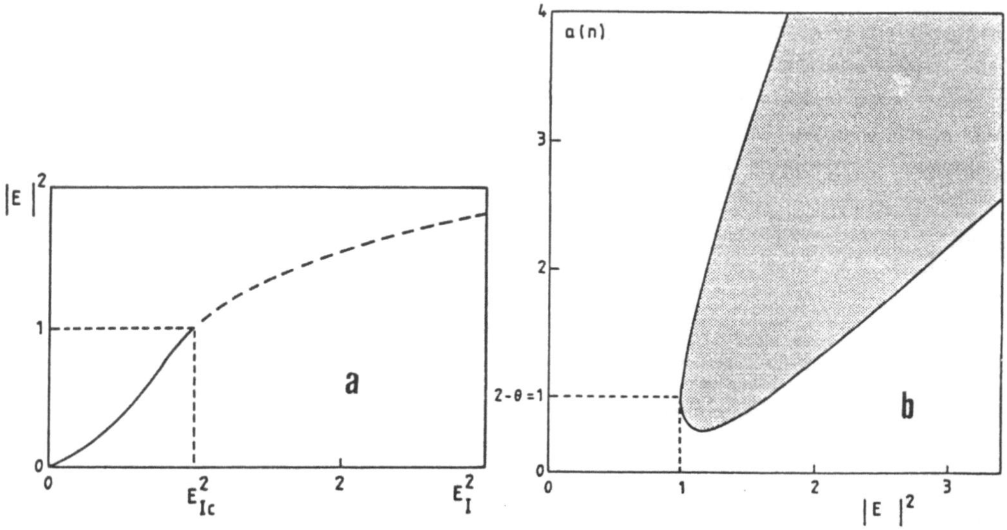

Fig.10. Self-focusing case, θ = 1. Fig. 10a displays the unstable (broken) part of the steady state curve when condition (33) is satisfied (See. Fig. 10b). In Fig. 10b the shaded region corresponds to the unstable domain in the plane of the variables $a(n) = a\pi^2 n^2$ and $|E|^2$, defined by Eqs. (33) and (34).

$$u(\bar{x}) = \pm\left\{\frac{E_I - E_{Ic}}{\Phi(\rho)}\right\}^{1/2}\cos(n_c\pi\bar{x})$$

$$+\left\{\frac{E_I - E_{Ic}}{\Phi(\rho)}\right\}\left[\, f_1(\rho) + f_2(\rho)\cos^2(n_c\pi\bar{x})\right] \qquad (37.1)$$

$$v(\bar{x}) = \pm\eta(1+\rho)(1-\rho)^{-1}\left\{\frac{E_I - E_{Ic}}{\Phi(\rho)}\right\}^{1/2}\cos(n_c\pi\bar{x})$$

$$+\left\{\frac{E_I - E_{Ic}}{\Phi(\rho)}\right\}\left[\, f_3(\rho) + f_4(\rho)\cos^2(n_c\pi\bar{x})\right] \qquad (37.2)$$

where

$$\rho = 1-\theta,$$

$$\Phi(\rho) = \frac{11+19\rho(1-\rho+\rho^2)-30\rho^4}{18(1+\rho^2)^{1/2}(1-\rho)^3} \qquad (38)$$

and $f_i(\rho)$, i=1,...,4 are rational functions whose explicit expressions are of no consequence for this discussion. The ± signs in front of the main term in Eqs (37) describe the bifurcation of the two inhomogeneous stationary solutions at the critical point. For values of θ such that $\Phi(\rho) > 0$ (or $\Phi(\rho) < 0$) the two solutions exist for $E_I > E_{Ic}$ (or $E_I < E_{Ic}$) and therefore are stable (or unstable) because the bifurcation is supercritical (subcritical) [33]. Note that the function $\Phi(\rho)$ shown in Fig. 11 is independent of η. In the self-focusing case, where $\eta=+1$, the bifurcation is supercritical for $\theta \leq 1.4$; in the self-defocusing case, where $\eta=-1$, the instability develops for $\theta>2$ and therefore is always subcritical.

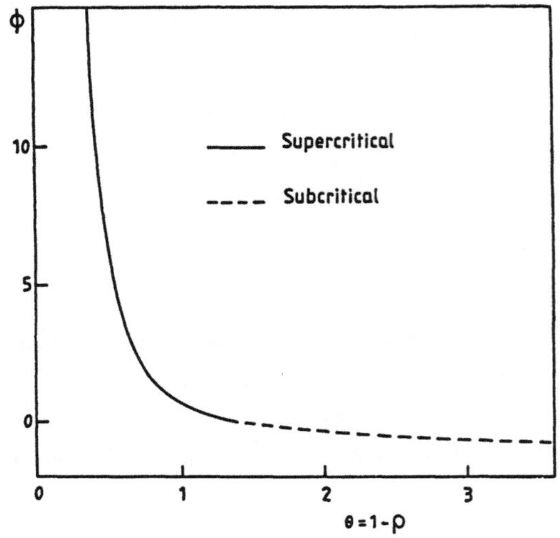

Fig.11. Plot of the function $\Phi(\rho)$ [see Eq. (38)]. The spatial structures that bifurcate at the critical point are stable (unstable) when $\Phi>0$ ($\Phi<0$)

Clearly, several additional investigations are in order; the most pressing would appear to be a numerical analysis to calculate the large amplitude stationary states that evolve far from the critical point, and a search for higher-order bifurcations. Our results predict that a plane-wave input field is spontaneously converted into a stationary beam that displays a transverse grating structure. An experimental observation of this phenomenon would be of great interest.

Acknowledgements

We dedicate this article to Professor Hermann Haken whose scientific and educational activities have always been a guideline and a source of inspiration for all of us. His constant encouragement was invaluable for our research.

We are also grateful to J.H. Kimble for lending us the originals of Figures 6 and 7 displaying experimental results obtained in his laboratory. This research was carried out in the framework of the EEC twinning project on Dynamics of Nonlinear Optical Systems, and it was also partially supported by the U.S. Army Research Office (Durham, N.C.).

1. H.M. Gibbs, S.L. McCall and T.N.C. Venkatesan, Phys. Rev. Lett. 36, 1135 (1976).
2. H. Haken, Laser Theory, Encyclopedia der Physik, Vol XXV/2c, Springer Verlag, Berlin, 1970.
3. H. Haken, Light 2 - Laser Light Dynamics, North-Holland, Amsterdam, 1985.
4. H. Haken, Synergetics - An Introduction, Springer Verlag, Berlin, 1977.
5. H. Haken, Advanced Synergetics, Springer Verlag, Berlin, 1983.
6. R. Graham and H. Haken, Z. Phys. 237, 31 (1970).
7. V. Degiorgio and M.O. Scully, Phys. Rev. A2, 1170 (1970).
8. R. Bonifacio and L.A. Lugiato, Opt. Comm. 19, 172 (1976).
9. R. Bonifacio and L.A. Lugiato, Phys. Rev. A18, 1129 (1978).
10. H. Haken, Z. Phys. 190, 327 (1966).
11. H. Risken, C. Schmid and W. Weidlich, Z. Phys. 194, 337 (1966).
12. R. Graham and H. Haken, Z. Phys. 213, 240 (1968).
13. H. Risken and K. Nummedal, J. Appl. Phys. 49, 4662 (1968).
14. H. Haken, Phys. Lett. 53A, 77 (1975).
15. E.N. Lorenz, J. Atmos. Sci. 20, 130 (1963).
16. R. Bonifacio and L.A. Lugiato, Lett. Nuovo Cimento 21, 510 (1978).
17. R. Bonifacio, M. Gronchi and L.A. Lugiato, Opt. Comm. 30, 129 (1979).
18. K. Ikeda, Opt. Comm. 30, 257 (1979).
19. L.A. Lugiato, L.M. Narducci, D.K. Bandy and C.A. Pennise, Opt. Comm. 43, 281 (1982).
20. L.A. Lugiato, R.J. Horowicz, G. Strini and L.M. Narducci, Phys. Rev. A30, 1366 (1984).
21. L.A. Orozco, A.T. Rosenberger and H.J. Kimble, Phys. Rev. Lett. 53, 2547 (1985).
22. L.A. Lugiato and R. Lefever, submitted for publication.
23. G. Nicolis and I. Prigogine, Self-Organization in Nonequilibrium Systems, Wiley, New York, 1977.
24. R. Bonifacio and L.A. Lugiato, Lett. Nuovo Cimento 21, 505 (1978).
25. H.M. Gibbs, Optical Bistability: Controlling Light with Light, Academic Press, New York, 1985;
L.A. Lugiato and L.M. Narducci, submitted for publication.
26. R. Bonifacio and L.A. Lugiato, Lett. Nuovo Cimento 21, 517 (1978).
27. L.A. Lugiato, Theory of Optical Bistability, in Progress in Optics, edited by E. Wolf, North-Holland, Amsterdam, 1984.
28. S.S. Hassan, P.D. Drummond and D.F. Walls, Opt. Comm. 27, 480 (1978).
29. E. Abraham and S.D. Smith, Rept. Progr. Phys. 45, 815 (1982).
30. S.L. McCall, Appl. Phys. Lett. 32, 284 (1978).
31. D.W. McLaughlin, J.V. Moloney and A.C. Newell, Phys. Rev. Lett. 54, 681 (1985).
32. N.B. Abraham, L.A. Lugiato and L.M. Narducci, Eds., Feature issue on Instabilities in Active Optical Media, J. Opt. Soc. Am. B2, Jan. 1985.
33. D. Sattinger, Topics in Stability and Bifurcations Theory, Springer Verlag, Berlin 1973.

Optical Instabilities in Semiconductors

H. Haug

Institut für Theoretische Physik, Universität Frankfurt,
Robert-Mayer-Straße 8, D-6000 Frankfurt, Fed. Rep. of Germany

In semiconductors the optical properties change strongly with the concentration of electronic excitations. Due to these nonlinearities various optical instabilities occur. As examples spatial, temporal and spatio-temporal instabilities due to induced absorption will be described.

1. Introduction

At the beginning of the sixties a young, inspiring Professor for Theoretical Physics came to the University of Stuttgart. His name was Hermann Haken. He suggested in 1962 to me that I should work for my master's thesis on the lineshape theory of excitons in connection with his treatment of the exciton-phonon interaction (see HAKEN [1]). In 1964 he asked me to elaborate his theory of semiconductor lasers (see HAKEN and HAKEN [2]). Maybe because I could not solve the proposed problems so well as a young student, I returned later after some experience in other fields of theoretical physics to these intriguing problems of the optical properties of semiconductors.
Naturally, Prof. Haken himself gave many important, early contributions to this field, and today it is stimulated again by the developments in the theory of synergetics (see HAKEN [3]).
In this contribution I will first discuss briefly, why the calculation of the nonlinear optical properties of semiconductors is a genuine many-body problem and indicate which techniques are most suited to solve it.
Next, I will concentrate on just one type of optical nonlinearity, namely the induced absorption, which is a rather common phenomena in semiconductors, but which so far has not received much attention in quantum optics.
The increasing absorption can cause an intrinsic optical bistability which does not require any resonator feedback. Because bistability is a first-order nonequilibrium phase transition (see HAKEN [3]), the phenomena of phase coexistence and spatial phase separation should exist (see HAUG and KOCH [4]). As a first example I will discuss that this phase separation occurs as a partial sample switching. KOCH et al. [5] found that due to the bistability the density kink which separates the switched-on and the switched-off region can move only in discrete jumps if the incident intensity is changed. These theoretical predictions have been verified quantitatively by the experiments of GIBBS et al. [6].

By putting an intrinsically bistable induced absorber in a ring resonator, oscillations can be obtained under the condition that the cavity round trip time is larger than the medium relaxation time. If the absorber is in the state of low absorption, the intensity in the cavity builds up until the medium switches into the state of high absorption. The high absorption causes the intensity to decrease until the medium switches back into the state of low absorption, whereupon the process repeats itself. The analysis of the oscillations according to LINDBERG et al. [7] and HAUG et al. [8] exhibits a strong locking behaviour: the ratios of the mean oscillation periods and the round trip time follow a Farey tree construction. All these predictions have been verified recently by experiments of WEGENER and KLINGSHIRN [9], where the feedback has been delayed electro-optically.
After the examples of a spatial and a temporal instability, I will discuss in a last example the influence of temporal chaos on a spatial structure. Chaos arises in a cavity if the induced absorber is in an intrinsically monostable state (see GALBRAITH and HAUG [10]). The influence of the chaos on the transverse beam profile will be discussed for an incident Gaussian beam. The diffusion of the excitation provides the spatial coupling. Large qualitative changes in the beam profile and in its temporal variations are found when the ratio of the diffusion length to the beam width is varied.

2. *The Photo-Electronic Nonlinearities in Semiconductors*

The elementary excitation process in a semiconductor is the excitation of an electron from the valence band into the conduction band, *i. e.* the creation of an electron-hole pair. At low excitation levels these electron-hole pairs have a hydrogen-like spectrum with an infinite set of bound states and an ionization continuum. At higher excitation levels one excites a dense system of these mobile electron-hole pairs which all interact with one another via the long-range Coulomb interaction. The calculation of the optical properties of such a strongly interacting system obviously requires the techniques of advanced many-body theory. The filling of the single-particle levels - also called phase space filling or band filling - and the screening of the Coulomb forces modify strongly the properties of the system of electronic excitations and thus also the optical properties of the semiconductor. In the low excitation regime excitons can often be considered as composite Bosons. In highly excited semiconductors it is more natural to work in the electron-hole picture in order to take the consequences of the Fermi statistics and the strong modifications of the relative motion of the electron-hole pairs properly into account. The effects of the long-range Coulomb interaction can best be treated by using Green's function techniques. Due to the rapid interparticle scattering processes the electron-hole system can for most experimental situations be treated in the quasi-equilibrium approximation. Under this assumption temperature-dependent one- and two-particle

Green's functions can be used to calculate the optical properties of laser-excited semiconductors as has been described in the reviews by HAUG and SCHMITT-RINK [11,12]. The most important effects obtained in such a treatment are the renormalizations of the single-particle energies due to intraband Coulomb scattering, which leads to a reduction of the band gap and a broadening of the energy levels as a function of the concentration of the excited carriers. The polarization can be calculated from an integral equation, called the Bethe-Salpeter equation. The results show that above a critical concentration of electron-hole pairs the bound electron-hole pair states no longer exist. Above this critical density, called Mott density, one has an electron-hole plasma. Naturally, the vanishing of the sharp exciton resonances from the spectra cause large optical nonlinearities.

Generally speaking, the electron-hole system is a nonequilibrium system, which cannot be rigorously described by equilibrium Green's functions. Therefore, nonequilibrium Green's functions have to be used, which have been introduced by KELDYSH [13], based on the earlier work of SCHWINGER [14]. Instead of a single Green's function (e.g. the time-dependent Green's function of an equilibrium system) one has to calculate a two-by-two Green's function matrix in nonequilibrium systems. Physically, this is due to the fact that one has to calculate both the spectrum and the population of the interacting particle states. In a two-band model of a semiconductor which is excited by a coherent light field, the Green's functions have a further two-by-two matrix structure in the band index. The diagonal elements are the electron and hole Green's functions, the nondiagonal (interband) elements describe the polarization. In this sense the nonequilibrium Green's function formalism is the natural extension of the Bloch equations of a noninteracting atomic system.

The nonequilibrium Green's function theory allows to treat in addition to the above discussed high-excitation effects also the high-field effects, i.e., the influence of the light itself on the spectral properties of semiconductors. Examples of such high-field effects are the ac-Stark effect seen on nonresonantly excited exciton levels and light-induced gaps in the spectrum of a plasma.

A systematic presentation of the nonequilibrium theory of laser-excited semiconductors has been given by HAUG [15] in a recent book on "Optical Nonlinearities and Instabilities in Semiconductors".

3. *Intrinsic Optical Bistability due to Induced Absorption and Partial Sample Switching*

From the many optical nonlinearities which can occur in a semiconductor depending on the detailed excitation conditions, I will select only the effect of induced absorption. While the nonlinearity of saturable absorption, which occurs already in a two-level model, is often used in theories of optical instabilities, the nonlinearity due to induced absorption has not received much consideration in quantum optics.

Induced absorption is a frequently observed phenomena in semiconductors. It arises, e.g., due to the above-mentioned band-gap renormalization. A simple estimate for the band-gap shrinkage can be obtained by noting that it is approximately determined by the change of the self-energy of a point-charge due to the presence of N photo-excited electron-hole pairs:

$$\Delta E_g \simeq \lim_{r \to 0} \left(V_S(r) - V(r) \right) , \tag{1}$$

where $V(r) = \frac{e^2}{\epsilon_0 r}$ is the Coulomb potential and $V_S = V(r) \exp(-\kappa(N) r)$ is the screened Coulomb potential. $\kappa(N)$ is the inverse screening length due to the electron-hole plasma. By linear expansion, we get from (1)

$$\Delta E_g = - \frac{e^2 \kappa(N)}{\epsilon_0} = - 2E_0 \, (a_0 \kappa(N)) \ . \tag{2}$$

Actually, this simple result is the classical limit (Debye limit) to the RPA result for the band-gap renormalization. Eq. (2) shows that the band-gap shrinks (- sign) with increasing plasma density N and that the effect scales with the exciton Rydberg E_0, which is the natural measure of the strength of Coulomb forces in a given semiconductor. a_0 is the exciton Bohr radius.

Suppose now, that one excites at elevated temperatures, where excitonic effects are not very important, a semiconductor in the band tail of the band-to-band transitions. The absorption is low, but with increasing laser intensity an increasing concentration is created, which reduces the band-gap according to Eq. (2).

Fig.1 shows the calculated density-dependence of the absorption for CdS at the frequency of $\hbar\omega = 2.548$ eV. It agrees well with experimental observations. The electron-hole concentration N is determined via a rate equation by the laser light intensity

Assuming a weak diffusion, the stationary solution of Eq.(3) is obtained from

$$\alpha(\omega, N) = \frac{N \hbar\omega}{\tau I} \ . \tag{4}$$

The geometrical solution of Eq.(4) which is given by the intersections of the straight line with $\alpha(N)$ in Fig.1, yields a bistable S-shaped curve $N = N(I)$. Thus, one gets an intrinsic bistability which does not require any resonator feedback. The system can for the same intensity be locally either in the state of low density N and therefore low absorption or in the state of high N and $\alpha(N)$. Suppose a crystal of length L is excited with an intensity high enough to switch the sample into high absorption at least at the front

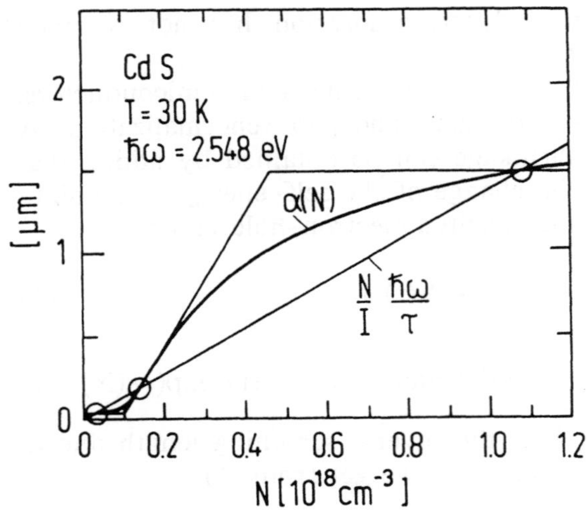

Fig. 1 Absorption coefficient α(N) for CdS versus electron-hole density, together with a piecewise linear approximation [5,16]. The intersections with the straight line are the solutions of Eq. (4).

edge. Due to the high absorption the intensity drops in beam direction. If the crystal is long enough and the diffusion is not too strong (i.e. $D\tau/L^2 \ll 1$) the end part of the crystal will remain in the state of low density and absorption. In between there must be a kink in the density profile. Numerical solutions of Eq.(3) together with Beer's law

$$\frac{dI}{dz} = - \alpha(\omega,N)\, I \tag{5}$$

(the intensity can be treated quasi-statically, because the density is the slow variable) confirm these expectations.

The density and the intensity just behind the kink correspond to the lower critical values on the bistable curve N = N(I) . If the incident intensity is now slowly increased, the kink remains in its position and the density behind the kink increases on the lower branch of the N(I) curve until it reaches the upper critical value of the bistable regime. Dynamically, one expects that, due to the critical slowing-down, one has to overshoot the critical values before the density kink will jump deeper into the sample. Numerical solutions by KOCH et al. [5] and analytical investigations by LINDBERG et al. [16] yield indeed such a stepwise motion of the kink profile for a slowly increasing incident intensity. The predicted highly non-linear kink dynamics was investigated experimentally by GIBBS et al. [6] using the slow photo-thermal induced absorption of a colour filter in which CdS_xSe_{1-x} microcrystallites are embedded in the glas matrix. The band-gap

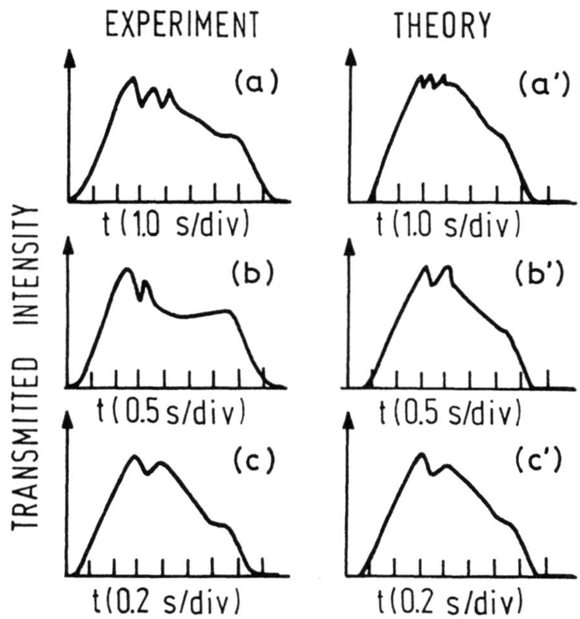

Fig. 2 Transmitted intensity versus time for a triangular incident pulse. Three different pulse lengths are shown [6].

reduction is here due to the electron-hole self-energies in the field of thermal phonons, which stem from the nonradiative decay of the photo-excited electron-hole pairs. A triangular pulse has been used. Each stepwise kink progression adds a highly absorbing layer and is thus seen as a sudden decrease of the transmitted intensity. Fig.2 shows the resulting sawtooth structure of the transmitted beam and the near perfect agreement between theory and experiment.

4 Frequency-Locked Oscillations of an Induced Absorber in a Resonator

Next, I will discuss a temporal instability. In order to simplify the situation only whole-sample switching will be treated, which occurs in the limit $D\tau/L^2 >> 1$. If an intrinsically bistable absorber is placed in a ring resonator with a round trip time τ_R equal or longer than the medium relaxation time τ, oscillations of the transmitted intensity can be expected. For suitably chosen parameters the intensity in the resonator will build up as long as the crystal is weakly absorbing. If the intensity is high enough it will switch the sample into the state of high absorption, which will cause a successive intensity decrease and finally a switch-down of the crystal, from whereon the process repeats itself.

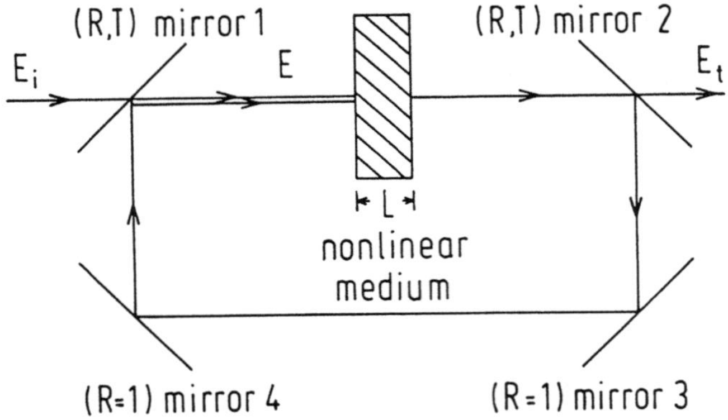

Fig. 3 Ring resonator

The resonator introduces - say for the field in front of the short sample - the following boundary condition

$$E(t+\tau_R) = E_i\sqrt{T} + RE(t)e^{-\alpha(N)\,L/2}, \tag{6}$$

where I assumed zero detuning and neglected nonlinear dispersive effects for simplicity. For a more complete treatment see LINDBERG et al. [7]. T and R are the coefficients of transmission and reflection of the semitransparent mirrors of the resonator shown in Fig. 3. Important in Eq.(6) is the fact that the feedback is retarded by a round trip time τ_R. Eq.(6) has to be solved together with the rate equation (3) (without the diffusion term, because N is assumed to be homogeneous throughout the sample). To make further analytical investigations possible, $\alpha(N)$ is replaced by a piecewise linear model as indicated in Fig.1.

The stationary stability analysis shows indeed that for a whole range of incident field values no stable stationary solutions exist. Fig.4 shows the oscillations which are obtained from the numerical solution of Eqs.(3) and (6) for three different ratios of τ_R/τ . Particularly, for $\tau_R = 100\tau$ (Fig. 4c), a discrete dynamics appears in form of stepwise variations of the intensity after each round trip. Before a state becomes unstable, there is some ringing on the time scale of the relaxation time τ.

In the limit $\tau_R >> \tau$, we can treat the density in the stationary limit Eq.(4), but one cannot eliminate N due to multiple solutions of Eq.(4). In this limit Eqs.(4) and (6) become a discrete nonlinear map, where $E(t) = E_n$ and $E(t+\tau_R) = E_{n+1}$ are the field values after n and n+1 round trips. Fig.5 shows examples of oscillations obtained by numerical iteration for various incident field values. Note, that e.g. for the incident field values E_i = 0.130 and 0.131 oscillations with higher periods occur. Furthermore, one sees that the oscillations become rather slow for the incident field values E_i = 0.06 and 0.319 which are close to the borders of the oscillating regime.

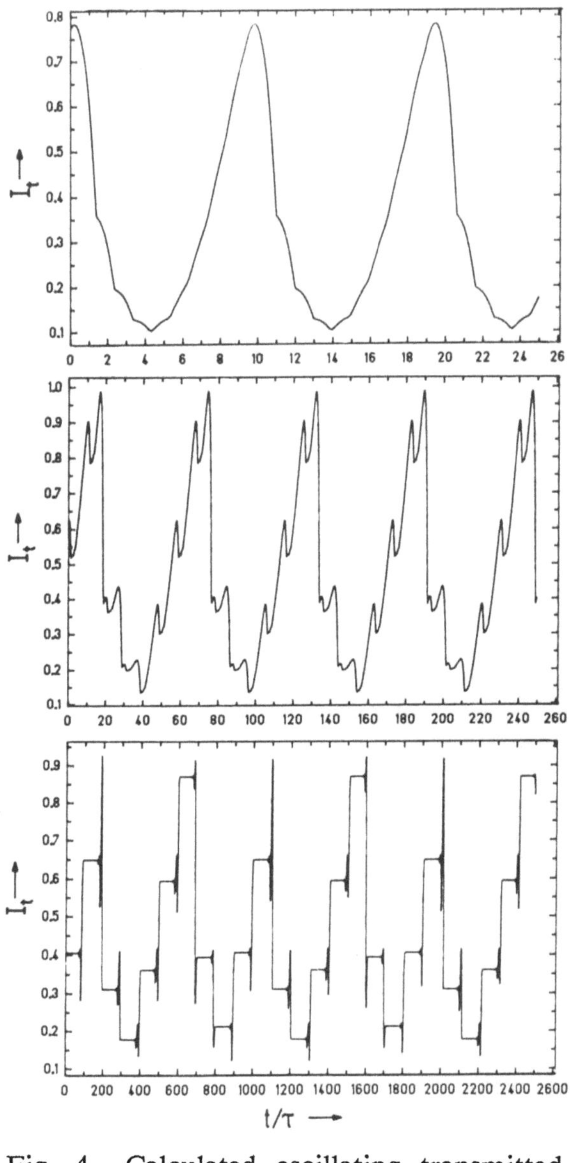

Fig. 4 Calculated oscillating transmitted intensity due to Eqs.(4) and (6) for three different ratios of the round trip time τ_R to the relaxation time τ: (a) $\tau_R/\tau = 1$; (b) = 10; (c) = 100. [7]

More insight can be obtained by plotting the ratio of the mean oscillation period T to the roundtrip time τ_R versus E_i (Fig. 6). One sees the critical slowing down at the boundaries of the oscillation regime and the locking of T/τ_R to simple ratios. The motion of the system in phase space is the

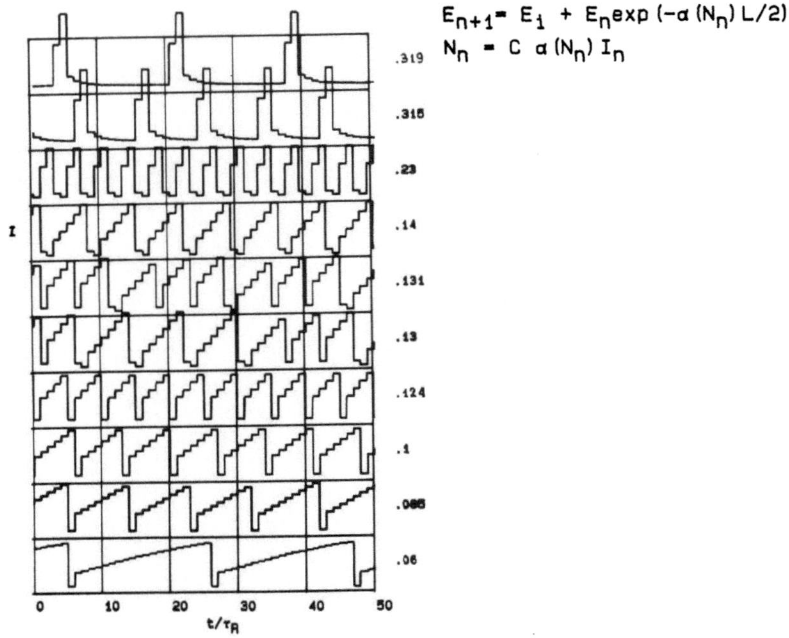

Fig. 5 Calculated oscillating transmitted intensity due to Eqs. (5) and (6) for various values of the incident field E_i [17]

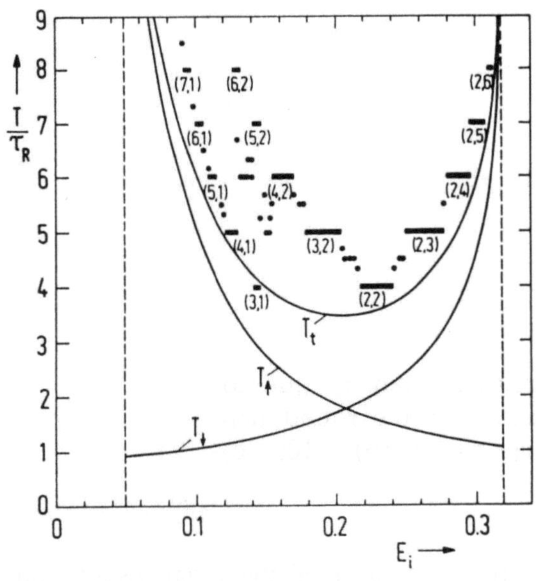

Fig. 6 Ratio of the mean oscillation period T to τ_R versus incident field [7]. The solid lines are analytic approximations.

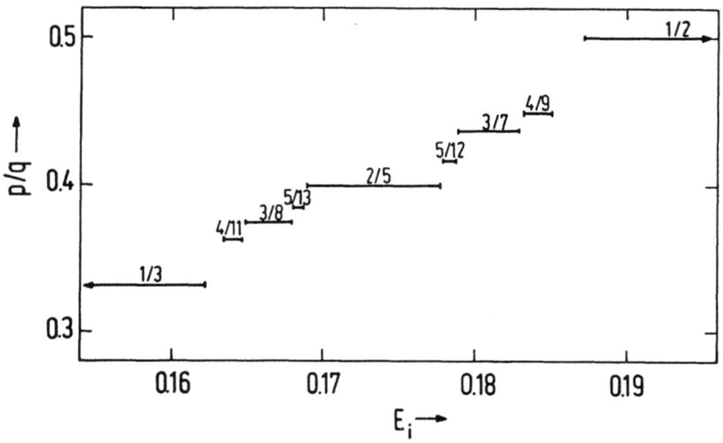

Fig. 7 The ratio τ_R/T versus E_i [8]

motion on a two-torus. The ratios of the two periods follow a Farey-tree construction as can be seen in detail from Figs. 7 and 8. The parameters used for the calculation of Fig. 7 are different from those of Figs. 5 and 6 (see HAUG et al. [8]). In between the rather stable oscillations with τ_R/T ratios of 1/2 and 1/3 one finds a daughter oscillation with a period ratio of 2/5, which is obtained by adding numerators and denominators of the parent oscillation.

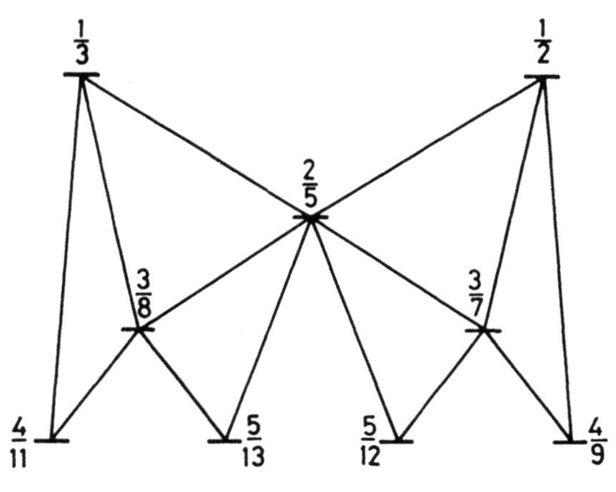

Fig. 8 Section of the Farey tree corresponding to Fig. 7[8]

Applying the same construction rule over and over, one finds increasingly complicated oscillation ratios with a rapidly decreasing range of stability. The oscillation ratios τ_R/T versus incident field E_i (Fig.6) form therefore a devil's staircase.

In Fig.6 the simple oscillations are characterized by (n,m), giving the number of round trips on the increasing and decreasing parts of the oscillation. There exist two families namely (m,1) and (m,2) together with (2,m). Around incident field values of $E_i \simeq 0.14$ the two families of modes coexist, one can show (for details see HAUG et al. [8]), that this mode coexistence occurs because the map becomes noninvertible in this parameter range. Chaotic solutions with irrational period ratios occur only at a set of incident field values of measure zero, and can thus not be observed.
The above-described frequency-locked oscillations have been verified experimentally by WEGENER and KLINGSHIRN [9], using the rather slow photo-thermally induced absorption of a CdS_xSe_{1-x} colour filter. The thermal relaxation time due to heat diffusion was in the order of ms. Therefore, an electro-optically delayed feedback was used. Fig. 9 shows the experimental setup, where PD denotes a photodiode and EOM an electro-optical modulator. Fig. 10 gives an example of an observed oscillation.

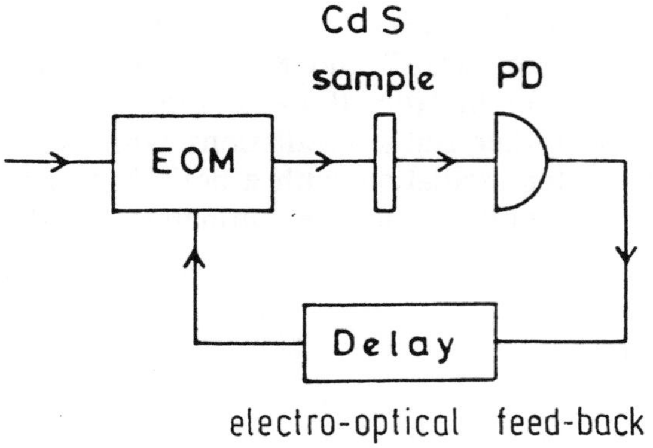

Fig. 9 Experimental setup of WEGENER and KLINGSHIRN [9]

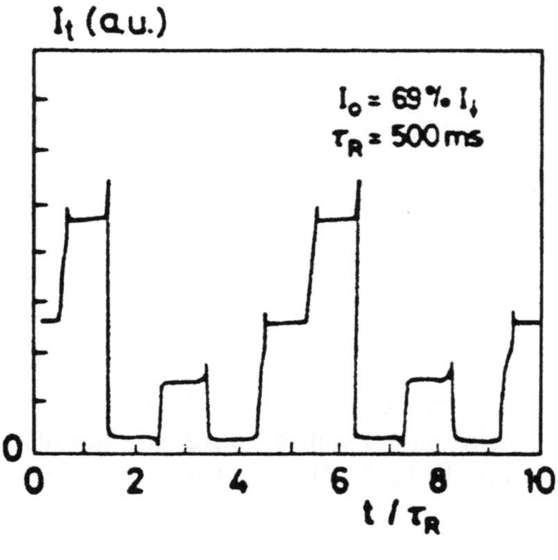

Fig. 10 Detected intensity oscillation [9]

The stepwise intensity variation due to the discrete dynamics of an iterative map and the frequency stability of the simple oscillations against fluctuations in the incident field due to the pronounced locking suggest that this device could be used as an optical clock in optical computing (see HAUG [17]).

5. *Spatio-Temporal Chaos in a Cavity*

After the examples of the formation of spatial structures in section 3 and of the occurrence of frequency-locked self-oscillations in section 4, I will finally present an example in which the influence of temporal chaos on spatial structures is studied following a recent investigation by GALBRAITH and HAUG [10]. Again, the the induced-absorption nonlinearity is used, but it is assumed that the function $\alpha(N)$ increases slowly enough so that the crystal remains monostable for all light intensities. The nonlinearity $\alpha(N)$ is modelled by

$$\alpha(N) = A - \frac{B}{1-e^{k(N-N_0)}} \quad . \tag{7}$$

In a resonator, the discrete map (6) has now a quadratic maximum and exhibits subharmonic bifurcations and chaos. Its attractor is shown in Fig. 11. The calculated Liapunov exponent is shown in Fig. 12 as a function of the input intensity. The period 4 and 2 oscillations at higher intensities do not longer belong to the Feigenbaum scenario due to the large oscillation amplitudes, as can be seen from the strange attractor of this map (see Fig.

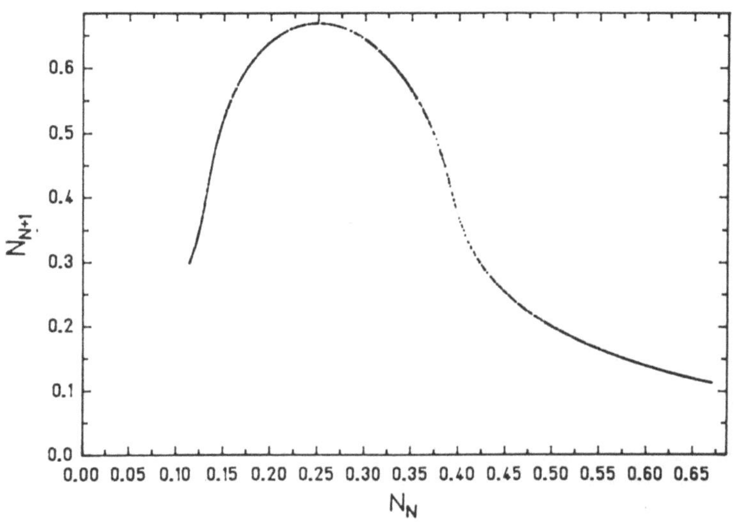

Fig. 11 Attractor of the map (6) together with Eqs. (4) and (7) according to Ref. [10]

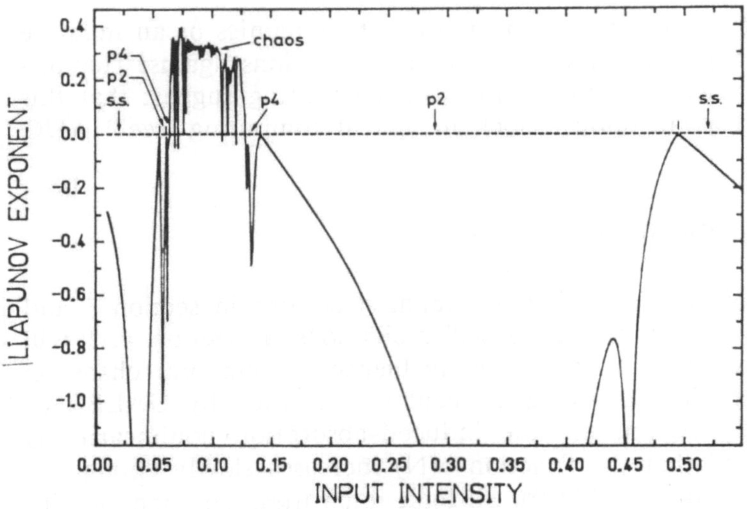

Fig 12 Calculated Liapunov exponent versus imput intensity [10]

11). This scenario too has been found experimentally by WEGENER and KLINGSHIRN [18].
In order to study the influence of the temporal chaos on spatial structures, an incident Gaussian beam profile is assumed. Then, the map (6) takes the form

$$E_n(x) = \sqrt{T\, I_{in}}\; e^{-x^2} + R\, e^{-\alpha(N)L/2}\, E_n(x) \quad , \tag{8}$$

while the excitation concentration obeys a rate equation of the type

$$D\,\frac{\partial^2 N}{\partial x^2} - \frac{N}{\tau} + \frac{\alpha(N)\, I_n(x)}{\hbar\omega} = 0 \quad , \tag{9}$$

where only one transverse coordinate x is taken into account. The diffusion provides the transverse coupling, while the diffraction coupling has been neglected. For the numerical solutions an incident peak intensity $I_{in} = 0.6$ (in normalized units) was chosen, which covers for various radii the whole oscillating and chaotic regime, as can be seen from Fig. 12. The resulting profiles of the transmitted beam are shown in Figs. 13 (a-h) for increasing ratios of the diffusion length $L_d = \sqrt{D\tau}$ to the beam width $w_0 = 1$. A number of successive iterations are plotted in each graph. For L_d=0 (Fig. 13(a),top left) the beam profile simply follows the prescription given by Fig. 12, i.e., the beam centre is constant while the neighbouring region is undergoing period two oscillations. Further out, we move through the chaotic region and the period doubling sequence to the wings, where again

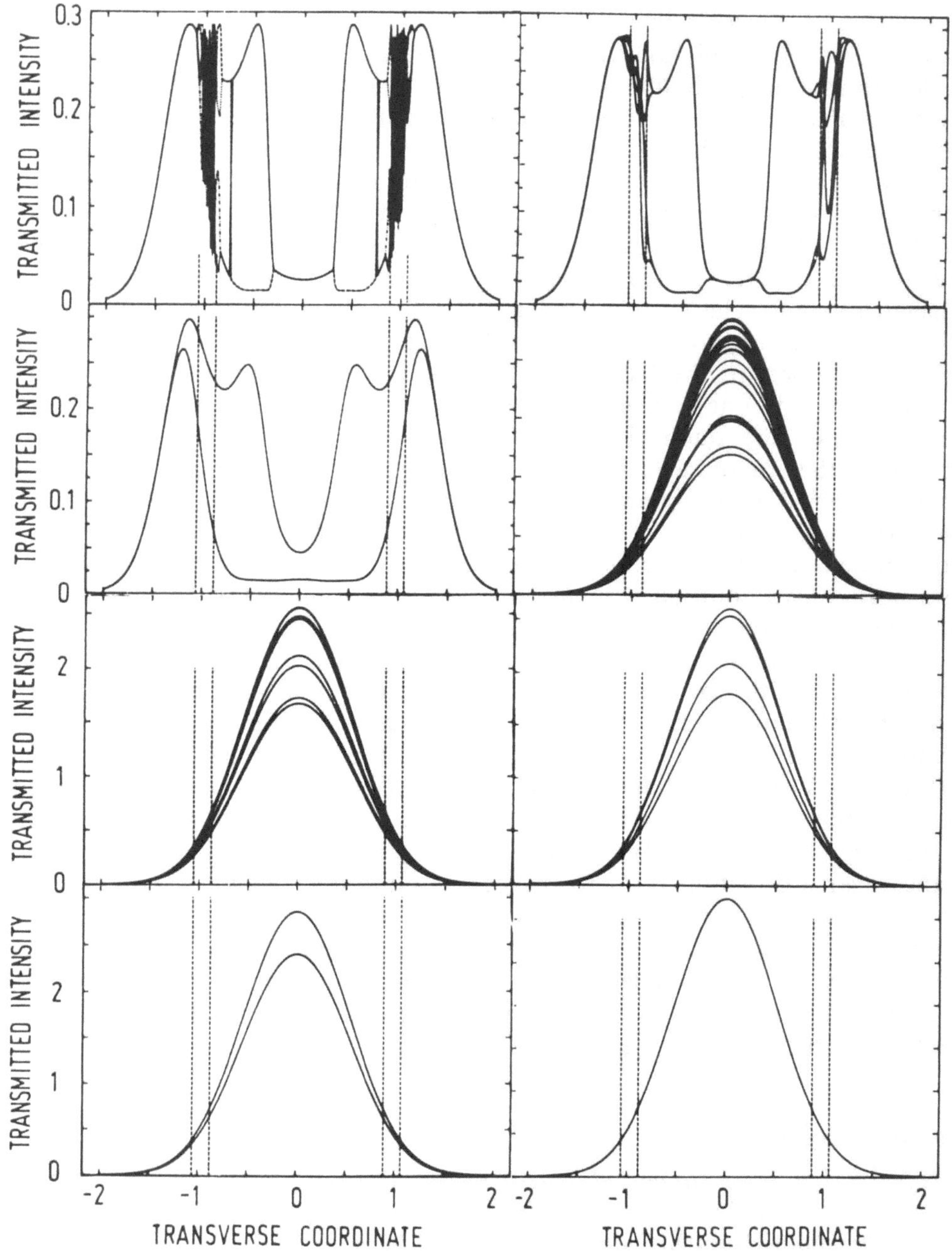

Fig. 13 Calculated beam profile for an inputintensity $I_i = 0.6$ and for various diffusion lengths in units of the beam width [10].(a) 1st row left $L_d = 0$; (b) 1st row right $L_d = 0.001$; (c) 2nd row left $L_d = 0.1$; (d) 2nd row right $L_d = 5.0$; (e) 3rd row left $L_d = 5.4$; (f) 3rd row right $L_d = 5.5$; (g) 4th row left $L_d = 6.0$; (h) 4th row right $L_d = 7.0$.

the solution is constant in time. At $x \simeq 0.75$ the period two solution flips phase between two neighbouring points because the maps at the two different locations, being slightly different, may approach the limit cycle with different phases. Clearly, this cannot physically occur as we may bring these two points infinitely close to one another and this is merely another anomaly of the local approximation. Introducing the diffusive coupling with a very small diffusion length L_d=0.01 - Fig. 13(b), top right - we see that little is changed except that the characteristic size of the spatial structure in the chaotic region structures is now many diffusion lengths in transverse extent. The correlation length in this chaotic region is $L_c \simeq 7L_d$. It is also interesting to note that the symmetry in this case remains broken, as it trivially was in the local case, but this can be at best indicative as we have treated only one transverse dimension. Note too that now the diffusion has eliminated the unphysical flip in phase at $x \simeq 0.75$ seen in the local case. A further increase in the diffusion length to 0.1 and the system shows a steady period two, symmetric oscillation as shown in Fig. 13(c), second row left. In this case the broad period two window (see Fig.12) dominates and through diffusion forces the rest of the unstable point to follow suit. This period two solution remains over a large range of diffusion lengths until eventually the loss in the excitation density from the beam region due to diffusion causes a drop into the chaotic region for L_d=5, Fig. 13(d), second row right. In this case the diffusion is too strong to allow any transverse structure ($L_c \gg w_0$) and the chaos is seen as a whole-beam phenomenon. At L_d=5.4 we see the end of the period doubling sequence with a period eight oscillation (see Fig. 13(e),third row left). The two highest oscillations are not distinguishable in this graph but can be seen clearly in the original numbers. With higher and higher diffusion lengths one goes in the reverse direction through the period doubling sequence and finally to the steady state as can be seen from Figs. 13 (f-h).
These results show that the diffusion enforces correlations in the beam profile over many diffusion lengths. A significant variation in the qualitative behaviour occurs as the diffusion length increases due to the reduction of the excitation density in the beam region.

Acknowledgements

This work has been supported partly by the Council of the European Communities and partly by the Deutsche Forschungsgemeinschaft through the Sonderforschungsbereich Frankfurt-Darmstadt.

1. H. Haken: In **Z. Physik 155**, 223 (1959)
2. H. and E. Haken: In **Z. Physik 176**, 421 (1964)
3. H. Haken: **Synergetics, An Introduction**, 3rd ed. Springer Ser. Synergetics, Vol.1 (Springer, Berlin 1983)

4. H. Haug and S.W. Koch: In **IEEE-QE 21**, 1385 (1985)
5. S.W. Koch, H.E. Schmidt and H. Haug: In **J. Luminesc. 30**, 232 (1985)
6. H.M. Gibbs, G.R. Olbright, N. Peyghambarian, H.E. Schmidt, S.W. Koch and H. Haug: In **Phys. Rev. A32**, 692 (1985)
7. M. Lindberg, S.W. Koch and H. Haug: In **JOSA B3**, 751 (1986)
8. H. Haug, S.W. Koch and M. Lindberg: In **Physica Scripta 34**, in print
9. M. Wegener and C. Klingshirn: In **Phys. Rev. A**, in print
10. I. Galbraith and H. Haug: In **JOSA B**, in print
11. H. Haug and S. Schmitt-Rink: In **Progr. Quantum Electronics 9**, 3 (1984)
12. H. Haug and S. Schmitt-Rink: In **JOSA B2**, 1135 (1985)
13. L.V. Keldysh: In **Sov. Phys. JETP 20**, 1018 (1965)
14. J. Schwinger: In **Proc. Nat. Acad. Sci. U.S. 37**, 452 (1951)
15. H. Haug: In "**Optical Nonlinearities and Instabilities in Semiconductors**", ed. H. Haug (Academic Press, New York), in print
16. M. Lindberg, S.W. Koch and H. Haug: In **Phys. Rev. A33**, 692 (1985)
17. H. Haug: In SPIE-Conference Proceedings "**Optical Bistability in Digital Optical Computing**", SPIE Washington to be published
18. M. Wegener and C. Klingshirn, preprint

Part III

Fluctuations in Quantum Optics

The Polarization Symmetric Laser *

S. Großmann and W. Krauth

Fachbereich Physik der Philipps-Universität, Renthof 6,
D-3550 Marburg, Fed. Rep. of Germany

1. Introduction

Since those days in 1964, when Hermann Haken gave the essentials of modern laser theory /1/, a comprehensive understanding of the laser physics has been developed (cf./2,3/). Besides the high photon intensity I the most striking feature is the extremely small linewidth λ_u of the laser's electric field u, measured with any interferometer. This field linewidth decreases with increasing pump rate p as $\lambda_u \propto p^{-1}$ in common single mode (SM) lasers. In contrast to the field linewidth λ_u the photon number fluctuates with a rather short correlation time λ_I^{-1}, resulting in a large intensity linewidth $\lambda_I \propto p$. It is only near threshold (p=0) that the intensity-intensity correlation function shows slowing down. The physics behind this markedly different behaviour of λ_u and λ_I is the different origin of the fluctuations /1/: while the induced emission and absorption mechanism provides a strong restoring force to fix I at p, the restoring time being all the smaller the larger p is, the phase of u is completely free and undergoes a Brownian diffusion stimulated by the spontaneous emission processes. Since the available phase space for the diffusion is of order I, the decorrelation time is ~I, hence $\lambda_u \propto p^{-1}$.

The described features, whose physical explanation has by now become commonplace, seem to be so uniquely realized in a laser that the question sounds silly, whether one could build an apparatus which produces light in which *also the photon linewidth* is sharp. After a first hint /4/ there is a remark in the literature that this might be possible in multi-mode lasers with high symmetry, see /5,6/. It is based on numerics and on a general argument. So the possibility of a small intensity linewidth remained unclear. Only recently it was pointed out again /7/ that a symmetric two-mode laser, namely, the polarization symmetric (PS) laser should show large photon correlation times λ_p^{-1} for each of the *partial intensities* of chosen polarization. The proof was on the same line of argumentation (mean field type) as Haken's original argument for λ_u.

We thought it to be worthwhile to explore this in more detail, to understand better the physical origin of the *small photon linewidth*, to evaluate it quantitatively even in the nonlinear threshold region, and to clarify the chance to realize such a device experimentally by calculating the influence of polarization symmetry-breaking effects on λ_p. The details of our findings are in /8,9/. While the ideal PS-laser can be treated /8/ as completely and as exactly as it has been done for the SM-laser /10,2/, namely, by solving a 1-dimensional Fokker-Planck eigenvalue problem, there is no hope to achieve this also for the much more complicated perturbed PS-laser. We therefore used for the latter the continued fraction approximation (CFA) to derive closed though approximate expressions for λ_p as a function of the pump rate p as well as of the perturbation strength. This method has been introduced to laser physics in /11/ and had turned out to describe the dynamical laser properties to a surprisingly good accuracy even in the nonlinear regime. We calculated the *exact* solution as well as the CFA for the *ideal* PS-laser

---

* Dedicated to Hermann Haken on his 60th birthday.

and found excellent agreement in this case too. We therefore hope that the following findings for the symmetry-broken PS-laser based on the CFA are sufficiently reliable.

In this paper we keep the emphasis on this approximation, which combines the advantages of being simple to use, of being applicable to all the laser systems of interest to us, and of providing us with physical insight into the basic dynamical features. We support it by a numerical solution of the laser's Langevin equation and its spectral resolution. Only the results are stated and commented; the detailed derivations can be found in /8,9,11/.

2. Ideal PS-laser

The common SM-laser emits light of a fixed polarization. Imagine now a laser, in which neither the atomic transitions nor the cavity single out a preferred polarization. This is the PS-laser. The laser field will have any polarization with equal probability. Slow changes of the polarization are caused by quantum noise, i. e., by the spontaneous processes. In the time mean all polarizations occur.

The electric field can be described by the slowly time varying real amplitudes u_1,u_2 and u_3,u_4 of two perpendicularly polarized (either circularly or linearly) states with the same frequency. $I_x=u_1^2+u_2^2$ and $I_y=u_3^2+u_4^2$ are the partial intensities of chosen polarization, $I=I_x+I_y=\sum u_i^2$ is the total intensity irrespective of the photon polarization. The equations of motion for the two components $u_x=u_1+iu_2$ and $u_y=u_3+iu_4$ of the general elliptically polarized wave are obtained along wellknown arguments /1,2,3/.

$$\partial_t u_i = (p - I)u_i + \Gamma_i \quad , \quad i = 1,2,3,4. \tag{1}$$

The noise Γ_i is supposed to be δ-correlated in time, $\langle\Gamma_j(t)\Gamma_k(t')\rangle = 2\delta_{jk}\delta(t-t')$. All quantities are expressed in reduced units as usual (cf./10,11,.../).

The stationary distribution ρ_s is derived from the Fokker-Planck equation equivalent to (1),

$$\partial_t\rho(u,t) = [-\partial_i(p-I)u_i + \partial_i\partial_i]\rho(u,t), \tag{2}$$

as the $t\to\infty$ solution

$$\rho_s(u) \propto \exp\ (\tfrac{1}{2}\, pI - \tfrac{1}{4}\, I^2)\ . \tag{3}$$

The mean total intensity I behaves as that of a SM-laser, $I\cong p$ for large pump rate, $I\cong 4/|p|$ far below threshold, with a rather sharp onset at threshold.

One may analyze the electric field transmitted from the laser using polarization filters outside the cavity. In contrast to the SM-laser it is a physically interesting information for the PS-laser how large the two independently polarized partial intensities I_P are. The probability distribution for $I_P=I_x$ or $=I_y$ is, from (3),

$$\rho(I_P) \propto 1 + \mathrm{erf}\ [\tfrac{1}{2}(p-I_P)]. \tag{4}$$

According to the sigmoid shape of the error function $\mathrm{erf}\ x = (2/\sqrt{\pi})\int_0^\infty \exp\ (-t^2)dt$ the partial density (4) is nearly p-independent for $I_P\leq p$ and drops to zero around $I_P\cong p$, cf. Fig.1. This is in remarkable contrast to the total intensity I, which is sharply peaked at p and has a limited variance $\langle(\delta I)^2\rangle=2$ in the lasing state, i.e. a decreasing relative r.m.s. $\sqrt{2}/p$. The partial intensity has an increasing variance $\langle(\delta I_P)^2\rangle\cong p^2/12$ and a non-decreasing constant relative r.m.s..

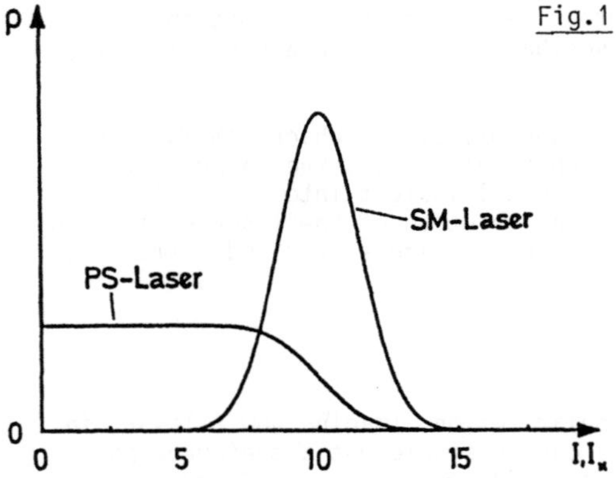

Fig.1 Comparison of the stationary photon distribution ρ_s vs I for the SM-laser and vs partial intensity I_x for the PS-laser. It is p=10 in both cases

This large variance clearly disqualifies the PS-laser for various applications, as the partial intensity of the chosen filtered polarization temporally is not stable. There is no restoring force that prevents the exchange of photons with one polarization to others with another polarization at constant total intensity I. Since this change in polarization is due to the weak fluctuations, it lasts rather long until I_p has varied. The motion of the state of polarization is diffusion-like on the hypersphere I=const and therefore is expected to lead to a long time scale in the I_p-correlation, i.e., to a small linewidth λ_p.

3. Partial Intensity Linewidth

Let us study how the correlation function $\langle \delta I_p \delta I_p(t) \rangle$ of the partial intensity decays. The average $\langle ... \rangle$ as usual is the mean in the stationary distribution ρ_s and the time dependence of $I_p(t)$ is governed /11/ by the Liouvillian L of the stochastic process, i.e., by the adjoint of the Fokker-Planck operator (2),

$$I_p(t) = e^{Lt} I_p \quad , \quad L = (p-I)u_i \partial_i + \partial_i \partial_i \ . \tag{5}$$

It is convenient to consider the Laplace transform of the temporal correlations

$$C^{(0)}(\omega) = \langle \delta A | (\omega - L)^{-1} \delta A \rangle \langle \delta A | \delta A \rangle^{-1} \tag{6}$$

for any set of dynamical laser variables A_k. This can be expressed /11/ by a continued fraction

$$C^{(0)}(\omega) = \cfrac{1}{\omega + \Lambda_o^{(0)} - \cfrac{N_1 N_o^{-1}}{\omega + \Lambda_1^{(0)} - \ddots}} \ . \tag{7}$$

Here $\Lambda_o = -\langle \delta A | L \delta A \rangle$, $\Lambda_1 = -\langle QL\delta A | LQL\delta A \rangle$, etc., $N_o = \langle \delta A | \delta A \rangle$, $N_1 = \langle QL\delta A | QL\delta A \rangle$, etc., are the successive relaxation rate and static correlation matrices. Q=1-P, and P projects onto the chosen set of variables $\delta A_k \equiv A_k - \langle A_k \rangle$. The upper index $^{(0)}$ denotes normalized quantities, $\Lambda_m^{(0)} = \Lambda_m N_m^{-1}$. The CFA consists in studying the truncated continued fraction (7). If the set of variables is "properly" chosen, a few steps are sufficient, two for the ideal PS-laser. The choice is "proper" if the observables δA_k are all of the same, longest time scale.

For the PS-laser the intensities I_x, I_y or $I = I_x + I_y$ and $\Delta = I_x - I_y$ are assumed to be a proper choice. We were able to verify /8/ this assumption by comparing the results of the CFA with those of the exact solution of the Fokker-Planck equation.

In the lowest order CFA one obtains

$$N_0 = \begin{pmatrix} \langle(\delta I)^2\rangle & 0 \\ 0 & \langle I^2\rangle/3 \end{pmatrix} , \qquad (8)$$

$$\Lambda_0 = \begin{pmatrix} 4\langle I\rangle & 0 \\ 0 & 4\langle I\rangle \end{pmatrix} , \qquad (9)$$

and consequently for the decay rate matrix in lowest order

$$\Lambda_0^{(o)} = \begin{pmatrix} 4\langle I\rangle/\langle(\delta I)^2\rangle & 0 \\ 0 & 12\langle I\rangle/\langle I^2\rangle \end{pmatrix} . \qquad (10)$$

The 1-1-element is the total intensity decay rate. It is large below threshold, slows down near threshold and increases above threshold like $2\langle I\rangle \cong 2p$ since $\langle(\delta I)^2\rangle \cong 2$. The 2-2-element is the Δ decay rate, characteristic for the partial intensity correlations. It is also large below threshold and slows down near threshold. But above threshold, although its numerator also increases, the denominator increases even stronger, $\propto p^2$. Hence $\lambda_{p,0} \cong 12/p$ becomes small well above threshold. In the next order one gets $\lambda_p \cong 8/p$, which is already indistinguishable from the effective linewidth obtained /8/ from the eigenvalue analysis of the Fokker-Planck equation as displayed in Fig.2 for *all* p; the expressions just given are valid asymptotically, for $p \gtrsim 5$.

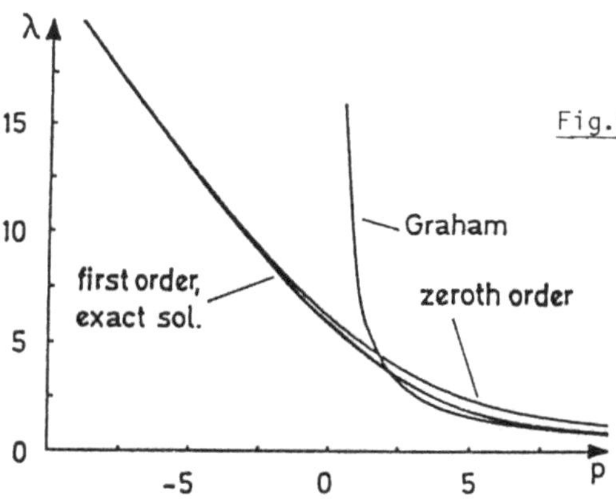

Fig.2 The linewidth λ_p of the partial intensity vs pump rate p in zeroth order and in first order, for which the deviations from the exact λ_p is less than 0.1. Graham's /7/ mean field solution 8/p above threshold is also shown. Since it is equivalent with Haken's /1/ treatment of the SM-laser one observes how well this early, first theory explained the point. - The intensity linewidth λ_I (not drawn) *increases* beyond threshold $\cong 2p$

It may be useful to give a physical interpretation to the lowest order matrix (10) in addition to being the mean value of the Liouvillian, $-\langle\delta A|L\delta A\rangle/\langle\delta A|\delta A\rangle$. Namely, the diagonal matrix elements of (10) can be rewritten as

$$\langle(\delta A_k)^2\rangle = 4\langle I\rangle \tau_k , \qquad (11)$$

with $\tau_k \equiv \lambda_k^{-1}$ the decorrelation times. $\langle I\rangle$ can be understood as the diffusion constant D of the Brownian motion for the intensity variables by the following argument: the u-equation (1) has an additive noise coupling, hence the $I_{x,y}$ equations have a coupling $\propto u$, hence the corresponding I-diffusivity is $\propto |u|^2$ or I. There is no difference between the I-motion and the I_p-motion, or even to the intensity motion of a SM-laser.

Equation (11) tells that decorrelation of the variable A_k is achieved, if the diffusive spreading has reached the order of the stationary variance, a very reasonable estimate. For the total intensity variable I (also in the SM-laser) the variance is $\cong$const=2. Therefore, with increasing diffusivity $D \cong p$ the decorrelation time τ_I shrinks accordingly. But for the partial intensity variable I_p we have learnt that its variance *increases* $\propto p^2$, even faster than the diffusivity ($\propto p$). Therefore τ_p has to increase also, evidently $\propto p$. This argument also precisely meets the reason for the small electric field linewidth of the SM- as well as of the PS-laser: for the u-motion the diffusivity is constant, and the variance is $2\pi\langle I\rangle$ (namely the circumference of the circular valley of $|u|^2$=const), hence $\tau_u \sim p$.

The drawback of the non-stability of the partial intensity I_p due to the Brownian diffusion in the whole range $0 \leq I_p \leq I \cong p$ turns out to be the very source of the small correlation decay linewidth, cf.(11). Since the ρ_s-plateau in Fig.1 implies a variance of order p^2, τ_p must be $\propto p$.

To make this explanation more transparent we integrated /8/ the Langevin equation (1) numerically for the SM-laser and for the PS-laser, cf.Fig.3. Note the same small-scale fluctuations but the additional slow meandering of I_p which would scan the whole range between 0 and I. The slow large-scale motion is dominant in the spectrum as an extremely small peak on the noisy background. (In Fig.3 the PS-laser width is limited in fact by the periodicity-interval T instead of λ_p^{-1}.) To state it alternatively: the range of different time scales is much broader in the PS-case, from the slow diffusion scales to the fast noise scale. They all add up to a rather narrow Fourier transform, in contrast to the SM-laser where only the fast noise scales are present, whose Fourier transform is correspondingly broader.

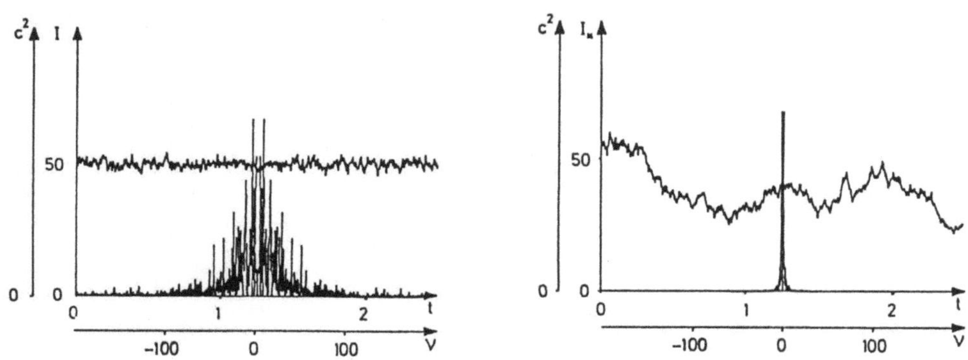

Fig.3 I(t) vs time t for the SM-laser with p=50 (l.h.s.) and $I_p(t)$ vs t for the PS-laser at p=100 (r.h.s.). Also plotted is the spectral distribution (1024 Fourier coefficients, periodicity interval T=2.5), scaled to have the same maximum height in both cases

4. Perturbed PS-laser

Two causes of breaking the polarization symmetry can be expected. Either the cavity prefers certain polarizations or the laser-active material within the cavity does it, in particular by the selection rules of the atomic transitions. While the resonator anisotropies will affect the loss and gain properties directly, describable by an additional linear term $A_{ij}u_j$ in (1), the preferential amplification by the atoms act via the material polarization and the stimulated emission, describable by a modified interaction in (1), an additional $B_{ijkl}u_j u_k u_l$. This has been discussed in detail by van Haeringen /12/. Resonator frequency and atomic transition line are always supposed to coincide.

We restricted ourselves /9/ to the case of *intensity coupling* of the anisotropy producing disturbances. It is well known /13/ that phase coupling is much more complicated; we do not contribute in the following to *such* symmetry-breaking schemes. For intensity coupling it is possible /9/ to obtain the stationary distribution ρ_s of the Fokker-Planck equation even in the presence of the perturbations $\sim A_{ij}$,

$\sim B_{ijkl}$. We discuss the particular cases of three types of symmetry breaking, which are controlled by three respective parameters a_c, a_ℓ, and b, which can be related to the physical quantities discussed by van Haeringen /12/.

$$\rho_s \propto \exp\,[\tfrac{1}{2}pI - \tfrac{1}{4}I^2 + \tfrac{1}{4}a_c\,(I_+ - I_-) + \tfrac{1}{4}b\,(I_+ - I_-)^2 + \tfrac{1}{4}a_\ell\,(I_x - I_y)]. \tag{12}$$

To fix the notation let us briefly recall /14/: $u_1+iu_2=u_+=E_+e^{i\phi+}$, accordingly E_-, ϕ_- from u_3+iu_4. $|u_+|^2=E_+^2=I_+$ is the intensity of the clockwise circular polarization if one faces the incoming light (in physical terms: negative helicity, i.e., left-hand screw of the electric field vector rotation and the direction of light propagation). I_- denotes the intensity of anticlockwise circular polarization facing the light (so positive helicity). It is $I=I_x+I_y=I_++I_-$. The electric field traces generally an ellipse in the x-y-plane of polarization, whose semimajor axis is $(E_++E_-)/\sqrt{2}$, whose opening (ellipticity) angle is $tg\chi=(E_--E_+)/(E_-+E_+)$ with $-\pi/2\leq 2\chi\leq\pi/2$, and whose orientation (tilt) angle is $2\theta=\phi_--\phi_+$ with $0\leq 2\theta\leq 2\pi$. The light is linearly polarized if $\chi=0$ and circularly polarized if $E_x=E_y$ and $2\chi=\pi/2$ (positive helicity, counter-clockwise if facing the incoming beam, I_- preferred) or $2\chi=-\pi/2$ (negative helicity, clockwise, I_+ preferred). The variables I, 2χ, 2θ can be visualized by the Poincaré sphere, having radius I, colatitude $\pi/2-2\chi$ (i.e. $\chi=0$ is the equatorial plane), and longitude (i.e. azimuthal angle) 2θ. The north pole $2\chi=\pi/2$ corresponds to $I=I_-$, counter-clockwise circular polarization, while the south pole $2\chi=-\pi/2$ represents clockwise circular polarization $I=I_+$. One has

$$I_- - I_+ = I\sin 2\chi\ , \tag{13}$$

$$I_x - I_y = I\cos 2\chi\cos 2\theta\ . \tag{14}$$

Having this in mind we observe that b>0 leads to the preference of circularly polarized light, $I\cong I_+$ or $\cong I_-$, while b<0 induces preference of linear polarization, $I_+\cong I_-$ or $\chi=0$. The reason for $b\neq 0$ is the saturation anisotropy of the lasing atoms. The resonator anisotropy (e.g. $\partial_t I_+=a_c I_+$ and $\partial_t I_-=-a_c I_-$) is represented by a_c, the circular absorption anisotropy parameter of the cavity and by a_ℓ, the linear absorption anisotropy parameter of the cavity. Evidently,positive or negative a_c, a_ℓ are equivalent (corresponding simply to an exchange of +/- or x/y), whereas b>0 and b<0 correspond to physically different situations.

Let us consider now three cases in order to study, how with increasing perturbance the ideal PS-laser loses its prominent feature of small partial intensity linewidth. Details are published in /9/.

i) $a_c>0$, $b=a_\ell=0$: circular asymmetry due to the cavity. From (12) we conclude that $I=I_+$ and $I_-=0$ has maximum probability. ρ_s has a peak at the Poincaré sphere's south pole. Clockwise rotation, i.e. negative helicity is preferred. The tangential spread of ρ_s is measured by

$$\left\langle(\delta\sin 2\chi)^2\right\rangle_{I=p} = (4/pa_c)^2 + 1 - \coth^2(pa_c/4)\ . \tag{15}$$

While for $pa_c/4\cong 0$ this is 1/3, for $pa_c/4\gg 1$ it decreases as $(4/pa_c)^2$. Hence we expect that for small $pa_c/4$ we will essentially have PS-laser behaviour, while at $pa_c/4\cong 1$ some switchover will occur.

Indeed we find this in the dynamics. Choosing again the variables δI and δD with $D=I_+-I_-$ the leading order $\Lambda_0^{(0)}=\Lambda_0 N_0^{-1}$ of the CFA is now nondiagonal already. But after diagonalization $C^{(0)}(t)$ again is a superposition of two exponentials:

$$\left\langle\delta I_+\delta I_+(t)\right\rangle \propto N_I e^{-\lambda_I t} + N_D e^{-\lambda_D t}\ . \tag{16}$$

The behaviour of the weights and decay rates is given in Fig.4 for $a_c=1$. We also evaluated /9/ the correlation function in the parameter range $0.1\leq a_c\leq 10$ and found

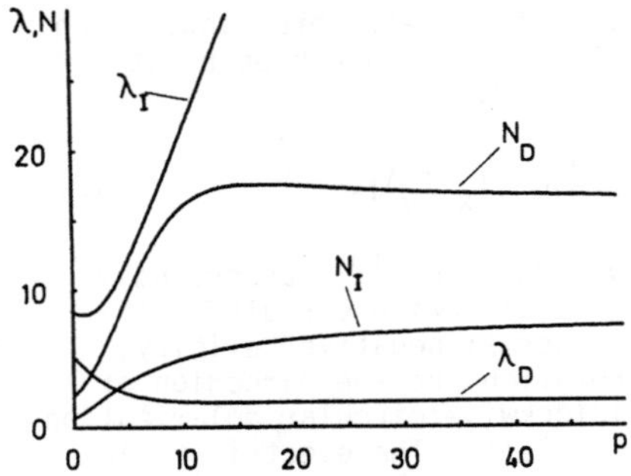

<u>Fig.4</u> Weights and decay rates of the two components of the correlation function for the partial intensity plotted vs pump rate p. It is $a_c=1$, $b=a_\ell=0$

$$\left\langle \delta I_+ \delta I_-(t) \right\rangle \propto 8e^{-2pt} + \frac{16}{a_c^2} e^{-2a_c t}, \tag{17}$$

provided $pa_c/4 >> 1$. While for $a_c=0$ we have $\lambda_D \cong 8/p$ (large p), which decreases to zero with p, λ_D stays constant at a *finite* value if the asymptotic range $pa_c/4 \gtrsim 1$ is reached. For small a_c the second term in (17) dominates in magnitude, so we have PS-like behaviour; for large a_c the first term wins, so we have SM-behaviour. For intermediate a_c there are appreciable deviations from a Lorentzian line shape.

<u>ii) $b>0$, $a_c=a_\ell=0$</u>: saturation anisotropy of the amplifying atoms in an isotropic resonator. ρ_s has two peaks at $2\chi=\pm\pi/2$, either $I \cong I_-$ or $I \cong I_+$. The laser stays close to circular polarization but changes irregularly its helicity under the influence of the quantum noise forces. The mean time τ_{tr} of this transition can be estimated by Kramers' rule to yield

$$\tau_{tr} \cong \pi \exp(bp^3/16), \tag{18}$$

steeply increasing with the pump rate p. For instance, if b=0.05 one has $\tau_{tr} \cong 4.6$ for p=5 but $\tau_{tr} \cong 70$ for p=10, larger than the other relevant time scales. We therefore may restrict ourselves to one Poincaré hemisphere, e.g. $-\pi/2 \leq 2\chi < 0$ or I_+ preferred. Again the tangential spread shrinks from O(1) to

$$\left\langle (\delta \sin 2\chi)^2 \right\rangle_{I=p} \cong 4/b^2p^4 \text{ if } bp^2/4 >> 1. \tag{19}$$

Note the p^{-4} behaviour instead of p^{-2} in (15) which is due to the *quadratic* $(I_+ - I_-)$ term $\propto b$ in ρ_s. Therefore soon a strong restoring force will fix the polarization to the south pole, implying an increase of the correlation decay rate with p, cf. Fig.5. Since N_D/N_I decreases with p the typical SM-behaviour results for stronger pumping, the mode having circular polarization.

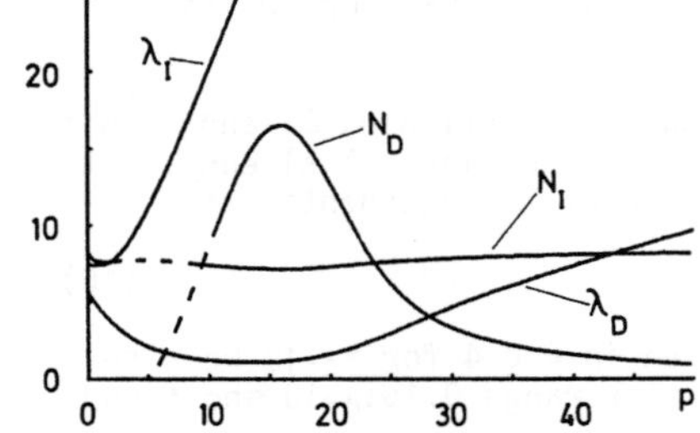

<u>Fig.5</u> Weights and decay rates of $\langle \delta I_+ \delta I_+(t) \rangle$ for b=0.05 but $a_c=a_e=0$. Since the truncated ρ_s (one-hump distribution) was used, the results are reliable only for not too small p ($\gtrsim 5$, thus $\tau_{tr} \gtrsim 5$)

In the parameter interval $0.02 \leq b \leq 0.2$ we obtained /9/ for $bp^2/4 >> 1$

$$\langle \delta I_+ \delta I_+(t) \rangle \propto 8e^{-2pt} + \frac{4}{b^2p^2} e^{-4bpt} \quad . \tag{20}$$

A PS-like behaviour only survives, if $4bp << 2p$ as well as $4/b^2p^2 >> 8$, i.e., if the following two conditions hold simultaneously.

$$2b << 1 \quad , \qquad p \lesssim (2b)^{-1} \ . \tag{21}$$

If either $2b \geq 1$, there is no slow partial intensity decay rate any more (due to the strong restoring force at the pole), or if $p > (2b)^{-1}$, then the residue of the slow motion is small, or if both happens, the PS-feature is gone. For positive b there is at best a p-window slightly above threshold in which PS-behaviour survives, otherwise a circularly polarized SM-laser behaviour is observed. In contrast to cavity anisotropies the PS-breaking by the laser-active material is very effective.

iii) $a_\ell > 0$, $b << -1$, $a_c = 0$: the variable linearly polarized laser. Strong negative b implies $I_+ \cong I_-$, i.e. $\chi = 0$ irrespective of θ. The maximum of ρ_S is at the equator but circular symmetric if $a_\ell = 0$. Such a laser is linearly polarized but with no preferred polarization direction. It might be realized /12/ if the atomic transitions are $j=1/2 \to j=1/2$ or $j \to j+1$ for $j>0$. The cavity still must be symmetric.

As long as $a_\ell = 0$ we still have the possibility of long time correlation in the partial intensities I_x or I_y due to slow quantum noise induced Brownian diffusion of the direction of the linear polarization. If a_ℓ is switched on, the polarization direction gets more and more fixed. From (14) with $\chi = 0$ the relevant spread now is

$$\langle (\delta \cos 2\theta)^2 \rangle_{I=p,\chi=0} \cong 17/2p^2 a_\ell^2 \ , \ \text{if} \ pa_\ell/4 >> 1. \tag{22}$$

As in the case of circular anisotropy of the cavity ($a_c > 0$) also for a linear anisotropy of the resonator ($a_\ell > 0$) we have the p^{-2} confinement of the polarization. The dynamical behaviour is therefore expected to be similar in both cases. This is verified in Fig.6. For the a_ℓ-dependence a fit analogous to (17) was found /9/ if $0.1 \leq a_\ell \leq 5$,

$$\langle \delta I_x \delta I_x(t) \rangle \propto 8e^{-2pt} + \frac{17}{2a_\ell^2} e^{-2a_\ell t} \ , \tag{23}$$

provided $pa_\ell/4 >> 1$. The physical discussion at the end of sect.4.i may be repeated correspondingly.

To sum up our findings in ref 9: Polarization symmetry breaking by the atomic amplification is very strong and leads directly from a PS- to a SM-laser except for a window slightly above threshold. In contrast, symmetry breaking by resonator anisotropies does not change the basic features of the PS-laser — its exceptionally small partial intensity linewidth, as long as the anisotropy is not too large. There

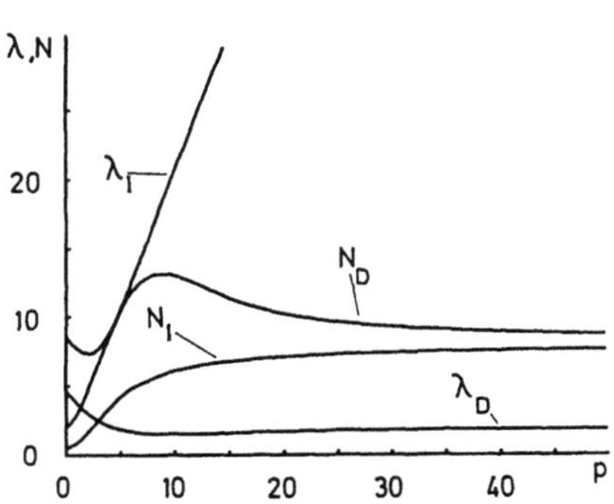

Fig.6 Amplitudes and decay rates of the I_x-correlation for $a_\ell = 1$, $a_c = 0$, $b << -1$. (The precise value of b is unimportant as long as it is big enough in modulus to guarantee linear polarization)

still is a remarkable difference between the large total intensity linewidth $\propto 2p$ and the small partial intensity linewidth $\cong 2a_{c,\ell}$, all the more pronounced the higher the pump rate is.

Needless to add that the field *amplitude* linewidth λ_u measured with an interferometer behaves qualitatively alike in both the SM- and the PS-laser. Only the linewidth factor is reduced from 2 to 1 in the former and from 4 to 3 in the latter case at threshold.

Also with respect to the total intensity I there is no interesting difference between the PS- and the SM-laser. The exciting distinction between them becomes important in an experiment that needs a small photon linewidth, which is provided by PS-laser light filtered for one chosen polarization.

While the original PS-laser does not seem to be a practical source of light with a small intensity linewidth (in consequence of the omnipresent saturation anisotropy), we showed that the partial intensity linewidth of a linearly polarized laser may well be very small. Unfortunately there presently seem to be no experiments in this direction.

Present address of W. K.: Laboratoire de Physique de l'Ecole Normale Superieure, 24, rue Lhomond, F-75231 Paris Cedex 05, France

References

1. H.Haken, Z.Phys.181,96(1964)
2. H.Haken, Laser Theory, Handbuch der Physik, Vol.XXV/2c, Ed.S.Flügge, Springer: Berlin, New York, Tokyo, 1970
3. H.Haken, Licht und Materie II, BI-Wissenschaftsverlag: Mannheim, Wien, 1981
4. S.Smirnov, A.M.Tumaikin, Opt.Spect.33,107(1972)
5. M.Tehrani, L.Mandel, Phys.Rev.A17,677(1978)
6. F.T.Hioe, J.Math.Phys.19,1307(1978)
7. R.Graham, Phys.Letters 103A,255(1984)
8. S.Grossmann, W.Krauth, Phys.Rev.A35, 2523 (1987)
9. W.Krauth, S.Grossmann, Phys.Rev.A35, (May 1987)
10. H.Risken, Fortschr.Physik 16,261 (1968)
11. S.Grossmann, Phys.Rev.A17,1123(1978)
12. W.van Haeringen, Phys.Rev.181,125(1969)
13. P.H.Richter, S.Grossmann, Z.Phys.255,59(1972)
14. M.Born, E.Wolf, Principles of Optics, Pergamon Press: London, 1959

Decay of Unstable Equilibria

F. Haake

Fachbereich Physik, Universität-GHS Essen,
D-4300 Essen, Fed. Rep. of Germany

1. Single-mode Superfluorescence and the Overdamped Pendulum

There is no simpler dynamical problem than that of the overdamped pendulum. Its equation of motion

$$\dot{\theta} = \frac{1}{\tau} \sin \theta \tag{1}$$

is solved by

$$\tan \frac{\theta(t)}{2} = e^{t/\tau} \tan \frac{\theta_0}{2} . \tag{2}$$

We recognize an unstable equilibrium at $\theta = 0$ and a stable one at $\theta = \pi$. The latter one is reached as $t \to \infty$ from any initial angle θ_0 except the one of unstable equilibrium. The time in which θ evolves from $\theta_0 < \pi/2$ to, say, $\theta = \pi/2$ is

$$t_d = - \tau \ln \tan \frac{\theta_0}{2} . \tag{3}$$

Somewhat less trivial is the quantum theory of the overdamped pendulum. Appropriate quantum observables are the components of an angular momentum vector $\vec{S} = (S_x, S_y, S_z)$, $[S_i, S_j] = i\varepsilon_{ijk} S_k$ and the dynamics is most conveniently described by a master equation for the density operator ρ [1],

$$\dot{\rho} = \frac{1}{2j\tau} \left\{ [S_-, \rho S_+] + [S_- \rho, S_+] \right\} , \tag{4}$$

where $S_\pm = S_x \pm i S_y$ are the usual raising and lowering operators. The evolution according to (4) conserves the squared angular momentum

$$\vec{S}^2 = j(j+1) \tag{5}$$

and can therefore be discussed separately for each integer or half-integer value of the quantum number j; for fixed j the density operator can be represented by a (2j+1) by (2j+1) matrix.

The classical dynamics (1) is contained in the quantum one as the limit $j \to \infty$. Indeed, by extracting equations of motion for expectation values from (4) and factorizing as

$$S_z = j \cos\theta \ , \quad S_\pm = j\, e^{\pm i\varphi} \sin\theta , \tag{6}$$

we recover (1) together with the conservation law $\dot\varphi = 0$.

One noteworthy difference between the quantum and the classical case remains even for arbitrarily large j: there is no unstable equilibrium quantum mechanically. Irrespective of its initial form the density operator $\rho(t)$ approaches the unique stationary solution, the projector on the eigenstate of J_z with the smallest possible eigenvalue, $S_z = -j$. The closest quantum analogue of the classical unstable equilibrium is the projector on the eigenstate of S_z with the maximum eigenvalue, $S_z = +j$. The nonstationary of that latter state is easily understood physically by looking at the relative variances of the transverse components of $\vec{S}$, i.e. at

$$\sqrt{\langle j|S_+ \ S_-|j\rangle / j(j+1)} = \sqrt{2/(j+1)} \ . \tag{7}$$

Even though small for large j the transverse variance is necessarily nonzero because of the noncommutativity of S_z and $S_\pm$: the transverse components $S_\pm$ cannot have sharp values in an eigenstate of S_z. Likewise, the initial state $|j\rangle$ cannot assign a vanishing value to the initial angle θ_0, but rather corresponds to a probability density $p(\theta_0, \varphi_0)$ which reads [2], for $j \gg 1$,

$$p(\theta_0, \varphi_0) = \frac{2j}{\pi}\, \theta_0\, e^{-2j\theta_0^2} \ . \tag{8}$$

The asymptotic (large j) treatment of the decay of the initial state $|j\rangle$ is thus obtained by using the classical trajectories (2), the correspondence (6) and the probability density (8) for the effectively random initial angle θ_0. Quantum mean values of operators like $S_+^m S_z^n S_-^p$ thus take the form of averages over an ensemble of classical trajectories. Moreover, for $j \gg 1$ and for times such that $\langle S_+ \ S_-\rangle \gg {}^1/j$ an individual classical trajectory can be associated with an experimentally realizable evolution of the pendulum. An ensemble of experimental trajectories will display fluctuations of, e.g., the decay time t_d according to a probability density defined by (3) and (8). The asymptotic (large j) behavior just discussed has corrections of relative weight ${}^1/j$ and should therefore prevail for moderate values of j, $j \gtrsim 100$, already.

HAROCHE and coworkers [3] have recently measured the decay of the unstable equilibrium of an overdamped "pendulum". The angular momentum $\vec{J}$ in that case is the collective Bloch vector pertaining to a pair of Rydberg states of $N = 2j$ Na atoms in resonance with an eigenmode of a nonideal microwave cavity. $S_z \triangleq ({}^N/2) \cos\theta$ then measures the degree of collective inversion of the N level pairs while $S_{\pm} \triangleq (N/2) \sin\theta \exp(\pm i\varphi)$ refer to the collective electric polarization. In the limit of the cavity damping constant exceeding the collective Rabi frequency the polarization operators are proportional to the electric field operators representing the radiation to the outside of the cavity. The radiated intensity is thus proportional to $S_+ S_- = (N/2)^2 \sin^2\theta \sim N^2$ and amounts to superfluorescence. The experiments in question allow an impressive study of the transition from quantum to classical behavior when the number of atoms is changed from one to up to several thousand. Fig. 1 shows the probability of finding m out of the N initially inverted atoms deexcited at time t [3],

$$W(m) = \int d\theta_0 \, d\varphi_0 \, p(\theta_0, \varphi_0) \, \delta(m - N \sin^2 \frac{\theta(t,\theta_0)}{2}) \quad . \qquad (9)$$

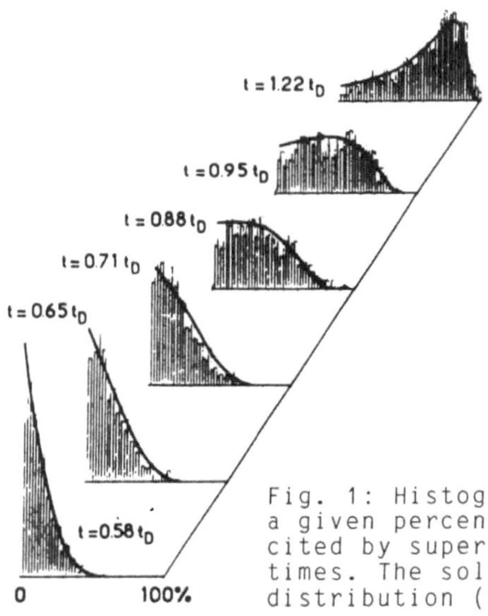

Fig. 1: Histograms giving the probability of having a given percentage of initially excited atoms deexcited by superfluorescence at time t, for various times. The solid lines represent the theoretical distribution (9).

2. Extended-Sample Superfluorescence

Classical deterministic trajectories originating from effectively random initial conditions, the randomness due to quantum uncertainties, are a quite universal phenomenon. It typically arises when the decay of unstable equilibria in macroscopic systems is triggered by quantum fluctuations. All types of superfluorescence are manifestations of this phenomenon. Of considerable interest to experimenters have been superfluorescent systems large in their spatial extension compared to a wavelength of the emitted radiation. A somewhat simplified version of the relevant Heisenberg equations of motion, representing the dynamics of the operators of the electric field $E^{\pm}(x,t)$, the polarization densities $S_{\pm}(x,t)$, and the inversion density $S_z(x,t)$, read [4]

$$\frac{\partial}{\partial t} S_\pm = S_z E^\mp$$

$$\frac{\partial}{\partial t} S_z = -\frac{1}{2}(S_+ E^+ + S_- E^-) \tag{10}$$

$$\left(\frac{\partial}{\partial x} + \frac{1}{c}\frac{\partial}{\partial t}\right) E^\mp = S_\pm \ .$$

The simplifications alluded to concern (i) the restriction to one spatial dimension and to unidirectional propagation of the radiated light, (ii) the neglect of antiresonant terms and the assumption of envelopes varying slowly in space and time (on the scales given by the wavelength and the optical period, respectively). Units have been chosen such that the various operators obey the commutation relations

$$[S_+(x,t),\ S_-(x',t)] = \frac{4}{N}\ \delta(x-x')\ S_z(x',t)$$

$$[S_z(x,t),\ S_\pm(x',t)] = \pm\frac{2}{N}\ \delta(x-x')\ S_\pm(x',t) \tag{11}$$

$$[E^+(x,t),\ E^-(x',t)] = \frac{4c}{N}\ \delta(x-x')$$

where N is the total number of initially excited atoms. The time t is measured in units of the "superfluorescence time" τ which is roughly the natural life time of the excited atomic state divided by the number of atoms; x is referred to the length of the active volume. Even though somewhat over idealized the model just sketched has reasonable validity for cylindrically shaped active volumes with a Fresnel number near unity, the symmetry axis of which then defines the direction of collective radiation. For collective radiation (peak intensity $\sim N^2$) to be possible in practice all competing mechanisms like incoherent relaxation of atomic excitation and polarization as well as dephasing phenomena must be rendered slower than the collective process described by (10).

The initial state of interest for superfluorescence is that of full inversion of the relevant atomic level pair, i.e. the eigenstate of S_z with maximum eigenvalue, $S_z = 1$; the electric field, on the other hand, must start in the vacuum state. Classically, that initial state is an unstable equilibrium. Its decay is initiated by spontaneous emission events, i.e. by quantum fluctuations.

For N >> 1 we can again treat all operators as c-number fields and take the Heisenberg equations (10) as classical (Maxwell-Bloch) equations. The appropriate initial condition is $E^\pm(x,0) = 0$, $S_z(x,0) = 1$ and $S_\pm(x,0) = S_-(x,0)^*$ random with a Gaussian probability density

$$p(S_+,S_-) \sim \exp\left(-\frac{N}{4}\int_0^1 dx\,|S_+|^2\right) \ . \tag{12}$$

The distribution (12) generalizes (8) to the case of spatially extended atomic excitation. In both cases the atomic polarization fluctuates on the very small scale $1/\sqrt{N}$; in the present case, no correlations occur initially between atoms at different locations x.

Due to the initially random polarization configurations $S_\pm(x,0)$ the radiation pulses will greatly fluctuate in shape from one run of the experiment to the next. Especially, the delay time t_d at which the radiation intensity reaches its maximum value will, as in the single-mode case, display a r.m.s. smaller than its mean only by a factor of the order $\ell n\ N$. An analytic expression for the probability distribution of t_d can be obtained [4,5,6] by linearizing the Maxwell Bloch equations (10) around the initial state, i.e. by setting $S_z = 1$. Such an approximation must yield good results (as in the single-mode case where $\sin\theta \to \theta$ in (1) gives $\tan(\theta_0/2) \to \theta_0/2$ in (3) which is quite acceptable since θ_0 is typically of the order $1/\sqrt{N}$) since the greatest part of the delay t_d is due to the slow traverse of the close neighborhood of the unstable equilibrium state where nonlinear effects are entirely negligible.

There have been numerous experiments on quasi one-dimensional superfluorescence. The first report of quantitative delay time statistics was given in [7]. Fig. 2 shows the probability density of the delay time difference of the pair of pulses radiated to the left and to the right of an active cylinder. The histogram is taken from [7] and the theoretical curve from [8] where the linearized theory sketched above was extended to include left- and rightgoing pulses. Highly interesting investigations have been reported of more complicated situations such as competition [9] and beats [10] between various transitions, three-dimensional effects [11,12,13], and the rivalry between collective radiation, incoherent relaxation, and dephasing [4b,8,14,15,16].

3. Onset of Raman Scattering

The linearized ($S_z = 1$) versions of the Maxwell Bloch equations (10) not only describe the early stages of superfluorescent pulses but also the onset of the Stokes wave in a Raman active medium irradiat-

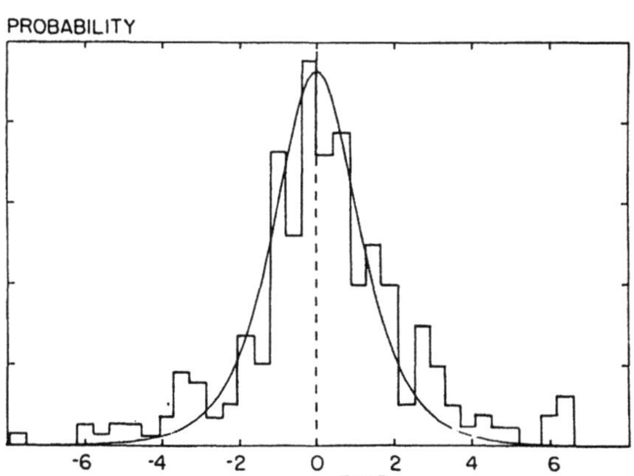

Fig. 2: Probability density of the delay time difference within a pair of pulses radiated to the left and to the right of an active cylinder. Experimental histogram from ref [7], theoretical curve from [8].

ed with a laser pulse of rectangular shape [17]. The field $E^{\pm}(x,t)$ then describes the Stokes wave while the polarization density $S_{\pm}(x,t)$ refers to the atomic level pair which picks up, for any elementary scattering act, the energy difference between the laser photon and the Stokes photon. The quantity of experimental interest in this case is the number of Stokes quanta appearing at the end face x = 1 of the active volume in the time interval $0 \leq t \leq \tau_L$ during which the rectangular τ_L laser pulse is on,

$$W \sim \int_0^1 dt\, E^-(1,t)\, E^+(1,t)$$

$$= \int_0^1 dx \int_0^1 dx'\, A(x,x')\, S_+(x,0)\, S_-(x',0)\ . \qquad (13)$$

The kernel $A(x,x')$ can be given in closed form on the basis of the analytic solution of the linearized version of (10) [17,18]. By again taking $S_-(x,0) = S_+^*(x,0)$ as random c-number variables we can find the probability distribution of the number of Stokes quanta as

$$P(W) = \langle \delta(W - \int_0^1 dx \int_0^1 dx'\, A(x,x')\, S_+(x,0)\, S_-(x',0))\rangle\ . \qquad (14)$$

The average in eq. (14) is over the function space accessible to $S_+(x,0)$ and involves the weight (12). The corresponding functional integral has been found to be closely approximated by an exponential [18,19]

$$P(W) \approx \frac{1}{\langle W\rangle} e^{-W/\langle W\rangle}\ . \qquad (15)$$

An experimental confirmation [20,21] of this theoretical result is shown in Fig. 3.

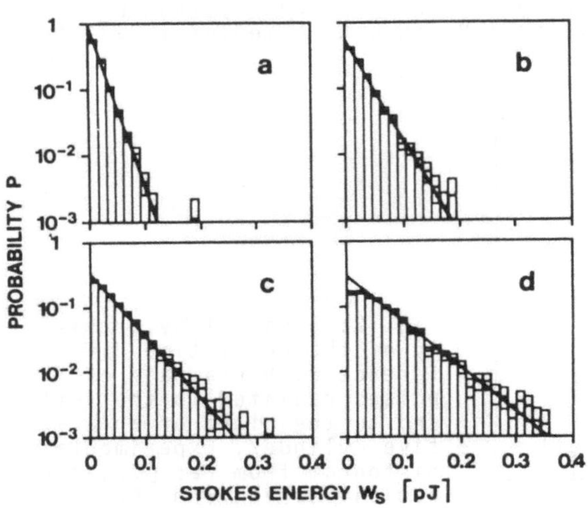

Fig. 3: Stokes energy statistics for various laser pump energies, the latter increasing from (a) to (d); taken from ref [21].

4. Quantum Measurements

Consider an harmonic oscillator with the free Hamiltonian

$$H_0 = \frac{1}{2M} P^2 + \frac{1}{2} M \omega_0^2 Q^2 \tag{16}$$

and let it be coupled to a "heat bath" of N other oscillators. The free bath Hamiltonian is

$$H_{bath} = \sum_{n=1}^{N} \left(\frac{1}{2m} P_n^2 + \frac{1}{2} m \omega_n^2 Q_n^2 \right) \tag{17}$$

and the coupling

$$H_{int} = \sum_{n=1}^{N} \varepsilon_n \sqrt{mM}\, Q Q_n \ . \tag{18}$$

This model is quite a popular one in the theory of damping [22,23, 24]. I shall here put the model to a slightly less conventional use.

The extremum of the potential energy in the Hamiltonian (16,17, 18) can be found from the extremum of the effective potential

$$V_{eff}(Q) = \frac{1}{2} M \left(\omega_0^2 - \sum_n \frac{\varepsilon_n^2}{\omega_n^2} \right) Q^2 \ . \tag{19}$$

We immediately infer from (19) that the bath tends to weaken the restoring force of the central oscillator. For sufficiently strong coupling

$$\omega_0^2 < \sum_n \frac{\varepsilon_n^2}{\omega_n^2} \tag{20}$$

the stability of the extremum at $Q = 0$ is lost. The Hamiltonian (16, 17,18) is then no longer bounded from below and has one imaginary eigenfrequency $i\Gamma_-$ in addition to N real ones. In the most interesting limit in which the unperturbed bath frequencies grow infinite in number and merge to a continuum the N real eigenfrequencies of the complete Hamiltonian H go dense, too, and give rise to decaying contributions to expectation values like $\langle Q(t)^n \rangle$, $\langle P(t)^n \rangle$, ... [22 - 24]. The imaginary eigenvalue $i\Gamma_-$ of H, however, remains isolated even in the continuum limit and clearly contributes a growing exponential to expectation values [25,26]. That growing exponential signals, as was the case in the problems discussed in the preceeding sections, the decay of an unstable equilibrium [27]. I now propose to show that the amplification process in question finds an interesting application in the quantum theory of measurement.

It is convenient to work with the reduced Wigner function $W(p,q,t)$ of the central oscillator. If the bath is assumed in thermal equilibrium initially,

$$\rho_{bath}(0) \sim e^{-\beta H_{bath}} \ , \tag{21}$$

the Wigner function can be shown to obey an equation of motion of the form $\dot{W} = \ell W$ with a generator of infinitesimal time translations ℓ containing first and second partial derivatives with respect to p and q [24]. If the continuum limit for the bath is taken with the spectral strength of bath frequencies

$$\gamma(\omega)\Delta\omega = \sum_{n}^{\omega \lesssim \omega_n \lesssim \omega+\Delta\omega} \varepsilon_n^2 \sim \frac{\omega^2}{\alpha^2+\omega^2}, \tag{22}$$

all coefficients in ℓ can be given rigorous closed-form expressions. For times much larger than the inverse bath correlation time, $t \gg {}^1/\alpha$, and high temperatures, $kT > \hbar\alpha$, the generator takes the form of the Fokker-Planck operator of the Ornstein Uhlenbeck process [24,26]

$$\ell = -\frac{\partial}{\partial q} p/M + \frac{\partial}{\partial p}\left\{(\Gamma_+ - \Gamma_-)\, p - M\Gamma_- \Gamma_+ q\right\}$$

$$+ M\, k\, T(\Gamma_+ - \Gamma_-) \frac{\partial^2}{\partial p^2}. \tag{23}$$

The amplification rate Γ_- and the damping rate Γ_+ are both assumed small compared to α and are determined by the quadratic equation

$$\Gamma_+ \Gamma_- = \int_0^\infty d\omega \frac{\gamma(\omega)}{\omega^2} - \omega_0^2$$

$$\Gamma_+ - \Gamma_- = \frac{1}{\alpha} \int_0^\infty d\omega \frac{\gamma(\omega)}{\omega^2}. \tag{24}$$

Their physical meaning becomes especially apparent in the behavior of the mean displacement [28]

$$\langle Q|t|\rangle = \dot{A}(t) \langle Q(0)\rangle + A(t) \langle P(0)\rangle/M \tag{25}$$

$$A(t) = \left(e^{\Gamma_- t} - e^{-\Gamma_+ t}\right)/(\Gamma_+ + \Gamma_-). \tag{26}$$

I shall henceforth consider times large compared to the inverse amplification rate and neglect the decaying transient in (26). In that limit the momentum of the central oscillator is in rigid adiabatic equilibrium with the displacement [29],

$$\langle q^m p^n\rangle(t) = (M\Gamma_-)^n \langle q^{m+n}\rangle(t)$$

$$\langle q^m\rangle(t) = \dot{A}_{as}(t)^m \langle (q+p/M\Gamma_-)^m\rangle(0)$$

$$\dot{A}_{as}(t) = \frac{\Gamma_-}{\Gamma_+ + \Gamma_-} e^{\Gamma_- t}. \tag{27}$$

For the large times in question the oscillator behaves classically. The uncertainty product $\langle(\Delta P)^2\rangle\langle(\Delta Q)^2\rangle$ has been amplified to a value so large compared to the minimum quantum size $\hbar^2$ that a "reading" of the displacement Q or of the momentum P can be thought of (i.e. performed so) as not amounting to a noticeable perturbation.

Moreover, the averages (27) can be interpreted as averages over an ensemble of classical trajectories, each of which corresponds to an experimentally realizable evolution; quantum effects are present only as an effective initial noise. An ensemble of many experimental displacement trajectories will thus reveal an effective initial displacement uncertainty characteristic of the initial quantum state of the oscillator.

If the amplifier in consideration is to be used as a meter recording the quantum state of some microscopic object it must undergo a coupling to that object before the amplification process is triggered. Consider therefore an object, an observable ξ of which is to be measured. Let the eigenvalues ξ_i of ξ be discrete,

$$\xi|\xi_i> = \xi_i|\xi_i> , \tag{28}$$

for the sake of simplicity, and let the density operator of the object before the measurement be an arbitrary mixture (or superposition) of the states $|\xi_i>$

$$\rho_{object} = \sum_{ij} \rho_{ij}|\xi_i><\xi_j| . \tag{29}$$

The object meter coupling should not change the occupation probabilities ρ_{ii}. The interaction Hamiltonian H' should therefore be designed so as to commute with ξ. On the other hand, H' must cause a shift of the probability density of the meter displacement and that shift must be different for different eigenstates of ξ which may be realized initially. Finally, there is no necessity to have the object involved in the amplification process of the meter. A suitable choice for the coupling is

$$H' = \varepsilon\ \xi\ P\ \ \delta(t) , \tag{30}$$

where ε is a coupling constant. Denoting the time immediately before the coupling as 0 and the time immediately afterwards by 0^+ we have for the joint density matrix of the object and the oscillator (the meter) in the representation of eigenstates of ξ and Q

$$<\xi_i,q|\rho(0^+)|\xi_j,q'> = \rho_{ij}\ <q - \varepsilon\ \xi_i|\rho_0(0)|q' - \varepsilon\ \xi_j> . \tag{31}$$

The initial condition for the meter must now be discussed. It is reasonable to assume the displacement and the momentum of the meter to vanish in the mean initially. The meter displacement brought about by the impulsive measurement is thus characterized by the following mean and variance,

$$\begin{aligned} <Q(0^+)> &= \varepsilon <\xi(0)> \\ \sigma(0^+) &= \sigma(0) + \varepsilon^2\ \sigma_{\xi\xi}(0) , \end{aligned} \tag{32}$$

where $\sigma_{\xi\xi}(0) = <\xi^2_{(0)}> - <\xi(0)>^2$ is the variance of ξ with respect to the object density operator (29). Clearly, with respect to measurements of ξ the contributions $\sigma(0)$ and $\varepsilon^2\ \sigma_{\xi\xi}(0)$ to $\sigma(0^+)$ represent noise and signal. For the oscillator to be useful as a meter we must require a small signal - to - noise ratio,

$$\sigma(0) \ll \varepsilon^2 \sigma_{\xi\xi}(0) . \quad (33)$$

If meter readings are to resolve the eigenvalues ξ_i of ξ on a scale $\Delta\xi$ the mean squared pointer displacement before the measurement must obey

$$\sigma(0) \ll (\varepsilon \Delta \xi)^2 . \quad (34)$$

With that condition met the probability density of meter displacements after the measurement,

$$\langle q|\rho_0(0^+)|q\rangle = \sum_i \rho_{ii} \langle q - \varepsilon \xi_i|\rho_0(0)|q - \varepsilon \xi_i\rangle , \quad (35)$$

will, as a function of q, display separated peaks at the values $\varepsilon \xi_i$ which are distinguished by the discrete eigenvalues ξ_i of the object observable ξ.

When the later amplification has gone on for times larger than $1/\Gamma_-$ the probability density $\langle q|\rho_0(t)|q\rangle$ will display peaks at the distinguished displacements $\varepsilon \xi_i \dot{A}(t)$ the separation of which will eventually be large enough to be easily readable. In an individual run of the measurement the meter displacement at a time $t \gg 1/\Gamma_-$ will most likely be close to one of those most probable values and thereby record the corresponding object eigenvalue ξ_i.

The reduced object density matrix immediately after the measurement follows from (31) as

$$\langle\xi_i|\rho_{object}(0^+)|\xi_j\rangle = \rho_{ij} \int_{-\infty}^{+\infty} dq \langle q - \varepsilon \xi_i|\rho_0(0)|q - \varepsilon \xi_j\rangle . \quad (36)$$

Whether or not the density operator of the object has collapsed to a diagonal mixture of eigenstates $|\xi_i\rangle$ obviously depends on the initial state of the meter. More specifically, the collapse will be the more pronounced the smaller offdiagonal elements of $\rho_0(0)$ with a skewness $\varepsilon(\xi_i - \xi_j)$ are.

The q-integral in (36) is easily recognized as the generating functional for expectation values of the initial meter momentum,

$$\int dq \langle q + \tfrac{1}{2} y|\rho_0(0)|q - \tfrac{1}{2} y\rangle = \sum_{n=0}^{\infty} \frac{1}{n!} (iy/\hbar)^n \langle p^n\rangle (0)$$

$$= 1 - y^2/2\lambda^2 + \ldots , \quad \text{where} \quad (37)$$

$$\lambda = \hbar/\sqrt{\langle p(0)^2\rangle} \quad (38)$$

is the initial de Broglie wavelength of the meter. A necessary condition for strong suppression of offdiagonal elements (36) therefore is that the initial de Broglie wavelength of the meter be small compared to $|\varepsilon(\xi_i - \xi_j)|$ for neighboring eigenvalues of ξ. For instance, any initial meter state (or mixture of states) the Wigner function of which is a Gaussian with a width $\hbar/\lambda$ such that $\lambda << |\varepsilon(\xi_i - \xi_j)|$ will entail the desired effective collapse, i.e. to a strong suppression of the offdiagonal terms of the density matrix (29).

To conclude, the model defined by the Hamiltonian (30,16,17,18) can describe a measurement of a quantum observable ξ provided the initial state of the oscillator (the meter) is chosen such that both the de Broglie wavelength and the root mean squared meter displacement are small in the sense

$$\sigma, \lambda^2 << \varepsilon^2 (\xi_i - \xi_j)^2 . \tag{39}$$

References

1. R. Bonifacio, P. Schwendimann, F. Haake: Phys. Rev. A4, 302 (1971); A4, 854 (1971)
2a. V. Degiorgio: Opt. Comm 2, 362 (1971)
b. F. Haake, R. Glauber: Phys. Rev. A5, 1457 (1972)
3. J. M. Raimond, P. Goy, M. Gross, C. Fabre, S. Haroche: Phys. Rev. Lett. 49, 1924 (1984)
4a. F. Haake, H. King, G. Schröder, J. Haus, R. Glauber: Phys. Rev. A20, 2047 (1979)
b. F. Haake, H. King, G. Schröder, J. Haus, R. Glauber: Phys. Rev. A23, 1322 (1981)
5. D. Polder, M. Schuurmans, Q. Vrehen: Phys. Rev. A19, 1192 (1979)
6. M. Gross, S. Haroche: Phys. Rep. 93, 302 (1982)
7. Q. H. F. Vrehen: In Laser Spectroscopy IV, ed. by H. Walther and K. W. Rothe, Springer Series in Optical Sciences, Vol. 21 (Springer, Berlin, 1979); see also H. M. Gibbs, Q. H. F. Vrehen, H. M. J. Hikspoors: Phys. Rev. Lett. 39, 547 (1977)
8. F. Haake, J. Haus, H. King, G. Schröder, R. Glauber: Phys. Rev. Lett. 45, 558 (1980)
9a. R. Florian, L. O. Schwan, D. Schmid: Phys. Rev. A29, 2709 (1984)
b. F. Haake, R. Reibold: Phys. Rev. A29, 3208 (1984)
10. Q. H. F. Vrehen, H. M. J. Hikspoors, H. M. Gibbs: Phys. Rev. Lett. 38, 764 (1977)
11. J. Mostowski, B. Sobolewska: Phys. Rev. A28, 2943 (1983); A30, 1392 (1984)
12. S. Prasad, R. Glauber: Phys. Rev. A31, 1583 (1985)
13. H. Ch. Fleischhauer, A. Schiller, L. O. Schwan, D. Schmid: Laser und Optoelektronik 3, 205 (1986)
14. M. F. H. Schuurmans, D. Polder: Phys. Lett. 72A, 306 (1979)
15. M. F. H. Schuurmans: Opt. Comm. 34, 185 (1980)
16. M. S. Malcuit, J. J. Maki, D. J. Simkin, R. W. Boyd: preprint (1986)
17. M. G. Raymer, J. Mostowski: Phys. Rev. A24, 1980 (1981)
18. M. G. Raymer, K. Rzazewski, J. Mostowski: Opt. Lett. 7, 71 (1982)
19. F. Haake: Phys. Lett. 90A, 127 (1982)
20. I. A. Walmsley, M. G. Raymer: Phys. Rev. Lett. 50, 962 (1983)
21. N. Fabricius, K. Nattermann, D. von der Linde: Phys. Rev. Lett. 52, 113 (1984)
22. P. Ullersma: Physica (Utrecht) 32, 27 (1966); 32, 56 (1966), 32, 74 (1966), 32, 90 (1966)
23. F. Haake, R. Reibold: Acta Physica Austriaca 56, 37 (1984)
24. F. Haake, R. Reibold: Phys. Rev. A32, 2462 (1985)
25. F. Haake, D. F. Walls: In Quantum Optics IV, ed. by J. D. Harvey and D. F. Walls, Springer Proceedings in Physics (Springer, Berlin, 1986)

26. F. Haake, D. F. Walls: to be published

27. It may be appropriate to remark that the phenomenon in question arises only for strong couplings between the central oscillator and the bath; a perturbative treatment of the coupling is therefore not possible.

28. There is an interesting strong-coupling anomaly in the effective initial condition for the Wigner function. The Fokker-Planck equation with the generator (23) holds only for $t \gg 1/\alpha$. In the time interval $0 \leq t \approx 1/\alpha$ the reduced Wigner function relaxes from $W(p,q,0)$ to $W_{eff}(p,q,0) = W(p + (\Gamma_+ - \Gamma_-)\, q,\, q,\, 0)$. This effective initial condition is accounted for in (25,26).

29. $\langle q^m\ p^m \rangle$ denotes a moment of the Wigner function and equals the expectation value of the symmetrized product $\{Q^m\ P^n\}$ of the corresponding operators.

Part IV

After-Dinner Speech

Synergetics and Weltanschauung

G. Caglioti

Istituto di Ingegneria Nucleare, CESNEF, Politecnico di Milano, Via Ponzio, 34/3, I-20133 Milano, Italy

1. Introduction

Last summer, when Professor Robert Graham invited me to deliver the after-dinner speech of this meeting, I accepted promptly, with gratitude and pleasure. Moved by enthusiasm, I did not realize that the organizers intended including a written version of my speech in the Proceedings. Only now do I realize how serious my task is: to tune what I am supposed to write during these winter days to the mood of the cheerful audience that is going to gather around Hermann Haken in a mid-summer evening next year.

After-dinner speeches are generally enlivened with witty remarks and jokes. The effectiveness of these witticisms, as measured by the favour with which the audience responds to them, is controlled mainly by the synergetic spirits of one's table-companions. But, as is usually the case for nonlinear systems removed far from equilibrium, the audience response depends to a large extent also on small erratic stimuli like the inflexion of the voice, as well as on extemporaneousness and unforeseeableness. In other, more common and less unforeseeable words, surprise turns out to be an important ingredient of success also in the context of such a convivial ceremony as that which we are about to attend. Perhaps that is why, all over the world, after-dinner speakers are reluctant to freeze well in advance a faithful written version of their talk, and to allow it to be distributed beforehand. But a proverb says: when in Rome do as the Romans do. And here I am - with my mind in Stuttgart, Germany, after dinner and my pen in Luvinate, Italy, before dinner - straining under tension. Moreover, the deadline fixed by Professor Graham is approaching. I must find a solution. Quickly.

For the reasons just presented, I propose to omit, in the present written version of my speech, most of what makes an after-dinner speech different from a before-dinner one. Without planning, of course, to do without the dinner.

The objective of this paper is to render homage to Hermann Haken on the occasion of his sixtieth birthday.

To find words adequate to the merits of this scientist is not easy.

Hermann Haken, the founder of synergetics, has turned our directions and our way of thinking. He has widened the domain of physics tremendously: the very definition of synergetics - the discipline concerning the cooperation of many subsystems such as atoms, molecules, cells, animals or humans that, under the action of external flows of energy, matter or information, may produce spatial, temporal or functional structures [1] - reminds us that, thanks to Hermann Haken, a method is now available enabling physicists not only to solve a variety of technical problems, but also to tackle a number of problems traditionally attributed to other disciplines, without betraying physics. According to synergetics, the behavior of open systems is not too sensitive to the nature of the subsystems and their supposedly nonlinear mutual interactions. It thus becomes possible to aim at a unified understanding of an ample variety of natural phenomena of direct interest to us: from physics to biology, from psychology to aesthetics, from economy to sociology, in short, from everyday life to Weltanschauung.

If you have not yet read Haken's book "Erfolgsgeheimnisse der Natur. Synergetik: die Lehre vom Zusammenwirken" [2] please do. There you will find a guide to the understanding of all kinds of structures and processes, and advice for disentangling all sorts of situations and impasses, including perhaps those of a sentimental genre.

This evening I am able to take my pick. The time allotted to the speech barely allows me to linger over just one of the topics considered in that book. I shall pick perception, a process of vital importance to all of us [3].

In synergetic terms, perception can be envisaged as a dynamic instability leading to the self-organization of an initially disordered heap of sensory stimuli into the final form of an ordered "visual thinking" [4]. Perception, as any dynamic instability, involves symmetry breaking: I shall take advantage of this circumstance in order to let you experience in vivo the importance of both symmetry and symmetry breaking in visual communication and, in general, in culture [5].

And now a few words on the structure of this contribution. We shall focus most of the time on images: the importance of images nowadays is paramount for several reasons which are indicated in Sect.2. A synergetic model of perception is outlined in Sect.3. In view of the nature of perception, I propose to concentrate attention on two concepts and to convey one message. Concept 1 consists of a pair of definitions of symmetry (Sect.4). Since these definitions require only at most four words each, it will be difficult to forget them immediately. Concept 2 correlates symmetry breaking and information (Sect.5). The message states that symmetry and symmetry breaking cooperate to make the language of art universal (Sect.6). A word of caution is in order at this point: Sect.6 deals with delicate systems belonging to the domains of aesthetics and psychology: therefore its validity is questionable. However, be that as it may, in order to illustrate the contents of Sects.4-6, in Sect.7 some instances of symmetry and symmetry breaking are presented in painting and in music, in design and in poetry. The last part of this final section is devoted to a case study: the role of symmetry and symmetry breaking in industrial trade marks.

2. Evolution of language and culture of images

In the 1950s Leslie Alvin White proposed that a measure of the degree of sophistication of a society could be provided by the per capita power consumption. The neoevolutionistic model of White has been severely criticized by several schools of cultural anthropology. Nevertheless today nobody would dare to argue against the evidence that the mutations that have occurred over the centuries and in the recent past in the dynamics of social relations have been made possible thanks indeed to the more and more conspicuous amounts of energy made available to people.

White's criterion could be formulated in synergetic terms by assigning the role of control parameter to the per capita consumption of power and the role of order parameter to language.

Like in the "society of photons" running in a laser cavity or in the "society of molecules" contained in a liquid heated from below (Fig.1), a hierarchy of dynamic instabilities develops also in the society of men: in all cases it will be sufficient that the flux of ingoing energy pumped in exceeds a threshold. Ultimately power input is the main cause of the dynamic instabilities undergone by society: we allude here to the transformations from agricultural to industrial and thereafter to post-industrial society. As these transformations occur, the objectives of human life change: from survival to consumerism and creation of options, and thereafter to creation tout court. And, of course, the means to achieve these more and more ambitious objectives also change: from procreation to industrial production and thereafter to industrial production and service supply assisted by robots and computers.

The objects around us become more and more numerous and the functions accomplished by these objects multiply together with the options created by them. While we are urged by the need to make up our mind, to make our choice, to remove uncertainty and to create information and knowledge in a world correlated and intricated to such an extent as to appear at times frustrating and chaotic, the number of new words increases. Languages expand too fast when compared to the cultural inertia of society. But "using the terminology of synerget-

Fig.1 Surface-tension-driven cells of convection beyond the Bénard instability in silicon oil. Light is reflected from aluminum flakes suspended in the oil. The oil flows upwards at the center of the hexagonal cells and moves towards their periphery, where it flows downwards. (Courtesy of M.Velarde)

ics, languages are the order parameters slaving the subsystems that are the human beings. A language changes only little over the life span of an individual. After his birth an individual learns a language, i.e. he is slaved by it, and for his lifetime contributes to the survival of the language" [1]. With the help of Hermann Haken we thus recognize in language the basis for "the system of the relations connecting all the aspects of the social life": a system that, according to Claude Lévi-Strauss, "in the transmission of the culture plays a more important role than that one played by any aspect considered separately" [6].

By definition, order parameters vary very slowly. No wonder then that a feeling of confusion may arise when a basis of reference such as language varies too rapidly with respect to the cultural inertia of society. Bigger and bigger vocabularies and treatises currently being published on languages are not sufficient to cope with this bewilderment. We must admit that in order to relieve this uneasy feeling a massive recourse is being made to images. Nowadays images as tools for communication and entertainment are being created at an increasing rate. Some people are literally overwhelmed by images.

It seems that we are on the way to recovering that ancient form of thinking and communicating (by pictograms, hieroglyphics and ideograms) that preceeded writing [7]: the culture of images explodes again.

The imbalance between the copious production of options and knowledge on the one hand and the cultural inertia of society on the other stimulates the utilization of images as a tool for communication. However this imbalance is not the only reason for the renewed importance of images. Additional reasons, of a technical nature, can be recognized in the large amounts of information as well as in the viability of the computer assisted production, animation and real time diffusion of images. But also important psychological reasons emerge. Images, while being perceived, propose themselves vividly to our fantasy and strike our sensitivity. Once perceived, they stick in our memory, ready to correlate themselves with our previous experience, and help us to evoke it. Images have a direct impact on our mood and may promote a sudden reorientation of our thinking, thus producing an emotion. At times, images procure us the illusion of understanding without effort. Some images, while we perceive them, make us believe that we are taking an active role in the process of creation of the object they illustrate. Occasionally, when we actively perceive an image, we are gratified by the feeling of creating. The reason for this exciting feeling lies in the universality of images. Like music, images don't need to be translated in order to be enjoyed.

What are the reasons for the universality of images and of visual communication? We shall reconsider this question in Sect.6, but first let us outline a model of perception, and discuss two ingredients contained in most images: symmetry and symmetry breaking.

3. Perception: an outline of a synergetic model

Images are attractive, they draw our attention.

An image can be envisaged as a sort of potential field of the attention. The minima of this potential are carved by signs and symbols: the sharper the symbols and the signs in the pattern, the deeper the corresponding minima.

The image acts as a source of sensory stimuli. These stimuli enter the memory during observation, but escape quickly, unless sufficient control is exerted on them by the mind. In order to accomplish this control efficiently, the mind correlates the stimuli, aggregating them in coherent schemes, precursors of the visual thinking. The control of the mind - in synergetic terms, the control parameter - is exerted especially on stimuli originating from the above minima: no control can be effective in practice on sensory stimuli suggested e.g. by a white canvas, whose potential of attention is irremediably flat.

When the control parameter exceeds a certain threshold, the initially disordered heap of uncorrelated sensory stimuli suddenly self-organizes into the final form of an ordered visual thinking. Hermann Haken writes: " the order parameters are ultimately the thoughts" [1]: his aphorism helps us to understand how the following of the birth and growth of a visual thinking during perception is indeed a way to experience in vivo the dynamics of an order parameter during a dynamic instability [5].

Once formed, the visual thinking behaves like a particle described by a nonstationary "quantum" state in the potential field of the attention, namely like a particle shaken across the field itself (this is a statement that should be taken with a grain of salt: of course no Planck constant is involved here). In simpler words, during its wandering, the visual thinking is bound to fall eventually into the minima of the potential of attention, like a river flowing along the bottom of the valley or rain water gliding downhill. The signs and symbols corresponding to these minima act as attractors of the visual thinking. We come back thus to the opening sentence of this section: images are attractive: they draw our attention. Thereafter they become attractors of our visual thinking.

Using the terminology of synergetics, so far we have schematically described a possible mechanism of perception. Perception is conceived as a dynamic instability produced cooperatively by the sensory stimuli. As such, it shares many features with dynamic instabilities of nonisolated physical or chemical systems. Perception leads from an uncertain and "symmetric" state of the mental system to an informed and ordered one. Perception involves a control parameter exerted by the mind on the stimuli, and an order parameter, the visual thinking, arising synergetically at the critical point of instability, when the image is first recognized. As we shall see later, symmetry breaking occurs at the instability. Beyond the instability, the visual thinking develops further, searching for the significance of the image. If I am allowed to use esoteric words only parenthetically on this occasion, the visual thinking behaves like a theoretical physicist when he analyzes a molecule or a quantum structure: for him the significance of the structure is its Hamiltonian or, equivalently, its energy. Energy is conserved: energy synthetizes all the symmetry elements leaving the structure invariant. Therefore energy and/or symmetry do indeed represent permanent reference values.

The following two sections are devoted to the concepts of symmetry, symmetry breaking and broken symmetries.

4. Symmetry

For the sake of simplicity, let us refer to an image containing identifiable geometrical elements. While exploring the significance

Fig.2 Corinthian crater. A rotation of the frieze of the horses by one module leaves the frieze practically identical to itself (from [5])

of the image, visual thinking operates essentially a screening of it. In other words visual thinking submits the elements of the image to a number of transformations, e.g. permutations, reflections, rotations, translations, inversions with respect to a point (for instance the geometrical center, if present).

Most of these mental transformations would produce a perceptible change in the structure of the image. However there might be transformations leaving the structure identical to itself: these transformations are called symmetry transformations (Fig.2).

Take, for instance, the wedding ring. The ring is left identical to itself if it is "transformed" by a rotation around its geometrical axis: that is very likely one of the reasons why the shape of this ring qualifies it naturally as the symbol of the invariance of conjugal love. To state (again parenthetically) that a rotation around the geometrical axis leaves the ring invariant, implies that the potential energy of the ring does not change if the angle of rotation changes. That is equivalent to saying that the moment of the external forces acting on the ring vanishes, and, consequently, that the rate of change of the angular momentum vanishes too. In other words, for an isolated axially symmetric ring there is only one choice: either the ring keeps still, or it rotates with constant angular velocity.

Therefore, symmetry is associated with invariance as well as the impossibility of measuring or perceiving a change: " the root of all symmetry principles in physics - according to T.D. Lee - is that it is impossible to measure certain fundamental quantities" [8].

In conclusion, symmetry is **indiscernibility of a transformation.** In three words, symmetry is **invariance against transformations.**

5. Symmetry breaking and information

If symmetry is present, a transformation, operated e.g. by the mind on the structure, produces imperceptible changes and leaves the structure invariant.

Conversely, if symmetry is broken or simply absent, such a transformation produces perceptible changes of the structure. These changes can be ascribed to an imbalance frozen in the configuration of geometrical, chromatic or morphological elements of the image (Fig.3). That is the case of **broken symmetries**.

However, sometimes these changes occur dynamically, promoted by the very process of perception. We allude for instance to ambiguous images, whose elements seem to possess e.g. a center of symmetry only at first, when they are considered superficially as belonging to a 2-dimensional surface. However, with reference for instance to the "condensed" binary structure of Fig.4, under closer observation the

Fig.4

Fig.3 Panathenaic amphora, showing two marathon runners and TWO MARATHON RUNNERS. Since the translational symmetry is broken, the image suggests relative change of momentum or velocity of the two athletes (from [5])

Fig.4 Graphical condensation (by Franco Grignani)

center of symmetry disappears suddenly: while the control parameter increases, the potential of attention changes producing two attractors, one on the left hand side and another on the right hand side. As soon as the critical stage is reached, a dynamic instability occurs, leading perception from flat to spatial: that is the case of **symmetry breaking**.

In order to correlate symmetry breaking and information, let us focus our attention again on the graphical condensation of the two cubic modules (Fig.4). Metaphorically, this figure can be interpreted as representative of the formation of e.g. the "hydrogen molecule" starting from two initially separated "hydrogen atoms" (the inner white square representing the "proton" and the outer black walls the delocalized "electrons").

Each of the six binary structures in this figure is composed of a pair of modules. By construction, each structure possesses a center of symmetry: by inverting any element of a binary structure with respect to its center, a structure identical to the original one is obtained.

It seems interesting to operate an introspective analysis of the perception of the "hydrogen molecule" at the bottom of Fig.4. During the observation of this structure the mind collects in the memory and interiorizes the sensory stimuli coming from the image. While controlling the initially scattered and disordered stimuli, the mind processes and correlates them: the aim is to build a collective and ordered pattern synergetically around the center of symmetry of the structure [9].

There are two possible patterns that the stimuli can form while the control exerted on them by the mind increases: these patterns are mutually in conflict, and yet each one separately is fully compatible and consistent with the logic and operative system of the mind itself. Both patterns assign the central wall 50% to the left hand side (lhs) module, and 50% to the rhs module. But while for one of the two patterns the center of symmetry is equally shared by the lhs and the rhs modules, for the other pattern the center of symmetry belongs neither to the lhs nor to the rhs module. As the control parameter increases further, a critical stage is reached finally where one could state: the center of symmetry is equally shared by the lhs and rhs modules **and, simultaneously,** the center of symmetry belongs neither to the lhs nor to the rhs module [10].

The mind is thus trapped in an ambiguous state where two conflicting, irreconcilable patterns coexist. This critical state of the mind is uneasy, the mind is anxious to escape through any possible way out. A small fluctuation, and it's catastrophe![11] A dynamic instability in the pattern of the stimuli occurs leading perception from flat to spatial, the center of symmetry disappears and, as symmetry breaking occurs, the order parameter - the visual thinking - starts to rise, driving the initially common wall between the two cubes toward the lhs or the rhs module. From an originally symmetrical or balanced situation the "center of symmetry" falls either **left** or **right**: in order to perceive symmetry we must destroy it: but, at the same time, a bit of information is created [5]. (Actually, visual thinking cannot remain fixed in any of these two positions: soon a dynamics of perpetual alternations from left to right to left prospects sets in, resembling the resonant behavior of the chemical bond during the spectroscopic observation of a diatomic molecule by the charge transfer spectrum. Details of this analogy are proposed in Ref [10].)

6. On the universality of images and visual communication

To identify **the reasons** for the universality of images is pretentious. To identify **reasons** for the universality of images and, more generally, of the language of art, is however a less demanding objective. In the present section we shall pursue this objective.

Images stimulate our state of mind. The stimulus produced by an image triggers a response, provided a component of it exists in the state of mind. Otherwise the stimulus is ignored. When the above triggering condition is satisfied, one can get the feeling of resonating on the image. In order to evaluate how likely this "resonance condition" is to occur, it would be necessary to know our state of mind and, in particular, its susceptibility to different stimuli, as measured by the amplitudes of the components of the image in the state of mind itself: large amplitudes imply that the state of mind resonates on the stimulus produced by the image.

A problem thus arises: to identify characteristic features of our state of mind that are susceptible to coupling with the stimulus produced by an image. Or, vice versa, to identify criteria by which images can be constructed in such a way as to tune in easily to the state of mind.

Without pretending to be exhaustive, one could observe that two basic, contradictory attitudes strive to balance in our state of mind.

Our state of mind longs for dream and reassurance, and, at the same time, for awareness and dynamics. On the one hand we feel an ancestral wish to rest in a reassuring repetitiveness such as that of the cradle song, metric in poetry, beating in dance, litany in prayer, slogans in the liturgy of political demonstrations. On the other hand, as if we were waiting for Godot, we are anxious to be involved in exceptional phenomena or to participate in extraordinary events, enabling us to break the monotony and the routine. Nietzsche would say that our aspiration is "to dream being aware of dreaming".

Now then, on the one hand dream and reassurance are susceptible to symmetry: symmetry is allied with homogeneity, isotropy, uniformity, repetitiveness, indifference, equilibrium, stasis, hypnosis, silence, invariance, immeasurability and imperceptibility. That is why symmetry is congenial to us. On the other hand, awareness and dynamics are susceptible to broken symmetries and symmetry breaking: broken symmetries are allied with gradients, anisotropy, variety, acceleration, differentiation, nonequilibrium, vivacity, anxiety, pause and variation; symmetry breaking is allied with the processes of measurement and perception or, all things considered, to information. But, according to Lila L.Gatlin, "life may be defined operationally as an information processing system [...] that has acquired through evolution the ability to store and process the information necessary for its own reproduction" [12]. Therefore information is a prerequisite of life. That is why symmetry breaking is congenial to us.

In conclusion, symmetry and symmetry breaking cooperate to make the language of art universal. After all, music is made of rhythm (symmetry) and melody (information). And perhaps no language is more universal than that of music.

7. Symmetry and broken symmetries in art and design

To put a sign or a symbol on a white canvas is perhaps the simplest way to break the symmetry characteristic of the homogeneity of space and to produce information.

However, signs and symbols may be assembled in patterns containing symmetry elements so that the original symmetry of the white canvas is at least partially restored. Since symmetry implies invariance, in principle the presence of symmetry elements in a pattern contributes to its stabilization.

Conversely, the absence of symmetry elements contributes to the animation of the pattern.

In this section several instances of symmetry and broken symmetries in both art and design are proposed.

Let us first consider translational symmetry. Translational symmetry evokes a constant momentum or a steady speed. Stepwise translational symmetry evokes a rhythmic or periodic motion, ultimately resulting in constant average speed. That is why, in music, the idea of a constant velocity, inherent to the conception of a fugue, arises from the sequential repetition of the theme throughout the score. That is why, in the Corinthian crater of Fig.2, the idea of a rhythmic run, typical of the gallop, arises from the periodic repetition of the horse image all around the frieze. That is perhaps the reason why, in order to evoke the same idea of rhythm and periodicity, Virgil, though not being aware of the solutions proposed by J.S.Bach and the unknown Corinthian decorator, imparts translational symmetry to the sequence of the first five meters in his famous hexameter [13]:

Quadrupe|dante pu|trem soni|tu quatit| ungula| campum

Fig.5 Metrical counterpart of the Corinthian crater of Fig.2 and of a baroque fugue (from [5])

From the point of view of rhythm, reading this pentadactylic hexameter metrically is tantamount to reading a version of it obtained from the hexameter itself by a "translation" of e.g. $2\pi/5$ (Fig.5):

Dantepu | tremsoni | tuquatit | ungula | quadrupe | campum.

This "hexameter" is literally nonsensical - indeed I am using here an old trick, played now and then by after-dinner speakers in order to recover the declining attention of the audience - : the significance of both the above hexameters lies in their onomatopoeia, namely in the invariance of their rhythm, as determined by the Hamiltonian - I beg your pardon, by the translational symmetry - of their structures.

The above examples show that in poetry, in painting and in music, translational symmetry (namely the indiscernibility of the translation) evokes the idea of a steady speed. Additional examples derived from both painting and music, more numerous than those one could possibly reproduce here, will be produced during the actual speech: they will help to emphasize the centrality of the role of symmetry and symmetry breaking in culture.

In what follows we shall expound on **trade marks**. These "hidden persuaders" [14], these idols of the technological olympus, these fruits of the science of communication and the art of persuasion, are devised specifically in order to accomplish a task: to bias the choice of the consumer so as to induce him to prefer a product or a service to other competitive products or services.

During the critical stage of the choice, the consumer is urged to make up his mind. He is on the alert and pays attention to any source of information.

On the one hand the consumer, being naturally suspicious, is sensitive to fixed or repetitive references evoking steadiness and inducing a feeling of reassurance. On the other hand, feeling entrapped in a labyrinth of uncertainties, he becomes impatient and will respond promptly to dynamic elements evoking new perspectives and capable of producing incitement to buy. That is why the consumer welcomes both symmetry and broken symmetries in a trade mark.

Take, for instance, the three-point star of Mercedes-Benz. Inscribed in a smooth circle, this image exhibits a ternary rotational symmetry, evoking the perspective of a wheel running forever at constant angular velocity. The potential buyer of a car feels reassured by the simple structure of this symbol, and looks at it as to a lucky star.

Take the double wing-stroke of the Citröen trade mark . Here the feeling of reassurance arises more subtly, most likely from the deep roots of the subconscious: the exchange symmetry, characterizing the structure of this symbol, is also a prerequisite of the stability of the structure of matter, all the way down to the atomic scale. In fact, should the indiscernibility of the permutation of two electrons (or the exchange symmetry) not be enforced, the Pauli exclusion principle could not be formulated, and the Mendeleev table at the basis of the structure and properties of the atoms would not be conceivable.

Take the FIAT trade mark (Fig.6). Its structure is the spatial counterpart of a musical phrase. Here the rhythm produced by the periodic sequence of the parallelograms is dressed by the melody of the acronym of the ancient factory of Torino.

Finally, take the image by Jean-Paul Goude , from a TV spot for the Volkswagen Polo (Fig.7). In this structure the palms follow one another regularly: in the absence of wind this fugue of palms would exhibit a rigorous translational symmetry. The red silhouette, mist-

Fig.6 The trade mark of FIAT. Here the stepwise translational symmetry produces the feeling of rhythm, while the acronym breaks the symmetry and provides "melodious" information

Fig.7 A fragment of a TV spot for the Volkswagen Polo, by Jean-Paul Goude. Symmetry and broken symmetry blend, suggesting the concept of a dynamic smoothness and the feeling of a dreaming awareness (from [15])

Fig.8 Saqqara, 2500 B.C. A bas-relief belonging to the ancient prince Ptah-hotep. As in Fig.7, symmetry and broken symmetry are the basic ingredients of this masterpiece. The vivacity of the scene relies on the ibis breaking the silence and warning its mates: "the path makes an elbow!" (from [5])

ed by the splashing of the car, breaks the symmetry of the composition and becomes the pole of our visual thinking.

Ever since I first saw this spot I have wondered whether Jean-Paul Goude drew his inspiration from the Egyptian bas-relief, presented in Fig.8, belonging to Ptah-hotep, an ancient prince of Saqqara. In this bas-relief the ibises follow one another regularly, exhibiting a translational symmetry. Nevertheless one of those ibises breaks the symmetry and becomes the pole of our visual thinking. 4500 years ago, that ibis has uttered a cry. This evening we hear its cry distinctly: that ibis warns its ibis-companions: "der Weg macht einen Haken!"

Acknowledgments

It is a pleasure to thank Mrs Dolores Cashman for her careful scrutiny of the manuscript and Mrs Carla Cattaneo for her patient and dedicated setting of this camera-ready version.

References

1. H.Haken, Synergetics. An Introduction, 3rd Edition, Springer-Verlag, Berlin, Heidelberg, New York, 1983
2. Hermann Haken, Erfolgsgeheimnisse der Natur. Synergetik: die Lehre vom Zusammenwirken, Deutsche Verlag-Anstalt, Stuttgart, 1981.
3. H.Haken, Pattern Formation and Pattern Recognition - An attempt at a Synthesis, in **Pattern Formation by Dynamic Systems and Pattern Recognition**, Ed. by H.Haken, Springer - Verlag, Berlin 1979, pp 1-13
4. R.Arnheim, Visual Thinking, Regents of the University of California, Berkeley, Los Angeles, 1969
5. G.Caglioti, Simmetrie Infrante nella Scienza e nell'Arte, Clup, Milano, 1983. German edition: Symmetriebruch in Wiessenschaft, Kunst und Design, to be published by Vieweg, Wiesbaden, 1987
6. Claude Lévi-Strauss, in **Enciclopedia del Novecento**, Istituto dell'Enciclopedia Italiana, Roma, 1975
7. Gillo Dorfles, Elogio della Disarmonia - Arte e vita fra logico e mitico, Garzanti, Milano, 1986
8. T.D.Lee, Symmetry Principles in Physics, in **Evolution of Particle Physics**, a volume dedicated to Edoardo Amaldi on his sixtieth birthday. Ed. M.Conversi, Academic Press and Periodici Scientifici, Milano, 1968
9. G.Caglioti and E.R.Caianiello, Biol. Cybernetics, 31, 205-208, 1978
10. G.Caglioti, The World of Escher and Physics, in **M.C.Escher: Art and Science.** Eds. H.S.M.Coxeter, M.Emmer, R.Penrose and M.L.Teuber, North-Holland, Amsterdam, 1986
11. G.Caglioti, La catastrofe percettiva: basta una fluttuazione e e' catastrofe, in Prometheus 1 , P.Bisogno, Director, **La Teoria delle Catastrofi**, AA. VV., Franco Angeli, Milano, 1985, pp. 227-242
12. Lila L.Gatlin, Information Theory and the Living System, Columbia University Press, New York, 1972
13. Virgil, Aeneid, VIII, 596
14. Vance Packard, The Hidden Persuaders, David McKay Company, Inc., New York, 1957
15. G.Caglioti, Finzione e realta' nelle simmetrie infrante, in **AD, Architectural Digest**, Edizione Italiana, Giorgio Mondadori, Milano, N.45, febbraio 1985, pp 20-24.

Part V

Methods of Synergetics

Boltzmann-Gibbs Entropy as a Measure of Order in Self-Organizing (Synergetic) Systems

Yu.L. Klimontovich

Department of Physics, Moscow State University, SU-117234 Moscow, USSR

This paper is dedicated to Prof. Hermann Haken, an outstanding scientist and, no less important, outstanding personality.

1. Dynamic and Statistical Description of Complex Behavior

The statistical theory of nonequilibrium processes was launched by Ludwig Boltzmann, and the theory of dynamic systems by Henri Poincaré. We know well how dramatic was the rivalry of these two schools in the beginning. And while the controversy has cooled down by now, these two directions have been developing more or less independently. The new theory of self-organization, synergetics, seems, however, to necessitate the merger of the two /1-5/.

The interest towards the study of complex dynamic behavior - the dynamic (deterministic) chaos - has been significantly stimulated by the works of E. Lorentz, D. Ruelle, F. Takens.

The main features of dynamic chaos are given different priority by different authors; as a rule, however, they can be listed as follows:

1) Exponential divergence of trajectories, often causing acute sensitivity to changes in the initial conditions. The measure of exponential instability is the K-entropy (Krylov-Kolmogorov or Kolmogorov-Sinai entropy).

2) Ergodicity. The existence of invariant measure.

3) Mixing. Damping (fading) of temporal correlations, continuous spectrum.

The first important step in the development of Boltzmann kinetic theory was made by M.A. LEONTOVICH /6/ in 1935, who constructed a kinetic equation for many-particle distribution function for Boltzmann gas, which later was studied independently by M. Kač, I. Prigogine, P. Resibois and others.

Unfortunately, Leontovich's paper passed almost unnoticed, and further development of the kinetic theory took a different route with the milestones associated with the names of A. Vlasov, L. Landau, N. Bogolyubov, M. Born and H. Green, J. Yvon, M. Green.

To be certain, the dynamic theory of complex behavior and the statistical theory of nonequilibrium processes do converge, but at an uneven pace. The latest (to author's knowledge) review /7/ on hydrodynamic instability, turbulent (stochastic) motion contains no references whatsoever to the works on the statistical theory of nonequilibrium processes.

What is then the reason for continuing separation of these two fundamental theories? Is this separation justified?

The answer to the latter question is definite: no. Independent (or nearly independent) pursuit of each is not justified. A way of bringing the two together must be found.

The former question cannot be answered as readily. Looking for a clue for merging the two approaches we find that the term dynamic chaos itself sounds wrong, since the dynamic instability can prove constructive as well: it can promote dissipative structures and provide a dynamic description of macroscopic processes.

2. Hierarchy of Equations of Fluid Mechanics. Dynamic Instability as a Constructive Factor in the Statistical Theory of Nonequilibrium Processes

Let us consider a model of rarefied gas, consisting possibly of chemically reacting components. We recognize four levels of the description of macroscopic processes in a gas:

1) Kinetic description, based on the Boltzmann equation, which is a dynamic equation for the mean density in the six-dimensional space of coordinates and momenta.

2) Gasdynamic description, in which the equations are derived from the Boltzmann equation and are examples of dynamic equations in mean density, velocity and temperature.

3) Chemical kinetic equations including spatial diffusion, which represent the basic dynamic equations in synergetics /1-5/. They can be derived from gasdynamic equations for a mixture of gases.

4) Chemical kinetic equations for spatially uniform distribution of components, which derive from the preceding case when the time scale is superior to the characteristic time of diffusion and have the form

$$\frac{dn_i(t)}{dt} = f_i(n_1, \ldots, n_N), \quad i = 1,2, \ldots, N . \qquad (1)$$

This brings us to the typical equations of the dynamic theory: a set of nonlinear dissipative first-order differential equations in $n_i(t)$.

Now let us return to the Boltzmann equation, which stands at the top of our hierarchy of dynamic equations. The way down the hierarchy from 1 to 4 is made possible by the existence of appropriate small parameters. Accordingly, the hierarchy of dynamic equations of continuous medium based on the dynamic instability of atomic motion can only be justified through the analysis of the assumptions of the initial Boltzmann equation. This brings us back again to the problems which fuelled the polemics in the early days of the kinetic theory and the dynamic theory of complex behavior. This time, however, we can benefit from the knowledge accumulated on both the nature of dynamic instability and the applicability of Boltzmann equation.

First we examine the assumptions made in deriving the Boltzmann equation from the equations of motion of gas atoms. This discussion will follow the lines drawn by Bogolyubov and Born-Green. Assumption 4 is treated as in Sect. 3.16 in /8/ and Ch. 9 in /9/.

We single out four assumptions, all based on the existence of the same small parameter, the density parameter $\varepsilon = nr_0^3 \ll 1$:

1. Approximation of paired collisions.
2. Neglect of variation of the distribution function on the atomic scale. This amounts to the neglect of the interactive contribution in thermodynamic functions /8/.
3. Damping of small-scale fluctuations.
4. Continuous medium approximation (neglect of discrete collisions).

The last of these is usually implicit; presently, however, we attach special importance to it.

Treating a gas as a continuous medium presumes smoothing of the microscopic phase density over physically infinitesimal time τ_{ph}

and volume V_{ph}. The scales of τ_{ph}, ℓ_{ph} for a rarefied gas are found from relations /8,9/

$$\tau_{ph} = \frac{\tau}{N_{ph}} \sim \frac{\tau}{n\,\ell^3_{ph}} \quad ; \qquad \tau_{ph} = \frac{\ell_{ph}}{V_{ph}} \quad , \tag{2}$$

where τ is the mean free path time for the chosen particle, and τ_{ph} is that for any particle from V_{ph} . Using the density parameter $\varepsilon = nr_0^3$ we find from (2) τ_{ph}, ℓ_{ph} and the average number of particles N_{ph} in the volume V_{ph}:

$$\tau_{ph} \sim \sqrt{\varepsilon}\,\tau \ll \tau; \quad \ell_{ph} \sim \sqrt{\varepsilon}\,\ell \ll \ell; \; N_{ph} \sim \frac{1}{\sqrt{\varepsilon}} \gg 1 . \tag{3}$$

The last of these implies that for a rarefied gas even the fluctuations on the minimum scale ℓ_{ph} are small. This allows to neglect in many cases the fluctuations of the distribution function and thus treat the Boltzmann equation as a dynamic equation.

Replacing in the above equations N_{ph} by the total number of particles N we obtain estimates for the corresponding minimum time scales:

$$(\tau_{ph})_{min} \sim (\tau_{rel})_{min} \sim \frac{\tau}{N} \quad . \tag{4}$$

Hence follows that in the thermodynamic limit the minimum scales of relaxation and smoothing are zero, which means that the concept of a reversible process in this approximation is not valid.

Stochastic processes in the kinetic theory can be described using the Boltzmann equation with Langevin source /8,9,10/. This equation can also top the hierarchy of equations similar to 1-4, now with appropriate Langevin sources.

We see that the present state of the statistical theory allows to justify the use of both dynamic and stochastic equations for macroscopic functions. The solutions of these equations enable to discern the more orderly behavior of macroscopic characteristics against the background of molecular motion. Such characteristics are appropriately termed the <u>functions</u> or <u>parameters of order</u>. Among these we can once again single out the processes with large relaxation times and build hierarchies of dynamic or stochastic equations in the parameters of order. This approach is much used in synergetics.

It turns out that a statistical theory of nonequilibrium processes can be constructed without the explicit use of such fundamental concepts as the dynamic instability, dynamic chaos, Krylov-Kolmogorov entropy and Lyapunov indices. The use of these concepts, however, allows us to better understand the causes of irreversibility and to describe irreversible processes by the relatively simple equations in functions of order. This approach has been much contributed to by I. PRIGOGINE /3/.

The motion of ball-like atoms in the Boltzmann gas, as well as in the "Sinai's Billiards", is not stable. For the chosen ball-like atom the time over which instability develops ($\tau_{inst} = 1/K$, K being the Kolmogorov-Krylov entropy) is of the order of the mean free path time:

$$\tau_{inst} \sim \tau \, . \tag{5}$$

Let us also define (cf. (2,3)) the time of development of instability for any single particle in physically infinitesimal volume V_{ph} which we denote by τ^{ph}_{inst}. It is N_{ph} times smaller than τ_{inst} and hence of the same order as τ_{ph}, ie

$$\tau^{ph}_{inst} \sim \tau_{ph} \quad ; \quad (\tau^{ph}_{inst})_{min} \sim (\tau_{ph})_{min} \propto \frac{1}{N} \, . \tag{6}$$

(Here we have made use of definition (2) and relations (5), (4)).

We see that the characteristic time of development of instability τ^{ph}_{inst} for a particle in V_{ph} is of the same order as the time τ_{ph} for Boltzmann gas. Because of this, the dynamic instability of atomic motion in the volume V_{ph} results in the mixing over the time τ_{ph}, considered infinitesimal in deriving the Boltzmann equation.

In other words, the estimates for the time of dynamic instability (and hence of mixing) support the validity of our choice in (2) of the time τ_{ph}, over which the microscopic distribution in the phase space is smoothed. It is exactly here that the constructive role of dynamic instability (exponential divergence of trajectories) is revealed, enabling us to use the kinetic Boltzmann equation for deriving the hierarchy of dynamic equations based on appropriate small parameters.

It is important that the motions described by the equations from this hierarchy can themselves become unstable. A classical example is the Lorentz set of equations for convective motion.

After the works of E. Lorentz, D. Ruelle, F. Takens, many authors have discovered and investigated (both in laboratory and numeric experiments) a broad variety of low-dimensional (with the number of degrees of freedom $N > 2$) systems displaying diverse kinds of strange attractors. This proved the existence of dynamic instability in a wide range of dynamic systems of type (1).

Complex behavior in "simple" systems with more than two degrees of freedom is associated with the possible escape of trajectories from the plane into space (in continuous-time processes).

In discrete-time processes the escape of trajectories is not a necessary precondition of complex behavior, which may result from additional "freedom of choice" in between the discrete time lapses. This opens another possibility for mathematical simulation of complex motions, which will be demonstrated in the following section for a generalized Van der Pol oscillator. The control parameters, apart from the feedback parameter (pumping a_p), will include the discrete time scale Δ.

Now the dynamic instability of atomic motion can play a constructive part, since the mixing allows us to proceed from the microscopic description to a dynamic equation for the distribution function smoothed over V_{ph} - the kinetic Boltzmann equation.

The kinetic Boltzmann equation in its turn generates a hierarchy of simpler dynamic equations which may themselves describe unstable motions with exponential divergence - this time not of individual particles, but rather of collective variables. Thus a hierarchy of dynamic instabilities becomes possible. A question arises though, whether these new instabilities can be constructive and lead to a perfected description of evolutionary dissipative structures, or they result in dynamic chaos? To handle this question we need a measure of chaoticity of nonequilibrium states. As a matter of fact, the required measure can be furnished by the renormalized Boltzmann-Gibbs entropy /11-15/. The efficiency of this criterion will be presently shown for the case of self-organization taking place in the generalized Van der Pol oscillator.

3. Generalized Van der Pol Oscillator with Two Control Parameters (Pumping and Discrete Time Scale) /11, 18/

Our treatment is based on the set of equations in charge and current (x,v) for the symmetrized with respect to nonlinearity Van der Pol oscillator /11/:

$$\frac{dx}{dt} + \tfrac{1}{2}(-a + bE)x = v \ ; \qquad \frac{dv}{dt} + \tfrac{1}{2}(-a + bE)v + \omega_0^2 x = 0 \ ;$$
$$E = \tfrac{1}{2}(v^2 + \omega_0^2 x^2) \quad ; \quad a = a_p - \gamma \quad . \tag{7}$$

Here E is the energy of oscillations, γ and b are the coefficients of linear and nonlinear friction, a_p is the feedback (pumping) parameter, ω_0 is the proper frequency. From (7) follows the exact (non-averaged) equation for the energy of oscillations:

$$\frac{dE}{dt} = (a - bE)E = 0 \quad . \tag{8}$$

For the set (7,8) exists an exact solution of the Cauchy problem, which at $a > 0$ describes a limit cycle with the frequency ω_0.

Let us consider along with (8) a corresponding equation in discrete time. We use dimensionless variables $t' = \omega_0 t$, $a' = a/\omega_0$, $E' = bE/\omega_0$, further dropping the prime. The discrete time is denoted by n. Then, with the step $\Delta = 1$, we find

$$E_{n+1} = (aE_n - E_n^2) + E_n \equiv F(E_n) + E_n \equiv f(E_n) \quad . \tag{9}$$

This iterative equation differs from the logistic equation $x_{n+1} = \alpha(1 - x_n)x_n$ only in notation ($a \leftrightarrow \alpha - 1$, $E \leftrightarrow 4x$).

Basing on (9), let us construct a sequence of difference equations with two control parameters a_p, Δ. The first of these follows from (9) and has the form

$$E_{n+\Delta} = \Delta F(E_n) + E_n \quad . \tag{10}$$

The k-th iteration of the initial equation (9) can be written as

$$E_{n+k\Delta} = \Delta(f^{(k)}(E_n) - E_n) + E_n \equiv \Delta F(E_n)F_{(k)}(E_n), \ k = 1,2\ldots \tag{11}$$

In particular, for k = 1,2 the function $F_{(k)}(E)$ is

$$F_{(1)}(E) = 1, \quad F_{(2)}(E) = a + 2 - (a + 2)E + E^2 \quad . \tag{12}$$

Let us also define the corresponding sequence of differential equations for the energy of oscillations:

$$\lim_{\Delta \to 0} \frac{1}{k\Delta}(E_{n+k\Delta} - E_n) \to \frac{dE}{dt} = \frac{1}{k} F(E)F_{(k)}(E), \ k = 1,2,3\ldots \quad . \tag{13}$$

The function $F_{(k)}(E)$ has the property

$$\frac{1}{k} F_{(k)}(E) \rightarrow 1 \quad \text{at } E, a \rightarrow 0 \quad , \tag{14}$$

which conforms equations (13) to the generating equation (9). By replacing $a - E \longrightarrow (a - E) \frac{1}{k} F_{(k)}(E)$ we obtain from (7) the corresponding generalized equations in x,v. In the stationary regime they describe single-mode self-oscillations (a limit cycle). The possible energy levels for different values of the control parameter a_p coincide with the stationary points of (10) at the k-th step of iteration.

In the presence of noise of preset strength D the dynamic equations (13) correspond to the sequence of Fokker-Planck equations for the distribution function f(E,t) /11/:

$$\frac{\partial f(E,t)}{\partial t} = D \frac{\partial}{\partial E} (E \frac{\partial f}{\partial E}) + \frac{\partial}{\partial E} \left[(-a + E) E \frac{1}{k} F_{(k)}(E) f \right] , \quad k = 1,2,3 \ldots \tag{15}$$

or, in discrete time,

$$\frac{1}{k\Delta} (f(E, n+k\Delta) - f(E,n)) = D \frac{\partial}{\partial E} (E \frac{\partial f}{\partial E}) + \frac{\partial}{\partial E} \left[(-a + E) E \frac{1}{k} F_{(k)}(E) f(E,n) \right] , \quad k = 1,2,3 \ldots \tag{16}$$

The general analysis of these equations is very complicated. It would be worthwhile to investigate first the two limits, corresponding to the dynamic and the stochastic approaches.

4. Dynamic (Reproducible) and Stochastic (Irreproducible) Motion

There is no universally recognized definition of dynamic chaos or of stochastic process. Sometimes these two concepts are combined, and then they speak of the stochasticity of dynamic systems. Let us base our classification on the reproducibility of the motion given the initial conditions. Then, by definition, dynamic motions are reproducible, and stochastic irreproducible.

Realistically, since the noise cannot be excluded, all processes are stochastic. We can speak of dynamic motion only modelwise, imposing certain conditions on the characteristic times and the observation time τ_{obs}.

This classification becomes, however, quite useful in the numeric experiment, since we can always repeat exactly the initial conditions fed into any computer. The reproducibility of the motion will then depend only on the mathematical model. If the equations contain no random sources (which include bugs in the computer), then the process is reproducible and the motion, however complicated, is dynamic. Otherwise we are dealing with stochastic behavior.

In order to see more clearly the distinctions between the equations for f(E,t) and f(E,n) in the cases of dynamic and stochastic processes, we shall consider the established states for those values of parameters which correspond to the regime of well-developed generation.

We introduce characteristic time parameters

$$\tau_a = \frac{1}{a} \; ; \quad \tau_b = \frac{1}{b\langle E\rangle} \quad ; \quad \tau_D = \frac{\langle E\rangle}{D} \quad ; \quad \tau_{obs} \; . \tag{17}$$

In the regime of well-developed generation, at $\langle E\rangle \backsim a/b$,

$$\tau_a \backsim \tau_b \; ; \quad \tau_a \ll \tau_D \quad (\tau_a/\tau_D = Db/a^2 \equiv \varepsilon_a \ll 1) \; . \tag{18}$$

We see that the time τ_a the limit cycle takes to establish is much smaller than both the diffusion time τ_D and the observation time τ_{obs}. Depending on the relative values of τ_D and τ_{obs} we can distinguish two established regimes:

1) Steady generation for stochastic processes with finite-valued τ_D occurs in the limit $\tau_{obs} \rightarrow \infty$. Away from bifurcation points we can find solutions in low-noise approximation. This corresponds to the following sequence of limit transitions:

$$1. \; \tau_{obs} \rightarrow \infty \qquad 2. \; \tau_D \rightarrow \infty \quad (\varepsilon_a \rightarrow 0) \; . \tag{19}$$

2) Established dynamic regime, which corresponds to the reversed sequence

$$1. \tau_D \rightarrow \infty \qquad 2. \; \tau_{obs} \rightarrow \infty \quad . \tag{20}$$

Equations describing the established dynamic regime can also be derived from the Fokker-Planck equations. Indeed, when carrying out the first transition of (20) in (15) we may drop the diffusive terms, ie set D ≡ 0. Then the solution of (15) can be presented in the form

$$f(E,t) = \delta(E - E(t)) \; . \tag*{(21)$_1$}$$

In order to obtain the equation for E(t) we multiply (15) by E and integrate with respect to E. This will result in the dynamic differential equation (13) whose stationary solutions give the possible energy values for the limit cycle.

The solution of (16) at $D = 0$ can be presented as

$$f(E,n) = \delta(E - E_n) \quad . \tag{21$_2$}$$

The function E_n satisfies the difference equation (11).

5. S-Theorem and S-Information for Van der Pol Oscillator

Let us consider the simplest stationary solution of the Fokker-Planck equation (15) at $k = 1$ and watch its evolution as the feedback parameter a_p increases. We shall use two criteria for assessing the relative order of different states. To make the picture clearer, we distinguish three regimes /12/:

1. Zero pumping (physical chaos)

$$f_{(1)} = \frac{\gamma}{D} \exp\left(-\frac{\gamma E}{D}\right) \qquad (\text{at } Db/\gamma^2 \ll 1) \quad . \tag{22}$$

2. Generation threshold ($a_p = \gamma$)

$$f_{(2)} = \sqrt{\frac{2b}{\pi D}} \exp\left(-\frac{bE^2}{2D}\right) \quad . \tag{23}$$

3. Well-developed generation ($\varepsilon_a = Db/a^2 \ll 1$)

$$f_{(3)} = \sqrt{\frac{b}{2\pi D}} \exp\left(-\frac{b(E - a/b)^2}{2D}\right) \quad . \tag{24}$$

The corresponding values of entropy and mean energy (for a "good" oscillator, $Db/\gamma^2 \ll 1$, $a/\gamma \ll 1$) satisfy the relations

$$S_{(1)} < S_{(2)} < S_{(3)} \ ; \ \langle E \rangle_{(1)} < \langle E \rangle_{(2)} < \langle E \rangle_{(3)} \quad . \tag{25}$$

Accordingly, as the generations develop (going from state 1 to 3), the entropy increases. The mean energy also goes up.

The rise of the mean energy is, of course, natural, while the increase in the entropy ought to be explained.

Let us first try the criterion of S-information, proposed by H. HAKEN /15,16/. Importance here is attached to the evolution of the Shannon information, defined in the same way as the entropy:

$$I = -\int \ln f(E)\, f(E)\, dE \quad . \tag{26}$$

For illustration let us quote the result for information gain between states 2 and 3:

$$I_{(3)} - I_{(2)} = \ln 2 > 0 \quad . \tag{27}$$

The positive gain here is due to the following: in state 2 $E > 0$, and only one-half of the distribution (23) comes to work, while in state 3 the range of possibilities is doubled, which is reflected by the information gain.

The criterion of S-information is certainly very useful. It does not, however, detect self-organization in the oscillator as a_p increases. Moreover, entropy (25) and information do not depend on the control parameter a_p.

Now let us use as the measure of order the Boltzmann-Gibbs entropy, renormalized to the given value of the mean energy /12/. Renormalization is carried out using the renormalized noise values in (22-24):

$$\tilde{D}_{(1)} = \gamma \frac{a}{b} \; ; \quad \tilde{D}_{(2)} = \frac{\pi}{2} \frac{a^2}{b} \; ; \quad \tilde{D}_{(3)} = D \quad . \tag{28}$$

The new entropy values

$$\tilde{S}_{(1)} = \ln \frac{a}{b} + 1 \; ; \quad \tilde{S}_{(2)} = \ln \left(\frac{\pi a}{2b}\right) + \tfrac{1}{2} \; ; \quad \tilde{S}_{(3)} = \ln\sqrt{\frac{2\pi D}{b}} + 1 \tag{29}$$

now arrange as (cf. (25))

$$\tilde{S}_{(1)} > \tilde{S}_{(2)} > \tilde{S}_{(3)} \quad , \tag{30}$$

which plainly illustrates the statement called in /12/ the S-theorem: If, as the control parameter increases, the entropy renormalized to the preset value of effective energy decreases, then the process of self-organization is underway.

There are other examples illustrating the S-theorem, which in its general form is formulated in /14/. A detailed account can be found in the review prepared for the Uspekhi Fizicheskikh Nauk . Here we shall consider just two cases of interest.

6. S-Theorem for the Generalized Van der Pol Oscillator

We have discussed a mathematical model of generalized Van der Pol oscillator in both continuous and discrete time.

In the continuous-time regime the increased feedback produces a cascade of bifurcations of the number of possible energy values in a stable limit cycle. Up to the critical Feigenbaum point this cascade follows a simple doubling pattern. Further increase in a_p gives rise to a more complicated, seemingly chaotic distribution of the possible energy values. Here the question of validity of purely dynamic description with $D \equiv 0$ remains open, except for some "gaps of order", corresponding to a limit cycle with, say, three possible energy values.

The calculation of Boltzmann-Gibbs entropy at each step of iteration is based on the stationary solution of the Fokker-Planck equation (15). Renormalization is carried out for the value of effective energy in the state of physical chaos ($a_p = 0$). As a result, we find the corresponding renormalized noise $\tilde{D}(a_p)$ and entropy $\tilde{S}$.

Our calculations indicate that as the cascade of bifurcations progresses, the renormalized entropy increases. According to the S-theorem, this means that increasing feedback in generalized Van der Pol oscillator promotes self-organization.

7. Self-Organization in a System Without Simple Attractors

Let us return to equations (7,9) for the symmetrized with respect to nonlinearity Van der Pol oscillator. Assume that nonlinear friction $b = 0$, but the energy range is restricted: $E \leqslant E_0$. We find the stationary solution of the corresponding Fokker-Planck equation for $f(E,t)$ and consider the low-noise approximation ($D/|a|E_0 \ll 1$). We are interested in two characteristic states:

1) Physical chaos ($a_p = 0$). In this case the dynamic system displays a simple attractor: rest point $E = 0$. The stationary distribution is given by

$$f(E) = \frac{\gamma}{D} \exp\left(-\frac{\gamma E}{D}\right) \quad . \tag{31}$$

2) $a = a_p - \gamma > 0$. When $E < E_0$ the dynamic motion is unstable (positive Lyapunov index), although cumulation is possible at $E = E_0$. In zero approximation with respect to $D/|a|E_0$ the stationary distribution is

$$f(E) = (1 - C)\frac{1}{E_0} + C\delta(E - E_0), \qquad \int_0^{E_0} f dE = 1 \;;$$

$$0 \leq C = 2\frac{\langle E\rangle}{E_0} - 1 \leq 1 \;; \qquad \langle E\rangle = \int_0^{E_0} E f dE \leq E_0 \;. \tag{32}$$

From the definition of C it follows that the cumulation at point E_0 occurs only when $\langle E\rangle > E_0/2$. First we set $\langle E\rangle < E_0/2$. Then the entropy is given by (cf. (26)) /17/

$$S_{(1)} = \ln\frac{D}{\gamma} + 1 \;;\; S_{(2)} = \ln E_0 \;;\quad S_{(1)} < S_{(2)} \;, \tag{33}$$

positive information gain from 1 to 2.

Now we renormalize the entropy with respect to the given value of mean energy $\langle E\rangle = E_0/2$. The renormalized noise is $\tilde{D} = \gamma E_0/2$, and renormalized entropy

$$\tilde{S}_{(1)} = \ln\frac{\tilde{D}}{\gamma} + 1; \quad \tilde{S}_{(2)} = \ln E_0, \quad \text{hence} \quad \tilde{S}_{(1)} > \tilde{S}_{(2)} \;. \tag{34}$$

We see that according to both the S-information criterion and the S-theorem state 2 is more ordered, and the transition $1 \rightarrow 2$ is an example of self-organization.

$$\text{If} \quad \langle E\rangle > E_0/2, \quad \text{then} \quad f(E) = \delta(E - E_0) \;, \tag{35}$$

which corresponds to the highest degree of self-organization. Accordingly, the oscillator with the limit cycle energy distribution (35) is the most efficient.

It turns out that both criteria (S-information and S-theorem) can give concerted estimates of the degree of self-organization in different states. Each of them, naturally, has its own advantages and shortcomings.

References

1. G.Nicolis, I.Prigogine: Self-organization in nonequilibrium systems (Wiley, New York 1977)
2. H.Haken: Synergetics (Springer, Berlin, Heidelberg, New York 1978)
3. I.Prigogine: From being to becoming (Freeman, San Francisco 1980)
4. H.Haken: Advanced Synergetics (Springer, Berlin, Heidelberg, New York 1983)
5. Yu.M.Romanovskii, N.V.Stepanova, D.S.Chernavskii: Matematicheskaya biologiya (Nauka, Moscow 1984)

6. M.A.Leontovich: Osnovnyye uravneniya kineticheskoy teorii gazov s tochki zreniya teorii sluchaynykh protsessov, ZhETF, 5, 211 (1936)
7. A.S.Monin: Gidrodinamicheskaya neustoychivost', UFN, 150, 61 (1986)
8. Yu.L.Klimontovich: Kinetic theory of nonideal gases and nonideal plasmas (Pergamon Press, Oxford 1982)
9. Yu.L.Klimontovich: Statistical physics (Harwood, New York 1986)
10. B.B.Kadomtsev: O fluktuatsiyakh v gaze, ZhETF, 32, 943 (1957)
11. Yu.L.Klimontovich: Dinamicheskaya i statisticheskaya teoriya obobshchennykh generatorov Van der Polya, Pis'ma v ZhTF, 11, 21 (1985)
12. Yu.L.Klimontovich: Umen'sheniye entropii v protsessakh samoorganizatsii. S-teorema, Pis'ma v ZhTF, 9, 1412 (1983)
13. Yu.L.Klimontovich: Entropiya i proizvodstvo entropii v laminarnom i turbulentnom potokakh. Pis'ma v ZhTF, 10, 80 (1984)
14. Yu.L.Klimontovich: S-Theorem. Z.Phys.B - Condensed Matter, 65 (1987)
15. H.Haken: Information, information gain and efficiency of self-organizing systems close to instability points. Z.Phys.B - Condensed Matter, 61, 329 (1985)
16. H.Haken: Information and information gain close to nonequilibrium phase transitions. Numerical results. Z.Phys.B - Condensed Matter, 62, 255 (1986)
17. Yu.L.Klimontovich; M.Bonits: Samoorganizatsiya v sistemakh bez prostykh attraktorov. Pis'ma v ZhTF, 12(22) 1356 (1986)
18. Yu.L.Klimontovich: Posledovatel'nost' bifurkatsii energii ustoichivogo predel'nogo tsikla. Pis'ma v ZhTF, 13 (1987)

On the Slaving Principle

A. Wunderlin

Institute of Theoretical Physics and Synergetics,
University of Stuttgart, D-7000 Stuttgart 80, Fed. Rep. of Germany

1. Introduction

Synergetics has been founded by Hermann Haken in 1970 as a new scientific discipline [1,2]. Synergetics introduces a unified point of view to open systems which are composed of many subsystems. These systems show the ability of spontaneous self-organization on macroscopic scales. As a result of self-organization processes we may observe highly ordered spatial, temporal, or spatio-temporal patterns, or coherent functioning of these systems. If, for example, we characterize the subsystems through a length scale ℓ and assume that a typical length of the composed system is provided by L, the hierarchy between different scales may be expressed in the form

$$\ell \ll L \quad . \tag{1.1}$$

When we need quantum mechanics to treat the subsystems we call the scale of ℓ microscopic, if it is possible to describe the subsystems by classical equations, ℓ corresponds to the so-called mesoscopic level. On the other hand the scale L characterizes the macroscopic level. (1.1) then immediately implies that the spontaneously occurring ordering processes in the system which are observed on macroscopic scales must be connected with the fact that many subsystems have to behave in a completely coherent fashion. In the realm of synergetics Hermann Haken has developed not only a unified point of view on the emergence of these processes but in addition he has developed a complete mathematical theory to understand and predict the spontaneous occurrence of order in such systems quantitatively.

The central aspects of the theory are introduced through the notions of instability, fluctuations, and slaving. Processes of self-organization in an open system always arise via an instability. Similarly, switching between self-organized states of a different macroscopic behaviour again is connected with an instability. The system "notices" the presence of an instability through the fluctuations. Fluctuations may be considered as always testing the stability of the present state of a system under externally given conditions. Eventually slaving guarantees that the macroscopically evolving ordered state may be characterized by only a few collective variables, the so-called order parameters, which are generated by the system itself. It is understood that these order parameters slave the subsystems to act in a coherent fashion and thus maintain the macroscopically ordered state. Furthermore they may be considered as a quantitative measure for the evolving order.

Here we shall be mainly concerned with one, however particularly important, aspect of the theory, the slaving principle. An early version and mathematical formulation has implicitly been presented by H. Haken in 1962 [3] in the context of laser theory where quasi periodic motion has been analyzed. A general mathematical formulation has been given by H. Haken in a series of papers since 1975 (compare the references given in [1,2]). The spirit of this general principle may be formulated as follows: In the vicinity of critical regions (which means near an instability) the macroscopic behaviour of a system is dominated by few collective modes, the order parameters. They are the only variables needed to describe the evolving self-organized state of the system on macroscopic scales exhaustively. Slaving in this context is understood as a general principle of nature.

Parallel to its mathematical formulation the slaving principle has been successfully applied to various systems from natural sciences up to problems of social sciences [1,2]. Laser action, for example, may be considered as a fundamental paradigm of slaving. The non-linear interaction of the laser active atoms, which here play the role of the subsystems, with the light field generates a highly ordered state which is connected with the coherent laser action. Laser action is a particularly important example because here the theory may be developed rigorously from first principles. In addition we note that systems of non-linear optics may be discussed along similar lines. Furthermore the mathematical formulation of the principle has been applied to hydrodynamic systems, e.g. the Benard and Taylor problem, solidification processes, flame instabilities etc. There are applications in magnetohydrodynamics where the discussion of the generation of macroscopic stellar magnetic fields becomes of special interest. Self-organizing processes have also been discussed and identified in the field of plasma physics.

Applications, however, are not only restricted to physical systems. Indeed, the interdisciplinary aims are one of the most appealing aspects of synergetics. In chemistry, the Belousov-Zhabotinski reaction, the Briggs-Rauscher reaction and model reactions like the Brusselator and Oregonator have been analyzed by using the slaving principle. It has been applied to biological problems, especially to problems of morphogenesis and is expected to become an important tool in the analysis of brain functioning.

We finally mention applications in social sciences, the problem of the formation of public opinion in sociology, for example. Synergetics has also been used to discuss quantitative and qualitative models in economics.

In the following we shall outline the general mathematical method of synergetics (sometimes called the traditional method) when all the details of the subsystems are known, and show how the macroscopically ordered state may arise. We then shall discuss to some extent the behaviour of the slaved modes which especially becomes important when we wish to construct the equations of the order parameters. In the last part we shall discuss aspects of slaving in the newly developing field of macroscopic synergetics, which plays a fundamental role when one discusses self-organization processes in systems where the subsystems are so complex that one has only partial knowledge of their detailed behaviour.

2. Synergetics - from the Mesoscopic Level to Macroscopic Order

In the following we shall present an overview on the general mathematical method of synergetics developed by H. Haken for the situation where the details of the subsystems are completely known on a mesoscopic scale. Complete knowledge means that we know all the variables of the subsystems which we denote by U_i (i = 1, 2 ...), the evolution of the single subsystems, as well as their mutual interactions. If we put together all the variables into a state vector $\underset{\sim}{U}$ in a state space Γ

$$\underset{\sim}{U} = (U_1, U_2, \ldots) \quad , \tag{2.1}$$

its time evolution is typically governed by an equation of motion of the following general type

$$\dot{\underset{\sim}{U}} = \underset{\sim}{N}(\underset{\sim}{U}, \nabla, \{\sigma\}) + \underset{\sim}{F} \quad . \tag{2.2}$$

Here $\underset{\sim}{N}$ denotes a nonlinear vector field in $\underset{\sim}{U}$ which may additionally depend on spatial inhomogeneities symbolized through the ∇ - symbol as well as a set of external parameters which have been abbreviated by $\{\sigma\}$. $\underset{\sim}{F}$ symbolically summarizes the presence of fluctuations which may be considered as a result of the action of the already suppressed microscopic degrees of freedom. We note that this typical form of the evolution equations may be generalized in various ways: We also may include time delay-effects, non-local interactions in space; we can formulate them for time discrete processes, etc.

At this stage the first step of the systematic method developed by H. Haken is completed. It is this step which has to be performed by the specialized scientific disciplines. The following examples may clarify this statement. In laser physics the state vector is built from the atomic variables, i.e. the polarization and the inversion of the single atoms and the amplitudes of the light field in a semiclassical approach. The equations of motion for these variables may, as already mentioned, be derived from first principles. In hydrodynamics we can consider the velocity field, temperature field, passive scalars etc. as further examples for variables constituting $\underset{\sim}{U}$ and the equations of motion are then governed by conservation laws like the conservation of mass, momentum, energy, particle densities etc. In chemistry we may consider concentrations of the chemicals as appropriate variables, and there interactions are provided through the special chemical reaction under consideration. The same remains true for morphological models in biology, where the concentrations and interaction of morphogenes may be considered along similar lines. It is worthwhile to mention that even in social sciences such state vectors can be identified. In the following we wish to consider all these different systems from the unified view which has been introduced through the development of synergetics.

The second step consists in the analysis of equations (2.2). Generally we shall not be able - at least by analytical methods - to analyze the complete global behaviour of the solutions of (2.2) even if we confine ourselves to the deterministic part represented by $\underset{\sim}{N}$. We therefore shall concentrate ourselves to more simple subregions of the state space Γ which still are of fundamental importance for an understanding of the underlying system: We consider stationary states. The most simple stationary state which we may expect as a solution describes a homogeneous situation in time and space. This state is usually referred to as the so-called thermodynamic solution branch of the system. However, the method can be considerably generalized, i.e. we may even consider stationary states which are not homogeneous in space or in time (periodic and quasi-periodic states). For the time being, however, we confine ourselves to the most simple stationary state which is homogeneous in time and space. We may construct this state as a solution of

$$\dot{\underset{\sim}{U}} = 0 \longrightarrow \underset{\sim}{U} = \underset{\sim}{U}_o \quad . \tag{2.3}$$

The third step then consists in the test of the stability of the state (2.3). Because we may alter the external conditions through a change in the external parameters, the stationary state $\underset{\sim}{U}_o$ may lose its stability. Mathematically this means that we are looking for instability regions in the parameter space spanned by $\{\sigma\}$. A general method to achieve this is provided by linear stability analysis. We look for the behaviour of small deviations $\underset{\sim}{q}$ from the stationary state $\underset{\sim}{U}_o$, i.e.

$$\underset{\sim}{U} = \underset{\sim}{U}_o + \underset{\sim}{q} \quad . \tag{2.4}$$

We then immediately find the equation of motion for $\underset{\sim}{q}$ from (2.2):

$$\dot{\underset{\sim}{q}} = L\underset{\sim}{q} + O(\|\underset{\sim}{q}\|^2) \quad , \tag{2.5}$$

where, for simplicity, we did not take into account the fluctuations. The elements of the linear matrix L are given through

$$L = (L_{ik}) \quad \text{and} \quad L_{ik} = \partial N_i/\partial U_k|_{\underset{\sim}{U}=\underset{\sim}{U}_o} \quad . \tag{2.6}$$

Obviously the matrix L is still a function of the stationary state $\underset{\sim}{U}_o$, may depend on gradients and depends on the external parameters $\{\sigma\}$. There exists a well-known algebraic procedure to solve (2.5). The ansatz

$$\underset{\sim}{q} = \underset{\sim}{q}_o \exp(\lambda t) \tag{2.7}$$

transforms (2.5) into a linear algebraic eigenvalue problem from which we may gain the set of eigenvalues λ_i and the corresponding eigenvectors which we denote by $\underset{\sim}{O}_i$. We assume that they form a complete set. Instabilities in the space of the external parameters are then announced through

$$\text{Re}\ \lambda_i = 0 \ . \tag{2.8}$$

We may now immediately summarize all the information we get from our linear stability analysis. First we find the regions of instability in the space of the external parameters. Secondly we find the eigenvectors which we interpret as the collective modes of the system, and finally we obtain locally the directions in Γ along which the stationary state may become unstable.

The fourth step of the systematic treatment consists of a complete non-linear analysis of (2.2). We note that we may present the solutions of the full nonlinear equation in the form

$$\underset{\sim}{U} = \underset{\sim}{U}_o + \underset{\sim}{q} \quad \text{and} \quad \underset{\sim}{q} = \sum_i \xi_i(t)\ \underset{\sim}{O}_i \ . \tag{2.9}$$

We therefore may transform the original equation (2.2) into an equation for the amplitudes of the collective modes $\xi_i(t)$:

$$\dot{\xi}_i(t) = \Lambda_{ik}\xi_k(t) + H_i(\{\xi_i\}) + \tilde{F}_i \ . \tag{2.10}$$

Here Λ is a diagonal matrix or at least of the Jordan canonical form, H_i contains the complete nonlinear terms, and $\tilde{F}_i$ denotes the correspondingly transformed fluctuating forces. We note that the set of equations (2.10) is still exact.

We now arrive at the central step 5, the application of the slaving principle. As already noted, slaving means that in the vicinity of a critical region in the space of external parameters only few modes, the so-called order parameters, dominate the behaviour of the complex system on macroscopic scales. Obviously, already the linear stability analysis indicates which amplitudes of collective modes finally will become the order parameters. Indeed, the linear movement yields a separation in time scales: The modes which become unstable will move on a very slow time scale because of $|\text{Re}\ \lambda_u| = 1/\tau_u$, where we have identified the index i with u to exhibit unstable behaviour. In the vicinity of the critical point τ_u becomes extremely large. On the other hand, the modes still remaining stable have a comparably short time scale which we denote by τ_s. Therefore in the instability region we find a hierarchy in the time scales

$$\tau_u >> \tau_s \ . \tag{2.11}$$

Taking into account the interaction of stable and unstable modes, we expect in the long-term behaviour of the system that the stable variables are moving in a way which is completely determined through the unstable modes, the order parameters. This observation may be put into mathematical terms as we shall indicate now. We may split the set of $\{\xi_i\}$ into a set of modes which will become unstable and denote it by the vector $\underset{\sim}{u}$ and the remaining set of stable modes which we collect into the vector $\underset{\sim}{s}$. Accordingly we split the set of equations (2.10) into

$$\dot{\underset{\sim}{u}} = \Lambda_u \underset{\sim}{u} + \underset{\sim}{Q}(\underset{\sim}{u},\underset{\sim}{s}) + \underset{\sim}{F}_u \tag{2.12}$$

$$\dot{\underset{\sim}{s}} = \Lambda_s \underset{\sim}{s} + \underset{\sim}{P}(\underset{\sim}{u},\underset{\sim}{s}) + \underset{\sim}{F}_s \ . \tag{2.13}$$

Slaving of the stable modes through the order parameters now means

$$\underset{\sim}{s} = \underset{\sim}{s}_1(\underset{\sim}{u},t) + \underset{\sim}{s}_2 \quad . \tag{2.14}$$

(2.14) expresses the fact that the values of the stable modes are - besides some corrections $\underset{\sim}{s}_2$ which arise from the fluctuations - completely determined by the instantaneous values of the order parameters $\underset{\sim}{u}$. With the result (2.14) the complete set of equations (2.12) and (2.13) can be drastically simplified. In fact we may use this result to eliminate the stable modes from the equations of the order parameters to arrive at the equation

$$\dot{\underset{\sim}{u}} = \Lambda_u \underset{\sim}{u} + \underset{\sim}{Q}(\underset{\sim}{u},\underset{\sim}{s}_1(\underset{\sim}{u},t) + \underset{\sim}{s}_2) + \underset{\sim}{F}_u \quad . \tag{2.15}$$

which is the final order parameter equation. We observe that the idea of slaving generally leads to a drastic reduction of the degrees of freedom of the system which, however, now yields a complete description of the system on macroscopic scales. The last step then consists in the solution of the order parameter equation (2.15) and in the identification of the macroscopically evolving ordered states which may correspond to definite patterns or functioning of the system. The central problem which still remains to be solved rests in the construction of (2.14).

3. On the Movement of the Slaved Modes

We are concerned with the construction of

$$\underset{\sim}{s} = \underset{\sim}{s}_1(\underset{\sim}{u},t) + \underset{\sim}{s}_2 \quad . \tag{3.1}$$

Our interest is devoted to the long-term behaviour of the slaved modes and we can formally integrate (2.13) in the following way

$$\underset{\sim}{s} = \int_{-\infty}^{t} \exp[\Lambda_s(t-\tau)]\{\underset{\sim}{P}(\underset{\sim}{u},\underset{\sim}{s}) + \underset{\sim}{F}_s\}_\tau d\tau \quad . \tag{3.2}$$

Through (3.2) we define the operator $(d/dt - \Lambda_s)^{-1}$:

$$\underset{\sim}{s} = (d/dt - \Lambda_s)^{-1} [\underset{\sim}{P}(\underset{\sim}{u},\underset{\sim}{s}) + \underset{\sim}{F}_s] \quad . \tag{3.3}$$

We mainly consider the purely deterministic case where $\underset{\sim}{F}_s$ vanishes and $\underset{\sim}{s}_2$ correspondingly (generalizations are given in [2]). We further introduce the operators

$$(d/dt) = \underset{\sim}{Q}\partial/\partial\underset{\sim}{u} \tag{3.4}$$

and in addition

$$(d/dt - \Lambda_s)^{-1}_{(o)}\underset{\sim}{P} = \int_{-\infty}^{t} \exp[\Lambda_s(t-\tau)]\ \underset{\sim}{P}(\underset{\sim}{u},\underset{\sim}{s})_t d\tau \quad . \tag{3.5}$$

A partial integration in (3.2) may now be performed by using the above defined operators in the following way:

$$\begin{aligned}(d/dt - \Lambda_s)^{-1} &= (d/dt - \Lambda_s)^{-1}_{(o)} \\ &- (d/dt - \Lambda_s)^{-1} (d/dt)\ (d/dt - \Lambda_s)^{-1}_{(o)} \quad .\end{aligned} \tag{3.6}$$

We now may use (3.6) to construct a systematic iteration scheme to find (3.1). We take the ansatz

$$\underset{\sim}{s} = \sum_{n=2}^{N} \underset{\sim}{C}^{(n)} \quad , \tag{3.7}$$

where $\underset{\sim}{C}^{(n)}$ is precisely of order n in $\underset{\sim}{u}$. Correspondingly we may define $\underset{\sim}{P}^{(n)}$ and $(d/dt)^{(n)}$. Inserting (3.6) and (3.7) into (3.2) we obtain the solution in the form

$$\underset{\sim}{C}^{(n)} = \sum_{m=0}^{n-2} (d/dt - \Lambda_s)^{-1}_{(m)} \underset{\sim}{P}^{(n-m)} \tag{3.8}$$

where we have defined

$$(d/dt - \Lambda_s)^{-1}_{(m)} = (d/dt - \Lambda_s)^{-1}_{(o)} \sum \prod_i [(-d/dt)^{(i)} (d/dt - \Lambda_s)^{-1}_{(o)}] \tag{3.9}$$

where the product over i has to be taken in such a way that $i \geq 1$ and $\sum i = m$. The sum means summing up over all different products. (3.9) provides us with a systematic method to construct (3.1) and it has been generalized in various ways, especially by taking into account the fluctuations. We shall here, however, draw our attention to another aspect, namely we now shall interpret the solution $\underset{\sim}{s}_1$ of (3.2) as the fixed point of

$$\underset{\sim}{s}^{(N+1)} = (d/dt - \Lambda_s)^{-1} [\underset{\sim}{P}(\underset{\sim}{u}, \underset{\sim}{s}^{(N)})] \quad . \tag{3.10}$$

We note that, if $(d/dt - \Lambda_s)^{-1} \underset{\sim}{P}(\underset{\sim}{u},.)$ defines a contraction on an open connected set in a normed space, (3.10) has a unique fixed point which is guaranteed by the fixed point theorem. This fixed point in our special case defines the so-called center manifold. We consider two examples to illustrate (3.10). In a first one we assume

$$\underset{\sim}{P}(\underset{\sim}{u}, \underset{\sim}{s}) = \underset{\sim}{P}_1(\underset{\sim}{u}) \quad . \tag{3.11}$$

The fixed point then is immediately obtained from

$$\underset{\sim}{s} = (d/dt - \Lambda_s)^{-1} \underset{\sim}{P}_1(\underset{\sim}{u}) \quad . \tag{3.12}$$

A nontrivial concrete example of (3.11) is provided by the Haken-Zwanzig model [2], where one considers just one unstable mode u and stable mode s.

As a second case we consider

$$\underset{\sim}{P}(\underset{\sim}{u}, \underset{\sim}{s}) = \underset{\sim}{P}_1(\underset{\sim}{u}) + 2\, A_{sus} : \underset{\sim}{P}_2(\underset{\sim}{u}) : \underset{\sim}{s} \tag{3.13}$$

where we used the notation introduced in [1,2]. The corresponding fixed point may now be obtained - at least formally - as the inhomogeneous solution of a linear difference equation from (3.10)

$$\underset{\sim}{s}^{(N+1)} = \sum_{n=0}^{N} (d/dt - \Lambda_s)^{-1} 2A_{sus} : \underset{\sim}{P}_2(\underset{\sim}{u}) \; \ldots\ldots \; [(d/dt - \Lambda_s)^{-1} \underset{\sim}{P}_1(\underset{\sim}{u})] \tag{3.14}$$

Here the dots indicate that one has to perform the operation n times and one has to take care about the chronological order. (3.15) can be transformed into a more elegant form by using the convention of time ordering. Denoting the time-ordered product as $\{\ldots\}_+$ we obtain

$$\underset{\sim}{s}^{(N+1)} = \sum_{n=0}^{N} (n!)^{-1} \{(d/dt - \Lambda_s)^{-1}\, 2A_{sus} : \underset{\sim}{P}_2\}^n_+ \, (d/dt - \Lambda_s)^{-1} \underset{\sim}{P}_1 \quad . \tag{3.15}$$

We find $\underset{\sim}{s}$ by taking the limit $N \to \infty$ and the series can trivially be summed up and yields the correct formal solution of the original equation. We now may draw some conclusions for a general $\underset{\sim}{P}$. If the mapping, defined through $\underset{\sim}{P}_2$ in (3.15) is a contraction we do not need to take into account the higher order terms in $\underset{\sim}{P}$ besides $\underset{\sim}{P}_{1,2}$ in order to construct (3.1). (3.13) may then be considered as linearly stable, and we expect to find a good approximation to (3.1) which generally exists only locally. The situation may change, however, when the map in some of its directions does no longer contract. This may be a result of the evolution of the u's. Then two possibilities might be considered. In a first case this direction might not be orthogonal to the inhomogeneity of the fixed point equation. The conclusion then would be the following. Besides the purely linear unstable modes in the order parameter equations, we have to take into account additional modes which are linearly stable but are now demanded as order parameters through the action of the nonlinearities. In the other case, where the inhomogeneity always remains orthogonal to the non-contracting directions of the mapping, we may apply the corresponding non-linear theory to construct (3.1). It is therefore suggested, through the discussion of (3.15), that the nonlinearities might become responsible that we have to take into account more order parameters than it is suggested from linear stability analysis. Finally we note that the discussion can be given at least formally along the same lines if fluctuations are included.

4. Macroscopic Synergetics

The origin of macroscopic synergetics (a review and complete list of references to the subject is given in [4]) results from the observation that even the subsystems of a complex system are complex units by themselves, so that it is difficult or even impossible to formulate the mesoscopic equations explicitly. When we go to the extreme and take the human brain as an example, the subsystems, i.e. the neurons, are by themselves not known to us in such detail that we could formulate mesoscopic equations. In this case the principle of Jaynes gives an important hint how to proceed. This principle has been formulated as follows: " If any macrophenomenon is found to be reproducible then it follows that all microscopic details which are not under the experimenter's control must be irrelevant for understanding and predicting it." Now let the macroscopic information be given by some state vector $\underset{\sim}{f}$ which is composed from measured mean values

$$\underset{\sim}{f} = (f_1, f_2, \ldots, f_N) \quad . \tag{4.1}$$

The index may distingish different variables but may also belong to the same variable but at different times. Following Jaynes we define the caliber through

$$i_{\underset{\sim}{f}} = \ln W_{\underset{\sim}{f}} \quad , \tag{4.2}$$

where $W_{\underset{\sim}{f}}$ denotes the number of microstates with the given values of $\underset{\sim}{f}$. Obviously this state will be realized in which the number of microscopic realizations is maximal. If this is translated into the notions of probabilities an extremal principle can be formulated for the caliber and we may be able to obtain the probability distribution from this principle. The crucial point, however, which remains is how to find the appropriate constraints. H. Haken has solved this problem for a large class of systems which are considered in synergetics. Here we are concerned with the slaving principle, and it appears that there exists a direct relation to Jaynes' principle. If we consider the caliber of a synergetic system in the vicinity of a critical point we may decompose it in a unique way into a part which corresponds only to the order parameters i_u and a second part which stems from the stable modes and their interaction with the order parameters

$$i = i_u + i_{us} \quad . \tag{4.3}$$

Under rather general conditions it may be shown that during the passage through a critical region the complete variation of the caliber is only due to i_u, i.e. the behaviour of the order parameters. Recalling the above formulation of Jaynes' principle we may immediately conclude that the slaving principle provides a microscopic derivation of Jaynes principle in regions close to nonequilibrium phase transition points.

5. Conclusion

I have just picked up one aspect from the fast developing field of synergetics. The idea of slaving has become one of the most important parts of the general theory as well as in its applications. The formulation of macroscopic synergetics on the basis of slaving opens for future research also the possibility to get a far deeper insight in the functioning of highly complex systems which are already composed of complicated and complex subsystems.

References:

[1] H. Haken: "Synergetics. An Introduction" 3rd. edition, Springer-Verlag Berlin-Heidelberg-New York (1983)

[2] H. Haken: "Advanced Synergetics", Springer-Verlag Berlin- Heidelberg-New York-Tokyo (1983)

[3] H. Haken: "Non-linear Interactions of Laser Modes", Talk given at the International Conference on Optical Pumping, Heidelberg (1962)

[4] H. Haken, A. Wunderlin: "A Macroscopic Approach to Synergetics" MIDIT Workshop TU of Denmark, Lyngby (1986), in press

Solutions of Fokker-Planck Equations in Terms of Matrix Continued Fractions

H. Risken

Abteilung für Theoretische Physik, Universität Ulm,
D-7900 Ulm, Fed. Rep. of Germany

1. Introduction

Synergetics deals with complicated systems consisting of very many subsystems. As one of HAKEN's famous examples [1-4] let us consider the laser systems. Here a few light modes are coupled to the many active atoms (subsystems) in the laser cavity. A total description of such a system would require very many variables. In general such a detailed microscopic description is not needed. Instead we are only interested in a few relevant variables, which describe the system on a macroscopic scale. In the laser the few light modes are such relevant macroscopic variables.

The equations of motion for these macroscopic variables are usually differential equations. In general the time scales in these differential equations are much slower than the timescales in the equations for the irrelevant variables. By applying the slaving principle of HAKEN [4], we can thus eliminate these irrelevant variables.

If the fluctuations of the irrelevant variables are not taken into account, we have to deal with a system of - in general nonlinear - differential equations for these macroscopic variables. If the differential equations have only one stable solution the usually small fluctuations may disturb the stable solution but without drastically changing the final result. If, however, the differential equations have multistable or unstable solutions fluctuations have a large influence on the behaviour of the solution. For multistable solutions one gets, due to the fluctuations, transitions between the different locally stable solutions. For the laser equation with one mode we have above threshold a stable solution with a finite intensity and an unstable solution with zero intensity. If the laser is switched on, the fluctuations force the unstable solution to grow and to fluctuate and thus determine the statistics of the laser light during the switching on process. Near laser threshold, where the stabilization is very small, the small fluctuations also force the light intensity to fluctuate appreciably. These fluctuations are taken into account by adding fluctuating forces to the differential equations of the macroscopic variables. The differential equations will then become stochastic differential equations.

For systems, where the equations of motion are nonlinear, it is in general impossible to solve the corresponding stochastic differential equations. For suitable assumptions of the noise forces (Gaussian distributed, δ-correlated), however, one can use a Fokker-Planck equation. This Fokker-Planck equation (FPE) is an equation of motion for the distribution function of the macroscopic variables.

For one variable this FPE is essentially equivalent to a Schrödinger equation. Using methods for solving Schrödinger equations one can obtain the solution of the one-variable FPE either analytically or numerically. For two or more variables, however, the FPE cannot generally be transformed to a (Hermitian) Schrödinger equation. A solution can then be obtained only in special cases, such as linear drift and constant diffusion coefficients (Ornstein-Uhlenbeck process) or for the detailed-balance condition. For various simple FPE with two variables we have calculated these solutions by using matrix continued fractions (MCF), see for instance Chaps. 9 and 11 of [5] for a review.

Ordinary continued fractions are well known in mathematics [6]. In the year 1572, according to [7], Bombelli wrote down the continued fraction for the square root of 13, i.e.

$$\sqrt{13} = 3 + \cfrac{4}{6 + \cfrac{4}{6 + \cfrac{4}{6\ldots}}} . \tag{1.1}$$

Many functions appearing in mathematical physics can be expressed by ordinary continued fractions, see [6-8]. In a MCF the inversions are performed with matrices. In Sect. 3 we will see the general structure of such a MCF.

In this article we want to show for a simple example, how one can solve the corresponding FPE in terms of MCF. The simple example is the overdamped motion of a classical particle in the simplest bistable potential (Landau potential) with noise. The equation of motion for a suitably normalized variable x and a normalized time t is given by([9], page 283ff)

$$\dot{x} = - \partial V/\partial x + \Gamma(t) = ax - x^3 + \Gamma(t) \tag{1.2}$$

$$\langle\Gamma(t)\rangle = 0, \quad \langle\Gamma(t)\,\Gamma(t')\rangle = 2\delta(t-t'), \quad \Gamma \text{ Gaussian distributed} . \tag{1.3}$$

The normalized Landau potential

$$V(x) = - ax^2/2 + x^4/4 = (x^2-a)^2/4 - a^2/4 \tag{1.4}$$

has a double well for $a>0$. It is shown in Fig. 1a. The equation (1.2) is the normalized one-dimensional classical model for the nonlinear laser operator equation valid near threshold, of HAKEN. (In (3.1) of [1] or (VI.4.23) of [2] the second derivative has to be neglected and the operator b^+ has to be replaced by the real variable x.) Though, as mentioned previously, we can obtain solutions of the corresponding one-variable FPE by other methods, we will nevertheless use this simple FPE to explain the MCF method. The more complicated FPE for the complex variable b of the laser light field can also be treated by this method [10]. In Sect. 7 we will discuss how more complicated FPEs having two variables can be solved by use of this method. Finally some other problems, where this method can be applied, are also mentioned.

2. Bistable System

As explained in the introduction we want to discuss the MCF method for the simple stochastic differential equation (1.2,3). The corresponding FPE takes the form

$$\frac{\partial W}{\partial t} = [\frac{\partial}{\partial x}(x^3 - ax) + \frac{\partial^2}{\partial x^2}]\, W . \tag{2.1}$$

The boundary conditions require that $W(x,t)$ goes to zero for $x \to \pm\infty$ in such a way that the normalization integral exists.

The main steps of the method are the following ones:

i) expansion of the distribution function into a complete set

ii) casting the coupled equations into a tridiagonal vector recurrence relation

iii) solving this tridiagonal vector recurrence relation in terms of MCFs

In the first step we expand the distribution function W(x,t) into a complete set obeying the boundary conditions. Here we assume natural boundary conditions, i.e. the integral over the distribution function should exist. As set we choose Hermite polynomials $H_m(\xi)$ times a suitable Gauss function, i.e. we use the expansion

$$W(x,t) = \sum_{m=0}^{\infty} c_m(t)\, H_m(\alpha x) e^{-\alpha^2 x^2} . \tag{2.2}$$

Here α is a scaling factor by which the width of the complete set can be adapted to the width of the distribution function. We insert (2.2) into (2.1) multiply the resulting equation with $H_n(\alpha x)$ and integrate it over x from minus infinity to plus infinity. By using [11]

$$\int_{-\infty}^{\infty} H_n(\xi)\, H_m(\xi) e^{-\xi^2} d\xi = \sqrt{\pi}\;\; 2^n n!\; \delta_{nm} \tag{2.3}$$

and repeatedly using

$$\frac{d}{d\xi} H_n(\xi) = 2n\, H_{n-1}(\xi) \tag{2.4}$$

$$\xi\, H_n(\xi) = \frac{1}{2} H_{n+1}(\xi) + n\, H_{n-1}(\xi) \tag{2.5}$$

we arrive at the following infinite system of differential equations for the expansion coefficients ($c_n = 0$ for $n<0$)

$$\begin{aligned} \dot{c}_n = &- \frac{1}{8\alpha^2} c_{n-4} + \left[\frac{a}{2} - \frac{3(n-1)}{4\alpha^2} + \alpha^2\right] c_{n-2} \\ &+ n\left(a - \frac{3n}{2\alpha^2}\right) c_n - \frac{n(n+1)(n+2)}{\alpha^2} c_{n+2} . \end{aligned} \tag{2.6}$$

As is seen in (2.6) only coefficients from c_{n-4} up to c_{n+2} are coupled. If such a finite coupling exists we can always cast the system into a tridiagaonal vector recurrence relation, see for instance [5] Sect. 9.1.2. Because of the symmetry of the problem even (odd) coefficients are only coupled to even (odd) ones. Introducing the two dimensional vectors $\mathbf{a}_n$ and $\mathbf{b}_n$ defined by

$$\mathbf{a}_n = \begin{pmatrix} a_n^1 \\ a_n^2 \end{pmatrix} = \begin{pmatrix} c_{4n} \\ c_{4n+2} \end{pmatrix} \;;\quad \mathbf{b}_n = \begin{pmatrix} b_n^1 \\ b_n^2 \end{pmatrix} = \begin{pmatrix} c_{4n+1} \\ c_{4n+3} \end{pmatrix}, \tag{2.7}$$

the recurrence relation (2.6) is equivalent to the following tridiagonal recurrence relations for vectors $\mathbf{a}_n$ and $\mathbf{b}_n$

$$\dot{\mathbf{a}}_n = \mathbf{Q}_n^- \mathbf{a}_{n-1} + \mathbf{Q}_n \mathbf{a}_n + \mathbf{Q}_n^+ \mathbf{a}_{n+1} \tag{2.8a}$$

$$\dot{\mathbf{b}}_n = \mathbf{P}_n^- \mathbf{b}_{n-1} + \mathbf{P}_n \mathbf{b}_n + \mathbf{P}_n^+ \mathbf{b}_{n+1} \tag{2.8b}$$

where the matrices $\mathbf{Q}$ and $\mathbf{P}$ are given by

$$\mathbf{Q}_n^- = \begin{pmatrix} -\dfrac{1}{8\alpha^2} & \dfrac{a}{2} - \dfrac{3(4n-1)}{4\alpha^2} + \alpha^2 \\ 0 & -\dfrac{1}{8\alpha^2} \end{pmatrix} \tag{2.9a}$$

$$
\mathbf{Q}_n = \begin{pmatrix} 4n(a-\frac{6n}{\alpha^2}) & -\frac{4}{\alpha^2}\, n(4n+1)(4n+2) \\ \frac{a}{2} - \frac{3}{4\alpha^2}(4n+1)+\alpha^2 & (4n+2\ [a-\frac{3}{\alpha^2}(2n+1)] \end{pmatrix} \tag{2.10a}
$$

$$
\mathbf{Q}_n^+ = \begin{pmatrix} 0 & 0 \\ -\frac{4}{\alpha^2}(4n+2)(4n+3)(n+1) & 0 \end{pmatrix} \tag{2.11a}
$$

$$
\mathbf{P}_n^- = \begin{pmatrix} -\frac{1}{8\alpha^2} & \frac{a}{2} - \frac{3}{\alpha^2}\, n + \alpha^2 \\ 0 & -\frac{1}{8\alpha^2} \end{pmatrix} \tag{2.9b}
$$

$$
\mathbf{P}_n = \begin{pmatrix} (4n+1)[a-\frac{3}{2\alpha^2}(4n+1)] & -\frac{2}{\alpha^2}(4n+1)(2n+1)(4n+3) \\ \frac{a}{2} - \frac{3}{2\alpha^2}(2n+1)+\alpha^2 & (4n+3)[a - \frac{3}{3\alpha^2}(4n+3)] \end{pmatrix} \tag{2.10b}
$$

$$
\mathbf{P}_n^+ = \begin{pmatrix} 0 & 0 \\ -\frac{4}{\alpha^2}(4n+3)(n+1)(4n+5) & 0 \end{pmatrix} \quad . \tag{2.11b}
$$

In the final step we have to solve the tridiagonal recurrence relations (2.8a,b). In the following section we explain the MCF method for obtaining stationary moments. In the next sections we apply the method to the calculation of eigenvalues and correlation functions.

3. Stationary Solution

For the simple FPE (2.1) the stationary solution is given by the Boltzmann distribution

$$
W_{st}(x) = N \exp[-V(x)] \tag{3.1}
$$

where $V(x)$ is the Landau potential (1.4). The distribution (3.1) is shown in Fig. 1b for the double-well potential of Fig. 1a. Any expectation value can be obtained by integration.

The normalization requires

$$
\int_{-\infty}^{\infty} W_{st}(x)\, dx = 1 \ . \tag{3.2}
$$

The normalization constant and the second moment read (the first moment is zero)

$$
N^{-1} = \int_{-\infty}^{\infty} \exp[-V(x)]dx = 2^{1/4}\sqrt{\pi}\ \ \exp(a^2/8)\ D_{-1/2}(-a/\sqrt{2}) \tag{3.3}
$$

$$
\langle x^2 \rangle = N \int_{-\infty}^{\infty} x^2 \exp[-V(x)]dx = N\ 2^{1/4}\sqrt{\pi}\ \ \exp(a^2/8)\ D_{-3/2}(-a/\sqrt{2}) \ . \tag{3.4}
$$

where $D_\nu(x)$ are the parabolic cylinder functions [11] page 337.

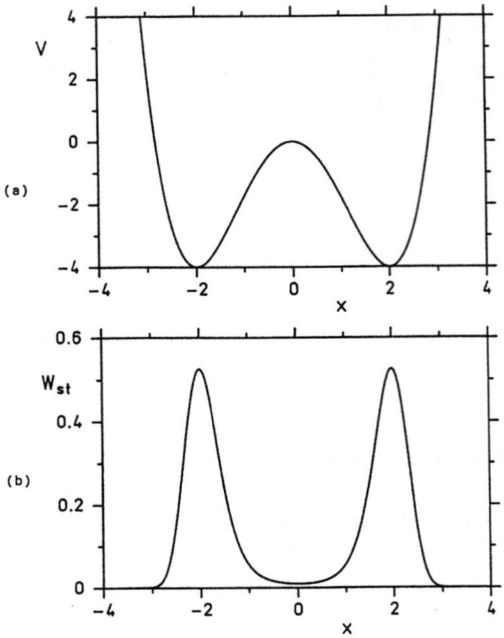

Fig. 1 The double-well potential (1.4) (a) and the stationary distribution (3.1) (b) for a=4

Here we want to illustrate the application of the MCF method to the calculation of (3.4). In the stationary state we have to solve (2.8a) with zero left-hand side, i.e.

$$\mathbf{Q}_n^- \mathbf{a}_{n-1} + \mathbf{Q}_n \mathbf{a}_n + \mathbf{Q}_n^+ \mathbf{a}_{n+1} = \mathbf{0} \ . \tag{3.5}$$

(The odd numbered coefficients in (2.2) are zero for the stationary solution.) Using (2.3) we get from the normalization (3.2)

$$c_0 = \alpha/\sqrt{\pi} \ . \tag{3.6}$$

Because of (2.3) and

$$x^2 = (H_2(\alpha x) + 2)/(4\alpha^2) \tag{3.7}$$

we can express the second moment by

$$\langle x^2 \rangle = \frac{2\sqrt{\pi}}{\alpha^3} c_2 + \frac{\sqrt{\pi}}{2\alpha^3} c_0 \ , \tag{3.8}$$

i.e. by the components of the vector

$$\mathbf{a}_0 = \begin{pmatrix} c_0 \\ c_2 \end{pmatrix}. \tag{3.9}$$

This vector is obtained from (3.5) by the following successive elimination procedure. Truncating (3.5) at n=N (i.e. $\mathbf{a}_{N+1} = \mathbf{a}_{N+2} = \ldots = \mathbf{0}$) we express $\mathbf{a}_N$ in terms of $\mathbf{a}_{N-1}$ according to (3.5) for n=N. Inserting this $\mathbf{a}_N$ into the equation for n=N-1 we then express $\mathbf{a}_{N-1}$ in terms of $\mathbf{a}_{N-2}$ and so on. The general expressions can be written in the form

$$\mathbf{a}_{n+1} = \mathbf{S}_n \, \mathbf{a}_n \tag{3.10}$$

where the $\mathbf{S}_n$ obey the relations ($n \leq N-1$)

$$\mathbf{S}_n = [-\mathbf{Q}_{n+1} - \mathbf{Q}^+_{n+1} \, \mathbf{S}_{n+1}]^{-1} \, \mathbf{Q}^-_{n+1} \tag{3.11}$$

starting with

$$\mathbf{S}_N = \mathbf{0} \, . \tag{3.12}$$

We continue this procedure till we reach the equation (3.5) with n=1, i.e. we obtain (3.10,11) with n=0. Inserting $\mathbf{a}_1 = \mathbf{S}_0 \, \mathbf{a}_0$ into (3.5) with n=0 we arrive at

$$\mathbf{R}_0 \, \mathbf{a}_0 = \mathbf{0} \quad \text{with} \tag{3.13}$$

$$\mathbf{R}_0 = -\mathbf{Q}_0 - \mathbf{Q}^+_0 \, \mathbf{S}_0 \, . \tag{3.14}$$

In components the relation (3.13) reads explicitly

$$R_0^{11} \, a_0^1 + R_0^{12} \, a_0^2 = 0$$

$$R_0^{21} \, a_0^1 + R_0^{22} \, a_0^2 = 0 \, . \tag{3.15}$$

As is easily seen from (2.10a, 2.11a) the coefficients R_0^{11} and R_0^{12} are zero and the first equation in (3.15) is fulfilled. From the second equation in (3.15) we obtain

$$a_0^2 = c_2 = - R^{21} \, a_0^1/R^{22} = - R^{21} \, c_0/R^{22} \, . \tag{3.16}$$

Therefore we finally have for the second moment (3.8)

$$\langle x^2 \rangle = - 2R^{21}/(\alpha^2 \cdot R^{22}) + 1/(2\alpha^2) \, . \tag{3.17}$$

Thus we see that the second moment can be expressed by the matrix elements of the matrix (3.14). For $N \to \infty$ this matrix is given by the infinite MCF

$$\mathbf{R}_0 = - \mathbf{Q}_0 - \mathbf{Q}^+_0 \cfrac{\mathbf{I}}{-\mathbf{Q}_1 - \mathbf{Q}^+_1 \cfrac{\mathbf{I}}{-\mathbf{Q}_2 - \ldots} \, \mathbf{Q}^-_2} \, \mathbf{Q}^-_1 \, . \tag{3.18}$$

In (3.18) the matrix inversions are written by fraction lines i.e. by

$$\mathbf{A}^{-1} = \frac{\mathbf{I}}{\mathbf{A}} \, , \tag{3.19}$$

in order to see the connection with an ordinary continued fraction more clearly. The MCF (3.18) can be easily translated into an algorithm for a digital computer. In order to do this one may use the recursion

$$\mathbf{R}_n = - \mathbf{Q}_n - \mathbf{Q}^+_n \, [\mathbf{R}_{n+1}]^{-1} \, \mathbf{Q}^-_{n+1} \tag{3.20}$$

starting with $\mathbf{R}_N = -\mathbf{Q}_N$ at $n = N$ and iterating (3.20) to lower n till we arrive at $\mathbf{R}_0$. The index N must be chosen so large, that the matrix $\mathbf{R}_0$ does not change within a given accuracy by increasing N. (For a theorem concerning the convergence of MCFs see [12].)

The average of x^2 as a function of a is shown in Fig. 2. The result for $\langle x^2 \rangle$ should be independent of the scaling factor α. By properly choosing the scaling

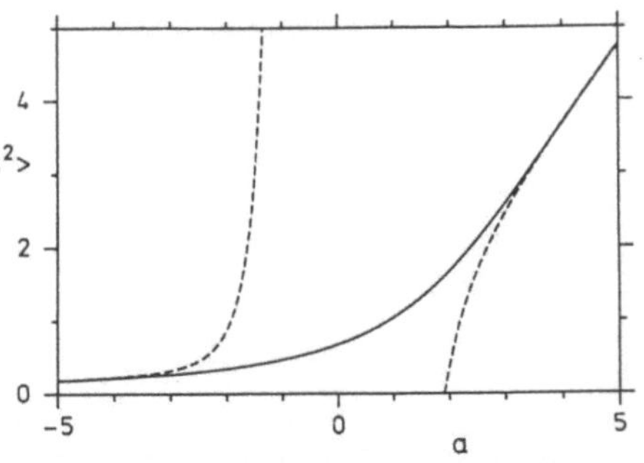

Fig. 2 The second moment $\langle x^2 \rangle$ (solid line) and its asymptotic expessions (3.21) (broken line) as a function of a

factor, however, the number of inversions to obtain accurate results can be minimized. Using a 14 digit arithmetic and choosing a=20 we obtain for instance with $\alpha=1.4$ the value $\langle x^2 \rangle=19.9496173$ using 300 inversions, whereas with $\alpha=0.1$ we had to use about 10,000 inversions to achieve the same accuracy. It is quite remarkable that the method works for such large a because the Hermite functions are in no way adapted to the double maxima of the stationary distribution (3.1,2).

For large a an asymptotic expansion of the integrals in (3.3,4) leads to

$$\langle x^2 \rangle = a - a^{-1} - 3a^{-3} - 24a^{-5} \quad ; \; a \gg 1$$

$$\langle x^2 \rangle = (1/|a|)\,(1-3a^{-2} + 24a^{-4}) \quad ; \; a \ll -1 \; . \tag{3.21}$$

The first expression agrees for a=20 up to 8 decimal places with the result just mentioned.

The stationary distribution function (2.2) can be obtained by calculating the vector $\mathbf{a}_0$ by the method described above and then, after calculating the MCFs $\mathbf{S}_n$ in (3.11), by iterating (3.10), i.e.

$$\mathbf{a}_n = \mathbf{S}_{n-1}\,\mathbf{S}_{n-2} \cdots \mathbf{S}_0\,\mathbf{a}_0 \; . \tag{3.22}$$

This method is numerically stable, whereas the up-iteration according to (3.5) starting with $\mathbf{a}_0$ and $\mathbf{a}_1=\mathbf{S}_0\mathbf{a}_0$ would be numerically unstable. (In the present simple example there is, of course, no need for calculating the distribution function in this way because we have the analytic expression (3.1,2)). Higher coefficients are also needed for the calculation of the higher order moments $\langle x^{2\mu} \rangle$ with $\mu \geq 2$.

4. Calculation of Eigenvalues and Eigenfunctions

For the calculation of the eigenvalues we can also use the MCF method. The most important eigenvalue in physics is the lowest nonzero eigenvalue. It is twice the transition rate from the left to the right well or vice versa. The eigenfunction belonging to this eigenvalue is antisymmetric for symmetric potentials. We therefore explain the method only for antisymmetric eigenfunctions. Inserting

$$\mathbf{b}_n(t) = e^{-\lambda t}\,\mathbf{b}_n \tag{4.1}$$

in (2.8b) we obtain the tridiagonal recurrence relation

$$\mathbf{P}_n^-\,\mathbf{b}_{n-1} + \hat{\mathbf{P}}_n\,\mathbf{b}_n + \mathbf{P}_n^+\,\mathbf{b}_{n+1} = \mathbf{0} \tag{4.2}$$

with

$$\hat{\mathbf{P}}_n = \mathbf{P}_n + \lambda \mathbf{I} \ . \tag{4.3}$$

The recurrence relation (4.2) has the same form as the relation (3.5). We can thus use the same succesive elimination procedure as done previously. The only difference will be that $\mathbf{Q}_n$, $\mathbf{Q}_n^{\pm}$ have to be replaced by $\hat{\mathbf{P}}_n$, $\mathbf{P}_n^{\pm}$. and $\mathbf{S}_n$ by $\hat{\mathbf{S}}_n$ The equation corresponding to (3.13) then reads

$$\hat{\mathbf{R}}_0 \, \mathbf{b}_0 = \mathbf{0} \tag{4.4}$$

with

$$\hat{\mathbf{R}}_0 = -\hat{\mathbf{P}}_0 - \mathbf{P}_0^+ \hat{\mathbf{S}}_0 \quad . \tag{4.5}$$

The determinant of $\mathbf{R}_0$ is no longer zero as was the case for the stationary solution. In order to have nontrivial solutions we have to require that the determinant of $\hat{\mathbf{R}}_0$ be zero, i.e.

$$D(\lambda) = \mathrm{Det}(\hat{\mathbf{R}}_0) = 0 \ . \tag{4.6}$$

This determinant depends on λ because the $\hat{\mathbf{P}}_n$ depend on it, see (4.3). Thus the eigenvalues follow as roots of the function $D(\lambda)=\mathrm{Det}(\hat{\mathbf{R}}_0)$. The matrix $\hat{\mathbf{R}}_0$ is obtained in analogy to the continued fraction (3.18). In Fig. 3 the lowest nonzero eigenvalue is shown as a function of a. As it is seen it decreases monotonously with increasing a. For large a it becomes extremely small. (Because the potential difference $a^2/4$ between the top of the potential at x=0 and its minima (see Fig. 1a) becomes very large, it is very unlikely that the fluctuations push the particle over the top at x=0 for large a.) The lowest nonzero eigenvalue is twice the Kramers rate [13], i.e the probability rate for a particle going from the left well to the right well or vice versa. Using the improved Kramers rate (see for instance Sect. 5.9 of [5]) we have the asymptotic result

$$\lambda_1 \approx \lambda_{as} = \frac{\sqrt{2}}{\pi} \, a \, (1 - \frac{3}{2a^2}) \, \exp(-\frac{a^2}{4}) \quad ; \quad a \gg 1 \tag{4.7}$$

which agrees very well with the numerical solution for large a, see Fig. 3.

To calculate the odd eigenfunctions we proceed as follows. For the eigenvalues λ the determinant of $\hat{\mathbf{R}}_0$ is zero. Up to a normalization constant the coefficient $\mathbf{b}_0$ follows from (4.4). Similar to the stationary case higher coefficients are obtained by repeated mutliplication with the matrices $\hat{\mathbf{S}}_n$, compare (3.22). The odd

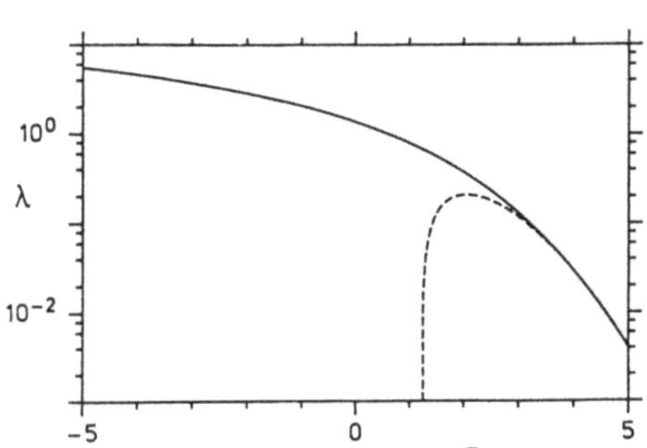

Fig. 3 The lowest nonzero eigenvalue (solid line) and its asymptotic expression (4.7) (broken line) as a function of a

eigenfunctions $\phi(x)$ belonging to the eigenvalues λ are then given by

$$\phi(x) = \sum_{m=0}^{\infty} [b_m^1 H_{4m+1}(\alpha x) + b_m^2 H_{4m+3}(\alpha x)]e^{-\alpha^2 x^2} . \quad (4.8)$$

5. Calculation of Correlation Functions

The calculation of correlation functions is also an important task. We illustrate this calculation for the stationary correlation function ($t \geq 0$)

$$K(t) = \langle x(t)\, x(0)\rangle = \int_{-\infty}^{\infty}\int_{-\infty}^{\infty} xx' G(x,x',t)\, W_{st}(x')dxdx' . \quad (5.1)$$

By using an eigenfunction expansion of the Green's function G one can obtain correlation functions, see for instance Sect. 5.4 of [5]. A method, which does not need the determination of eigenfunctions and eigenvalues is the following one. First we introduce the function ($t \geq 0$)

$$w(x,t) = \int_{-\infty}^{\infty} x'\, G(x,x',t)\, W_{st}(x')\, dx' . \quad (5.2)$$

It also obeys the FPE (2.1), because the Green's function G obeys it. The initial condition for $w(x,t)$ is given by

$$w(x,0) = \int_{-\infty}^{\infty} x'\, G(x,x',0)\, W_{st}(x')dx'' = \int_{-\infty}^{\infty} x'\, \delta(x-x')\, W_{st}(x')\, dx'$$

$$= xW_{st}(x) . \quad (5.3)$$

Because the initial condition is an odd function and the operator in the FPE (2.1) is even in the variables x, $w(x,t)$ is an odd function in x for all times. With the help of the two-dimensional vectors $\mathbf{b}_n$ of (2.7) the expansion of $w(x,t)$ is written in the form

$$w(x,t) = \sum_{n=0}^{\infty} [b_n^1(t)\, H_{4n+1}(\alpha x) + b_n^2(t)\, H_{4n+3}(\alpha x)]\, e^{-\alpha^2 x^2} . \quad (5.4)$$

Because of $x = H_1(\alpha x)/(2\alpha)$ and because of (2.3), (5.1,2) $K(t)$ is expressed by the first component of the vector $\mathbf{b}_0$, i.e.

$$K(t) = \int_{-\infty}^{\infty} xw(x,t)dx = \sqrt{\pi}\ \alpha^{-2}\, b_0^1(t) . \quad (5.5)$$

Using (2.5) the initial condition (5.3) transforms to the following initial condition for the expansion coefficients

$$\mathbf{b}_n(0) = \begin{pmatrix} b_n^1(0) \\ b_n^2(0) \end{pmatrix} = (1/\alpha) \begin{pmatrix} a_n^1/2 + (4n+2)a_n^2 \\ a_n^2/2 + 4(n+1)a_n^2 \end{pmatrix} \quad (5.6)$$

where a_n^1 and a_n^2 are the components of the vectors $\mathbf{a}_n$ of the stationary distribution, see (3.22). Thus the problem of calculation the correlation function (5.1) is transformed to the problem of finding a solution of the tridiagonal vector recurrence relation (2.8b) with the initial condition (5.6). Making a half-sided Fourier transform

$$\tilde{\mathbf{b}}_n(\omega) = \int_0^\infty \mathbf{b}_n(t) e^{-i\omega t} dt \tag{5.7}$$

we obtain the inhomogeneous recurrence relation

$$\mathbf{P}_n^- \tilde{\mathbf{b}}_{n-1} + \tilde{\mathbf{P}}_n \tilde{\mathbf{b}}_n + \mathbf{P}_n^+ \tilde{\mathbf{b}}_{n+1} = -\mathbf{b}_n(0) \tag{5.8}$$

with

$$\tilde{\mathbf{P}}_n = \mathbf{P}_n - i\omega \mathbf{I} . \tag{5.9}$$

For solving (5.8) we have to add vectors $\tilde{\mathbf{v}}_n$ to the ansatz (3.10), i.e we now use

$$\tilde{\mathbf{b}}_{n+1} = \tilde{\mathbf{S}}_n \tilde{\mathbf{b}}_n + \tilde{\mathbf{v}}_{n+1} \quad \text{for } n \geq 0 \quad \tilde{\mathbf{b}}_0 = \tilde{\mathbf{v}}_0 . \tag{5.10}$$

By inserting (5.10) into (5.8) and by truncating the system (5.8) at n=N ($\tilde{\mathbf{b}}_n=0$ for $n \geq N+1$) we have

$$\tilde{\mathbf{S}}_{N-1} = -\tilde{\mathbf{P}}_N^{-1} \mathbf{P}_N^-$$

$$\tilde{\mathbf{v}}_N = -\tilde{\mathbf{P}}_N^{-1} \mathbf{b}_N(0)$$

$$\tilde{\mathbf{S}}_n = -[\tilde{\mathbf{P}}_{n+1} + \mathbf{P}_{n+1}^+ \tilde{\mathbf{S}}_{n+1}]^{-1} \mathbf{P}_{n+1}^- \qquad \text{for } 0 \leq n \leq N-2$$

$$\tilde{\mathbf{v}}_n = -[\tilde{\mathbf{P}}_n + \mathbf{P}_n^+ \tilde{\mathbf{S}}_n]^{-1} [\mathbf{b}_n(0) + \mathbf{P}_n^+ \tilde{\mathbf{v}}_{n+1}] \qquad \text{for } 0 \leq n \leq N-1 . \tag{5.11}$$

Thus by iteration according to (5.11) we obtain the vector $\tilde{\mathbf{b}}_0 = \tilde{\mathbf{v}}_0$. Because of (5.5) the half-sided Fourier transform of the correlation function is expressed by the first component of this vector

$$\tilde{K}(\omega) = \int_0^\infty K(t)\, e^{-i\omega t} dt = \sqrt{\pi}\ \alpha^{-2}\, \tilde{b}_0^1 . \tag{5.12}$$

For obtaining susceptibilities we need this half-sided Fourier transform of the correlation fuction. By performing an inverse Fourier transform the correlation function K(t) itself can also be calculated.

6. Recurrence Relations for the Moments

A very simple ordinary tridiagonal recurrence relation for the moments

$$c_n(t) = \langle x^{2n} \rangle \tag{6.1}$$

is easily derived from the FPE (2.1). By multiplying (2.1) with x^{2n} integrating the resulting equation over x and using partial integrations one gets

$$\dot{c}_n = 2n(2n-1)\, c_{n-1} + 2na\, c_n - 2n\, c_{n+1} . \tag{6.2}$$

This relation is of course much simpler than (2.6) or (2.8a) and it would therefore be much easier to use it for the calculation of $\langle x^2 \rangle$. Let us apply the continued fraction method to this recurrence relation. For the stationary case (6.2) reduces to

$$(2n-1)\, c_{n-1} + a\, c_n - c_{n+1} = 0 \quad , \; n \geq 1 \; . \tag{6.3}$$

Using the same successive elimination procedure as before, the stationary second moment is expressed by the ordinary continued fraction ($c_0=1$)

$$\langle x^2 \rangle = \frac{c_1}{c_0} = \cfrac{1}{-a + \cfrac{3}{-a + \cfrac{5}{-a + \ldots}}} = -a\, \cfrac{1/a^2}{1 + \cfrac{3/a^2}{1 + \cfrac{5/a^2}{1 + \ldots}}} \; . \tag{6.4}$$

It is immediately seen from the continued fraction in the last form that the result cannot be valid for $a>0$ because $\langle x^2 \rangle$ would then become negative. Below threshold ($a<0$), however, the expression (6.4) is correct. The reason why (6.4) does not work for $a>0$ is the following. The ordinary recurrence relation (6.3) has two independent solutions. (Both starting values c_0 and c_1 in (6.3) for instance can be chosen arbitrarily and the other coefficients c_n with $n \geq 2$ then follow by upiteration of (6.3).) The continued fraction method, however, only leads to one solution. A closer investigation reveals that the continued fraction method singles out that solution where the coefficients $|c_n|$ decrease with increasing n in the fastest way (or increase with increasing n in the slowest way). This solution is called a minimal solution [7]. If the coefficients c_n are the expansion coefficients of a function as in (2.2) this would mean that the continued fraction singles out that solution, for which the expansion (2.2) converges in the best possible way. In expansions of the type (2.2) one is just looking for such a solution. If, however, the coefficients are some other quantities like the moments (6.1), it is not clear from the beginning whether the continued fractions lead to the correct result. For the example just discussed one may prove that for a large positive a the moments $\langle x^{2n} \rangle$ increase with increasing n whereas for large negative a the moments decrease with increasing n. Thus the ordinary continued fraction (6.4) is only applicable for $a<0$ whereas the MCF (3.18) is valid for all a i.e for $a<0$, $a=0$ and $a>0$.

The lowest nonzero eigenvalue of (2.1) was calculated from (4.2). Below threshold ($a<0$) it can also be obtained from the recurrence relation for the odd moments

$$d_n(t) = \langle x^{2n+1} \rangle \; . \tag{6.5}$$

This recurrence relation reads

$$\dot{d}_n = 2n(2n+1)\, d_{n-1} + (2n+1)\, d_n - (2n+1)\, d_{n+1} \; . \tag{6.6}$$

Insertion of $d_n(t)=d_n \exp(-\lambda t)$ in (6.6) leads to the ordinary continued fraction

$$D(\lambda) = a + \lambda + \cfrac{2}{a + \cfrac{\lambda}{3} + \cfrac{4}{a + \cfrac{\lambda}{5} + \cfrac{6}{a + \cfrac{\lambda}{7} + \ldots}}} = 0 \tag{6.7}$$

for determining the eigenvalues λ. For $a<0$ the eigenvalues agree with the MCF result of Sect 4. For positive a, however, the λ will become negative and thus the zeros of $D(\lambda)$ do not determine the eigenvalues of (2.1) any more.

7. Some Applications to Fokker-Planck Equations with two Variables

In Sects. 2-5 we have shown how the one-dimensional FPE (2.1) can be solved in terms of 2×2 MCFs. In particular all the formulas needed for determining $\langle x^2 \rangle$ have been given explicitly. As discussed in the introduction the FPE (2.1) can also be treated by other methods. It mainly served here as a simple example to illustrate the method. The MCF method was applied some time ago by VOLLMER and myself for solving the FPE describing the Brownian motion in a tilted cosine potential [14-15]

$$\frac{\partial W}{\partial t} = \left[- \frac{\partial}{\partial x}v + \frac{\partial}{\partial v}(\gamma v + d \sin x - F + \gamma\theta\frac{\partial}{\partial v})\right] W . \tag{7.1}$$

Here W is the distribution in x (position) and v (velocity) space, γ is the damping constant, F the external force and θ a normalized temperature. The equation (7.1) is applicable to such different devices as superionic conductors, Josephson tunneling junctions and phase-locked loops. Because x is usually an angle one has to find solutions of (7.1) which are periodic in x. Because detailed balance is not valid for $F \neq 0$ and because the drift coefficient is nonlinear it is very hard to obtain even stationary moments or distribution functions. In order to apply the MCF method to (7.1) the distribution function was expanded into two complete sets satisfying the periodic boundary conditions in x and the natural boundary condition in v. For x we have used the terms of the Fourier series and for v the Hermite functions defined by

$$\psi_n(v) = H_n(v/\sqrt{2\theta}) \exp[-v^2/(4\theta)] \Big/ \sqrt{n!2^n\sqrt{2\pi\theta}} \tag{7.2}$$

i.e. we made the expansion

$$W(x,v,t) = \psi_0(v) \sum_{m=-\infty}^{\infty} \sum_{n=0}^{\infty} c_n^m(t)\, \psi_n(v) \exp(imx)/\sqrt{2\pi} . \tag{7.3}$$

By inserting (7.3) into (7.1) one obtains a recurrence relation for the expansion coefficients $c_n^m(t)$. In these relations only terms with adjacent lower indices n are coupled. Truncating the expansion in m at ±M and introducing the (2M+1) dimensional column vectors

$$\mathbf{c}_n = \begin{pmatrix} c_n^{-M} \\ \cdot \\ \cdot \\ \cdot \\ c_n^M \end{pmatrix} \tag{7.4}$$

the recurrence relation for the expansion coefficients can be cast into the tridiagonal vector-recurrence relation

$$\dot{\mathbf{c}}_n = \mathbf{Q}_n^- \mathbf{c}_{n-1} + \mathbf{Q}_n \mathbf{c}_n + \mathbf{Q}_n^+ \mathbf{c}_{n+1} . \tag{7.5}$$

The stationary solution can thus be obtained in the same way as described in Sect. 3 for the simple FPE (2.1). In contrast to that example, however, the matrices $\mathbf{Q}_n$, $\mathbf{Q}_n^\pm$ have a much larger dimension here. In Fig. 4 a perspective plot of the distribution function is shown for those parameters, for which the deterministic part of the Langevin equation corresponding to (7.1) has two stationary stable soltions. As seen in this figure the maxima of the distribution function occur near these two stable solutions, the locked one (v=0) and the running one (v>0).

The Brownian motion in a double-well potential in position and velocity was treated similarly by this method [16]. The corresponding FPE takes the form

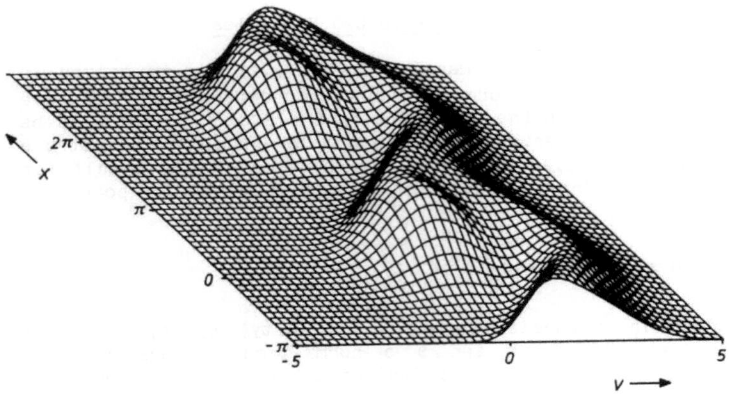

Fig. 4 Perspective plot of the stationary distribution of (7.1) for θ=1, d=1, γ=0.25 and F=0.5 (taken from [15])

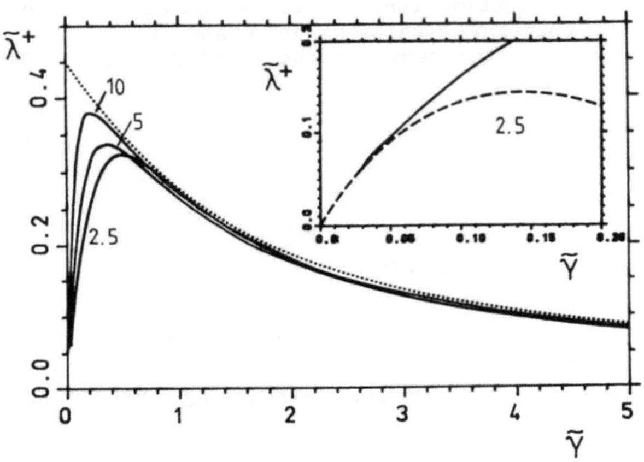

Fig. 5 The expression (7.7) scaled by $\sqrt{a}$ ($\tilde{\lambda}^+=\lambda^+/\sqrt{a}$) as a function of γ scaled by $\sqrt{a}$ ($\tilde{\gamma}=\gamma/\sqrt{a}$) for the three values of $a^2/(4\theta)=2.5$, 5, 10 (solid line). The asymptotic expansion (7.8) is shown by a dotted line, (7.9) by a broken line (taken from [16])

$$\frac{\partial W}{\partial t} = \left[- \frac{\partial}{\partial x} v + \frac{\partial}{\partial v} (\gamma v - ax + x^3 + \gamma\theta\frac{\partial}{\partial v})\right] W . \tag{7.6}$$

For the overdamped case $\gamma\to\infty$ (7.6) can essentially be reduced to the one-dimensional FPE (2.1) by applying the slaving principle and the elimination procedure of HAKEN [4,9]. In the general case we use the expansion (7.3) where the x-expansion in a Fourier series is replaced by an expansion in Hermite functions as in (2.2). In this way a tridiagonal vector recurrence relation of the form (7.5) can again be derived. The stationary solution is the well-known Boltzmann distribution (3.1) with an additional kinetic-energy term. The eigenvalues are more difficult to obtain. With the MCF method the first nonzero eigenvalue, describing the transition from the left to the right well and vice versa as a function of γ, was obtained in [16]. In Fig. 5 the lowest nonzero eigenvalue multiplied by the Arrhenius factor, i.e.

$$\lambda_1^+ = \lambda_1 \exp(\Delta E/\theta) \tag{7.7}$$

is shown as a function of the damping constants for several $\Delta E/\theta = a^2/(4\theta)$ ratios. For intermediate and large damping constants but large $\Delta E/\theta$ values there is good agreement with KRAMERS [13] result (a>0)

$$\lambda_1^+ = \frac{\sqrt{2}}{\pi}\left(\sqrt{a + \gamma^2/4} - \gamma/2\right) . \tag{7.8}$$

For very low damping constants the eigenvalue λ_1 has a dependence on γ of the form

$$\lambda_1 = \alpha\gamma + \beta\gamma^{3/2} . \tag{7.9}$$

The first term ($\sim\gamma$) was also derived by KRAMERS [13] whereas the second term ($\sim\gamma^{3/2}$) stems from a boundary layer theory [17-18]. As seen in Fig 5 the results obtained by the MCF method are in good agreement with (7.8,9). Fig. 6 shows the altitude charts of the eigenfunction belonging to the lowest nonzero eigenvalue. As seen the nodal line and the contour lines close to it twist around the stable points x = ±1, v = 0 of the deterministic equation.

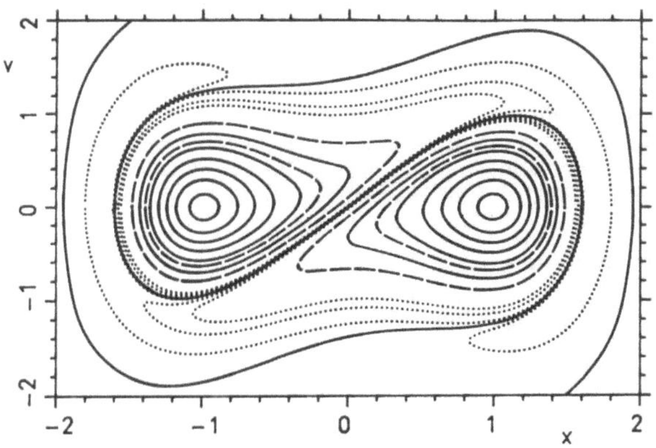

Fig. 6 Contour-lines of the antisymmetric eigenfunction of (7.6) belonging to to lowest nonzero eigenvalue for a=1, γ=0.25, θ=0.25. The lines ϕ=c=const are shown for c=0, ±0.1, ..., ±0.6 (solid lines) c=±0.05, ±0.15 (broken lines) c=±0.001, ±0.005, ±0.01 (dotted lines) (taken from [16])

A laser FPE with two variables (intensity and inversion) has been treated with the MCF method as well [19].

Colored-noise problems can also be handled by the MCF method. For the overdamped motion in the bistable potential driven by exponentially correlated noise we have the equation of motion

$$\dot{x} = ax - x^3 + \varepsilon(t) \tag{7.10}$$

$$\langle\varepsilon(t)\rangle = 0 \quad , \quad \langle\varepsilon(t)\,\varepsilon(t')\rangle = \gamma\, D\, e^{-\gamma|t-t'|} \tag{7.11}$$

where $1/\gamma$ is the correlation time and D the diffusion constant. If we now introduce ε as a new variable with the equation of motion

$$\dot{\varepsilon} = -\gamma\varepsilon + \gamma\sqrt{D}\,\Gamma(t) \tag{7.12}$$

and $\Gamma(t)$ being the white noise variable (1.3) we get the correlation function (7.11) in the stationary state. Thus we have to treat the two coupled Langevin equations (7.10,12). The FPE, corresponding to these Langevin equations, is given by

$$\dot{W} = \{\frac{\partial}{\partial x}(x^3 - ax - \varepsilon) + \gamma\frac{\partial}{\partial\varepsilon} + \gamma^2 D\frac{\partial^2}{\partial\varepsilon^2}\} W . \tag{7.13}$$

By a Hermite-function expansion in ε one again obtains the tridiagonal vector recurrence relation with the structure of (7.5) [20]. Problems with multiplicative colored noise can also be treated by the MCF method [21].

FPEs with timedependent terms have been investigated by the MCF method as well. Here I want to mention the harmonic mixing due to charged Brownian particles in a cosine potential [22] and the dithered ringlaser gyroscope [23] with additive noise. Some other applications are the dispersive optical bistability with an additive classical noise term [24] and the stochastic optical Bloch equations with multiplicative colored noise [25].

Master equations for the density operators are very important in quantum optics as has been shown by HAKEN in his famous encyclopedia article about laser theory [2]. It was further shown there that these master equations can be transformed to Fokker-Planck type equations for the Glauber-Sudarshan P-function or for the Q-function. For certain cases of operator equations these FPEs have a diffusion matrix which is no longer positive definite or positive semidefinite. Thus they can no longer be interpreted as describing the Brownian motion of particles in suitable potentials and no simple simulation is possible. (In some cases one can simulate these equations by doubling the variables [26].) These FPEs have been termed quantum-Fokker-Planck equations (QFPE). The Q-function is defined by

$$Q(\alpha) = \langle\alpha|\rho|\alpha\rangle/\pi , \qquad |\alpha\rangle = \text{coherent state} .$$

For the model of DRUMMOND and WALLS [27] describing dispersive optical bistability the QFPE for the Q-function reads

$$\begin{aligned}\frac{\partial Q}{\partial t} = &- \frac{\partial}{\partial\alpha}(-\kappa\alpha + i\Omega\alpha + 4i\chi\alpha - 2i\chi\alpha^2\alpha^* + iF)Q \\ &- \frac{\partial}{\partial\alpha^*}(-\kappa\alpha^* - i\Omega\alpha^* - 4i\chi\alpha^* + 2i\chi\alpha^{*2}\alpha - iF)Q \\ &+ i\chi\frac{\partial^2}{\partial\alpha^2}\alpha^2 Q + 2\kappa(n_{th}+1)\frac{\partial^2 Q}{\partial\alpha^*\partial\alpha} - i\chi\frac{\partial^2}{\partial\alpha^{*2}}\alpha^{*2}Q .\end{aligned} \tag{7.14}$$

In (7.14) Ω is difference between the frequency of the classical driving field F and the cavity frequency, χ is the imaginary part of the third order susceptibility, κ is the cavity damping constant and n_{th} is the number of thermal quanta. The QFPE (7.14) was also treated by the MCF method [28]. In Fig. 7 the contour lines of the stationary solution of (7.14) are shown. For the parameters of this figure the two-peak distribution is clearly visible. Three of the lowest nonzero real eigenvalues as a function of the driving field F are plotted in Fig. 8. In the bistable region the lowest nonzero eigenvalue becomes very small. This eigenvalue is connected with the tunneling rates between the classical two stable states. For $n_{th} = 0$ (as in Fig. 8) these tunneling rates are solely caused by quantum fluctuations.

Finally I would like to mention that the method is not only applicable to FPEs. One can, for instance, determine eigenvalues of a Schrödinger equation [29] or investigate the stability of an upside down double pendulum with harmonically moving support by this method [30]. ALLEGRINI et al. [31] used MCFs for solving the Bloch equations in connection with saturation effects in spin-1/2 radio frequency spectrocsopy, DURST et al. [32] applied them to two-level systems coupled to a vibrational mode.

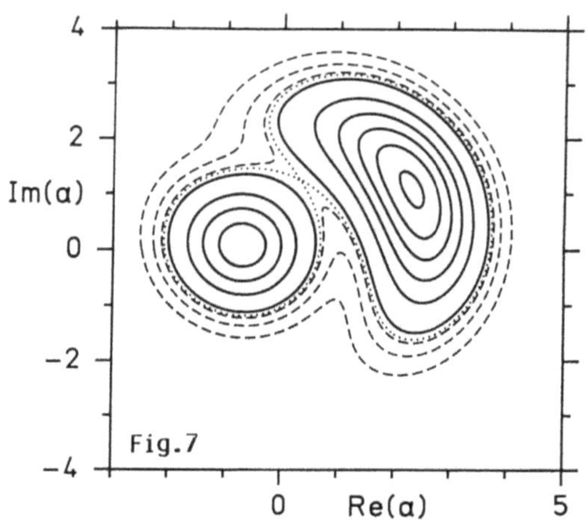

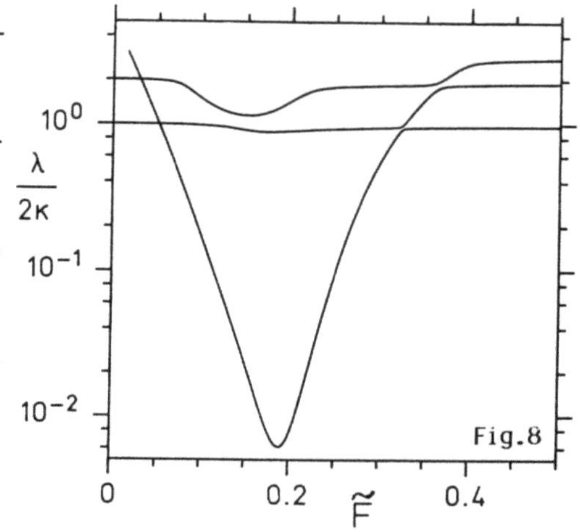

Fig. 7 Contour-lines of the stationary distribution Q=c of (7.14) for $n_{th}=0$, $\chi=0.1\Omega$, $\kappa=0.1\Omega$ and for the scaled driving field $\tilde{F}=\Omega^{-1}\sqrt{\chi/\Omega}$ F=0.2. Solid lines: c=0.02, 0.04, ...,0.12; broken lines: c=0.0005, 0.010, 0.015; dotted line: separatrix at c=0.01665 (taken from [28])

Fig. 8 Plot of the three lowest real nonzero eigenvalues as a function of the scaled driving field $\tilde{F}$ (see Fig. 7 for the scaling factor) for $n_{th}=0$, $\chi=0.1\Omega$ and $\kappa=0.1\Omega$ (taken from [28])

Thus quite a number of FPEs and other problems can be treated by the MCF method. By working with MCFs for eight years I am sure that they will become useful for solving still more problems in physics as well as in other areas of science.

References

1. H. Haken: Z. Physik **181**, 96 (1964)
2. H. Haken: Laser Theory, Encyclopedia of Physics, Vol. XXV/2c Springer, Berlin (1970)
3. H. Haken: Rev. Mod. Phys.**47**, 67 (1975)
4. H. Haken: Synergetics, An Introduction 3rd ed. Springer Series in Synergetics, Vol. 1, Springer, Berlin (1983)
5. H. Risken: The Fokker-Planck Equation, Springer Series in Synergetics, Vol. 18, Springer, Berlin (1984)
6. O. Perron: Die Lehre von den Kettenbrüchen, Vols. I and II, Teubner Stuttgart (1977)
7. W.B. Jones, W.J. Thron: Continued Fractions, Encyclopedia of Mathematics and its Applications, Vol. 11, Addison-Wesley, Reading, MA (1980)
8. P. Hänggi, F. Rösel, D. Trautmann: Z. Naturforsch. **33a**, 402 (1978)
9. H. Haken: Advanced Synergetics, Springer Series in Synergetics Vol. 20, Springer, Berlin (1983)
10. H. Risken, H.D. Vollmer: Z. Physik **B39**, 89 and 339 (1980)
11. I.S. Gradshteyn, I.M. Ryzhik: Tables of Integrals, Series and Products, Academic Press, New York (1965)
12. H. Denk, M. Riederle: J. Approx. Theory **35**, 355 (1982)
13. H.A. Kramers: Physica **7**, 284 (1940)
14. H. Risken, H.D. Vollmer: Z. Physik **B33**, 297 (1979)
15. H.D. Vollmer, H. Risken: Z. Physik **B34**, 313 (1979)
16. K. Voigtlaender, H. Risken: J. Stat. Phys. **40**, 397 (1985)

17. M. Büttiker, R. Landauer: Phys. Rev. **B30**, 1551 (1984)
18. H. Risken, K. Voigtlaender: J. Stat. Phys. **41**, 825 (1985)
19. M. Mörsch, H. Risken, H.D. Vollmer: Z. Physik **B49**, 47 (1982)
20. P. Jung, H. Risken: Z. Physik **B61**, 367 (1985)
21. P. Jung, H. Risken: Phys. Lett. **103A**, 38 (1984)
22. H.J. Breymeyer, H. Risken, H.D. Vollmer, W. Wonneberger: Appl. Phys. **B28**, 335 (1982)
23. W.Schleich, C.S. Cha, J.D. Cresser: Phys. Rev. A **29**, 230 (1984)
24. H. Haug, S.W. Koch, R. Neumann, H.E. Schmidt: Z. Physik **B49**, 79 (1982)
25. P. Zoller, G. Alber, R. Salvador: Phys. Rev. **A24**, 398 (1981)
26. M. Dörfle, A. Schenzle: Z. Physik **B65**, 113 (1986)
27. P.D. Drummond, D.F. Walls: J. Phys. A **13**, 725 (1980)
28. K. Vogel, H. Risken: in preparation
29. H. Denk: Ph. D. Thesis, Ulm (1984)
30. Th. Leiber: Diploma Thesis, Ulm (1985)
31. A. Allegrini, E. Arimondo, A. Bambini: Phys. Rev. **A15**, 718 (1977)
32. C. Durst, E. Sigmund, P. Reineker, A. Scheuing: J. Phys. C : Solid State Phys. **19**, 2701 (1986)

Part VI

Coherence and Structure in Physical Systems

Ultrafast Coherent Raman Spectroscopy

W. Zinth and W. Kaiser

Physik Department E11, Technische Universität München,
D-8000 München, Fed. Rep. of Germany

1. Introduction

During the past decade, time resolved coherent methods have attracted much interest for the study of fast dynamic processes. In particular, time resolved coherent Raman scattering allowed to measure - in the time domain - rapid dephasing processes of molecular vibrations in liquids and elucidated various line-broadening mechanisms /1-4/. It is the aim of this paper to focus attention on two more recent subjects. While time resolved Raman scattering is a method working in the time domain, it allows to obtain valuable information in the frequency domain /5,6/. The results of coherent experiments in the time domain have to be transformed by appropriate methods, e.g. by Fourier transformation, to show properties of the investigated system in the frequency domain. We present here two different approaches to relate time and frequency domain: (i) A time resolved coherent Raman experiment is performed and the spectrum of the coherently scattered light is recorded. (ii) A numerical transformation is applied to coherent data of high time resolution giving precise values of frequency differences of vibrational modes separated by several terahertz. Both techniques give - when applied in an appropriate way - results superior to steady-state spontaneous spectroscopy.

After a short theoretical description of the basic ideas of the two techniques we give experimental results. Measurements are presented where congested spectral regions consisting of several overlapping bands are resolved and the vibrational transition frequencies of the different components are determined. We also show that terahertz quantum beats can be measured with advanced experimental systems; they give precise values for the frequency differences between vibrational transitions separated by more than 200 cm^{-1}.

2. Theory

Time resolved coherent Raman scattering is commonly treated under the following assumptions /1/: Light fields are described by Maxwell's equation and the vibrational transitions are represented by two-level systems. Changes in the population of the two molecular levels are neglected (weak Raman interaction). Important is the coherent amplitude $\langle q \rangle$, i.e. the expectation value of the vibrational amplitude. First we discuss a single homogeneously broadened transition with dephasing time T_2. We use the ansatz of plane waves for the coherent amplitude $\langle q \rangle = (i/2)Q \exp(-i\omega_q t+ik_q x)$ + c.c. and assume an isotropic Raman tensor for the investigated vibration. The coherent amplitude of a Raman active mode at frequency ω_q is excited via transient stimulated Raman scattering by a pair of light pulses, the laser pulse $\tilde{E}_L$ and the Stokes pulse $\tilde{E}_S$ /1,7/. The electric fields are considered to be plane waves, e.g. $\tilde{E}_L = 1/2\ E_L \exp(-i\omega_L t+ik_L x)$ + c.c. with the wave vector k_L

and frequency ω_L for the laser field. The driving force $F(x,t)$ for the coherent amplitude Q is proportional to the product of the laser and the Stokes field, $F(x,t) \propto E_L E_S^*$; as a result the carrier frequency of the excitation force is $\omega_L - \omega_S$. The linear response theory applied to the excitation process leads to the following equation for the coherent amplitude Q:

$$Q(t) = \kappa \int_{-\infty}^{t} dt' \; E_L(t')E_S^*(t') \exp[(t'-t)/T_2 - i\Delta\omega t'] \,. \qquad (1)$$

The constant κ is defined in Ref./1/ and contains material parameters such as Raman cross section and vibrational frequency ω_q. T_2 stands for the dephasing time of the transition. Detuning between excitation frequency and vibrational mode is given by $\Delta\omega = \omega_L - \omega_S - \omega_q$. The coherent amplitude $\langle q \rangle$ under the action of a short exciting force evolves as follows: $\langle q \rangle$ rises to a maximum with the exciting force and, subsequently, decays exponentially with the dephasing time T_2. During the free exponential decay the molecules oscillate at their resonance frequency ω_q. In the experiments described here, the large bandwidth of the short driving force allows to excite several vibrational modes simultaneously. The additional modes may be treated with the same ansatz, introducing the individual resonance frequencies ω_{qj}, amplitudes Q_j, and phase factors ϕ_j.

In a time resolved coherent Raman experiment the amplitude $\langle q \rangle$, generated at time zero, is monitored by coherent Raman scattering of a delayed probing pulse E_{L2}. We investigate the anti-Stokes part of the spectrum, which is generated by the nonlinear polarisation P^{NL}_{AS}:

$$P^{NL}_{AS} = \gamma \langle q \rangle E_{L2} + \chi^{(3)}_{NR} \, E_L E_S^* E_{L2} \,. \qquad (2)$$

The coherent signal consists of two parts. The first part on the r.h.s. of Eq.(2) describes the coherent Raman signal scattered from the excited vibrational mode. The constant γ contains the Raman cross section of the mode. This part of the coherent signal is emitted at the resonant anti-Stokes frequency $\omega_{AS} = \omega_{L2} + \omega_q$ and contains information on the dynamic properties of the investigated mode. The second part on the r.h.s. of Eq.(2) is caused by the nonresonant nonlinear susceptibility $\chi^{(3)}_{NR}$, which is related to off-resonant modes and electronic contributions. Especially distant electronic transitions determine this signal /8/. The nonresonant part of the signal gives an instantaneous response, i.e. it follows the product $E_L E_S^* E_{L2}$ and decays very rapidly with the time resolution of the experiment. The resonant part, on the other hand, shows the slower response of the resonantly excited mode. The nonresonant part of the coherent signal does not provide any information on the investigated modes. It has to be separated from the resonant signal. This fact is not possible in the steady-state coherent anti-Stokes Raman spectroscopy (CARS). In the time-resolved coherent anti-Stokes experiments described here the separation is possible. Measuring at later delay times allows to obtain information on the resonant part of the signal alone, while the nonresonant part has vanished at that time.

The two time-resolved spectroscopic techniques discussed next apply the same transient excitation process, but differ in the way the probing process is performed.

2.1 Short Excitation and Prolonged Interrogation spectroscopy

We introduce the so-called short excitation prolonged interrogation (SEPI) spectroscopy. Performing this technique one uses a short and broad-band excitation process followed - at later delay times - by a coherent scattering process using a long probing pulse /5,9-11/. The frequency spectrum of the coherently scattered signal is detected. In order to gain optimum spectral resolution, a long and possibly Gaussian shaped probing pulse E_{L2} interrogates the coherent material excitation at a late delay time t_D after the peak of the exciting force. At that time the coherent amplitude decays exponentially and the nonresonant contribution to the coherent signal $\chi^{(3)}_{NR}$ has disappeared. The anti-Stokes spectrum becomes

$$S_{AS}(\Omega) \propto |E_{AS}(\Omega)|^2$$

$$\propto |\int_{-\infty}^{+\infty} dt\; e^{i\Omega t}\vec{E}_{AS}(t)|^2 = |\int_{-\infty}^{+\infty} dt\; e^{i(\Omega-\omega_{L2}-\omega_q)t} E_{L2}(t)\; Q(t)|^2. \qquad (3)$$

The anti-Stokes signal is centered at the anti-Stokes frequency $\omega_A = \omega_{L2}+\omega_q$, provided that the probing light field does not exhibit phase modulation. Eq.(3) represents the Fourier transform of the product (probing field E_{L2} times the coherent amplitude Q), i.e. the Fourier transform of the product of a Gaussian function times an exponential function. This product is again a Gaussian function of the same width, which is shifted in time. In other words, the width of the anti-Stokes spectrum is determined by the Gaussian shape of the probing field E_{L2}, which may become very narrow provided one works with long (i.e. spectrally narrow) probing pulses. When the probing pulses are sufficiently long, $t_P > 1.4T_2$, the observed SEPI line is narrower than the spontaneous Raman band with $\Delta\nu_{spont} = 1/\pi T_2$. As a consequence, the anti-Stokes frequency may be determined with high accuracy. Of special interest is the application of the SEPI technique to congested spectral regions made up of overlapping vibrational bands. In this case, the narrowing of the different transitions below the spontaneous width allows to resolve individual transitions buried under a broad inhomogeneous spectrum.

2.2 Fourier Transform Raman Spectroscopy

In a second approach we record the coherent signal for a long time period in a pure time-domain experiment. In order to have optimum temporal resolution pulses of very short duration for the exciting and probing laser fields are required. Under these conditions the experiment detects the time-integrated coherent signal as a function of the delay time between exciting and probing process:

$$S(t_D) \propto \int_{-\infty}^{+\infty} dt|E_{AS}(t,t_D)|^2$$

$$\propto \int_{-\infty}^{+\infty} dt|P^{NL}_{AS}(t,t_D)|^2 = \int_{-\infty}^{+\infty} dt|\gamma\langle q\rangle E_{L2} + \chi^{(3)}_{NR}\; E_L E_S{}^* E_{L2}|^2. \qquad (4)$$

For short exciting and probing pulses this experiment measures the component due to $\chi^{(3)}_{NR}$ at time zero and, most important, the time dependence of the coherent amplitude. When several modes are excited simultaneously, the signal decay is complex. It contains contributions of the different vibrational modes. It can be shown that for short driving and probing pulses the coherent signal is equivalent to the absolute square of the Fourier transform of the spontaneous Raman spectrum /5,6/. As a consequence, frequency domain information may be obtained from time resolved data by Fourier transformation.

The two techniques introduced here analyze the molecular system in a similar way. In the Fourier-transform spectroscopy the manipulation of the measured data prior to the transformation may lead to improved spectral resolution. In the SEPI experiment the handling of the information starts with the prolonged probing process using Gaussian shaped pulses. In this way, the exponential decay of the signal due to the dephasing process - corresponding to the homogeneous Lorentzian line shape in spontaneous spectroscopy - is removed prior to the spectral analysis. This procedure is equivalent to a numerical filtering of the time resolved data in the Fourier transform Raman spectroscopy /6/.

3. The Experimental System

An experimental set-up used successfully in various coherent Raman investigations is depicted schematically in Fig.1. A pumping laser source generates synchronized exciting and probing light fields at the frequencies ω_L, ω_{L2}, and ω_S in several dye lasers. The durations of the various laser pulses are determined by the special type of the laser. For the SEPI experiment we use a picosecond (synchronously pumped) dye laser emitting pulses E_L and E_{L2} with $\omega_L = \omega_{L2}$ of duration $t_p \approx 12$ ps and a second picosecond dye laser (pumped by the same mode-locked Ar-ion laser) to produce the Stokes pulse E_S ($t_p \approx 5$ ps) /12/. For the

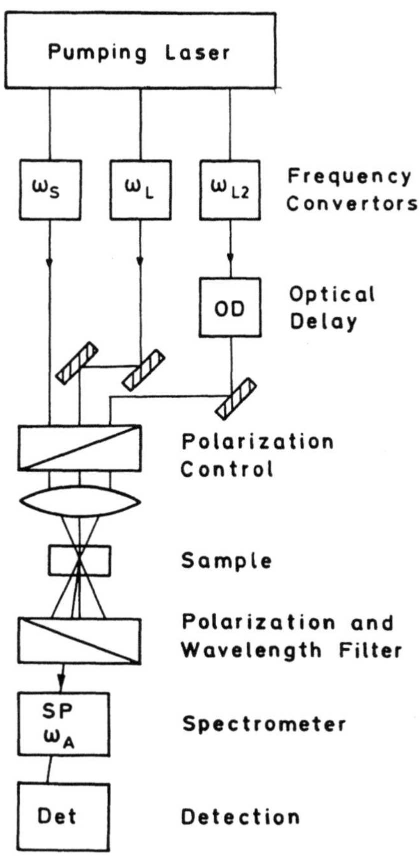

Fig.1 Schematic of the experimental system used to study time resolved coherent Raman spectroscopy. A mode-locked Ar-ion laser synchronously pumps dye lasers as frequency convertors. The coherently scattered signal is recorded by a highly sensitive detection system

Fourier-transform coherent Raman experiment one works with the same Stokes pulses, but the exciting and probing laser pulses are very short, namely 60 fs. The latter are produced in a femtosecond unidirectional ring dye laser /13/. The exciting Stokes and laser pulses are focused into the sample generating the coherent material excitation. The probing pulses are delayed relative to the exciting pulses in the optical delay OD and then traverse (at the phase-matching angle) the excited volume of the sample. The weak coherently scattered light is separated from the incoming pulses via diaphragms, polarizers, and appropriate spectrometers. In the case of SEPI spectroscopy a high-resolution spectrograph ($\Delta\lambda/\lambda \approx 10^{-5}$) serves to observe the anti-Stokes spectrum. The detection system is a highly sensitive cooled photomultiplier.

4. Short Excitation and Prolonged Interrogation Spectroscopy of Hydrogen-bonded Aggregates

With the SEPI technique mixtures of liquid pyridine and methanol were studied. In these mixtures hydrogen bonds between pyridine and methanol occur, strongly influencing the spontaneous Raman spectra. In Fig.2 the spontaneous Raman data of pyridine:methanol mixtures are depicted for the spectral range of 980 cm^{-1} to 1030 cm^{-1}. In neat pyridine (lowest trace in Fig.2) the ν_1 mode occurs at 990.5 cm^{-1}. When methanol is added to pyridine, an additional broad band appears around 1000 cm^{-1}, which shifts with increasing methanol concentration /14/. Shoulders in the spontaneous Raman spectrum indicate that this band is made up of several components. The spontaneous Raman data do not allow to determine the number, the frequency positions, and the amplitudes of the different components. In Fig.3b and 3c we present SEPI data taken of two mixtures. For each mixture a dozen spectra were recorded. The laser frequency ω_L was kept constant and different anti-Stokes spectra were measured, when tuning the Stokes frequency in steps of 1 cm^{-1}. Of special interest are the narrow SEPI spectra around 1000 cm^{-1}. One finds two distinct bands at 997.3 cm^{-1} and at 1000 cm^{-1}, and a weaker component at 1001 cm^{-1}. The three frequency positions are marked in Fig.3 by vertical dash-dotted lines. The relative amplitudes of the two major components change considerably with concentration. While the component at 1000 cm^{-1} is strongest at x_{py} = 0.33 molar fraction (Fig.3b), one finds the component at 997.3 cm^{-1} to dominate at x_{py} = 0.66 molar fraction (Fig.3c). A large number of SEPI spectra was taken for

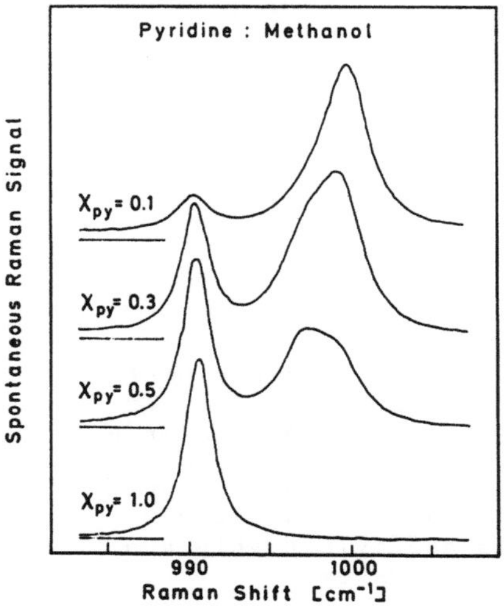

Fig. 2 Spontaneous Raman spectra of pure pyridine (lowest trace) and of mixtures of pyridine and methanol. In the mixtures a new band appears around 1000 cm^{-1}, which is due to complexes of pyridine and methanol. The band is resolved by SEPI spectroscopy into three lines (see Fig.3)

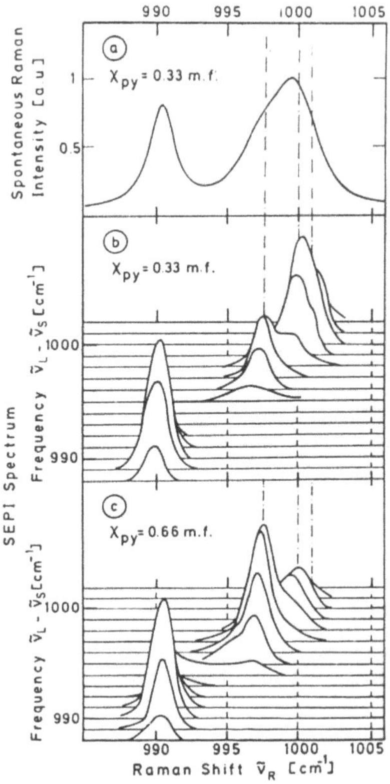

Fig. 3 Raman spectra of pyridine:methanol mixtures. (a) Polarized spontaneous Raman spectrum measured at a pyridine concentration of 0.33 molar fraction taken with a standard Raman spectrometer of spectral resolution of 0.5 cm^{-1}. (b) and (c) Short excitation and prolonged interrogation (SEPI) spectra obtained for a set of excitation frequencies $\tilde{\nu}_L-\tilde{\nu}_S$ at two pyridine concentrations, x_{py} = 0.33 molar fraction (b) and x_{py} = 0.66 molar fraction (c). The spectra, taken at excitation frequencies $\tilde{\nu}_L-\tilde{\nu}_S$ < 994 cm^{-1}, are drawn with reduced amplitudes [x0.25 in (b) and x0.02 in (c)]. The SEPI spectra resolve three Raman lines of different hydrogen-bonded aggregates hidden under the broad band of the spontaneous Raman spectrum (see dash dotted lines)

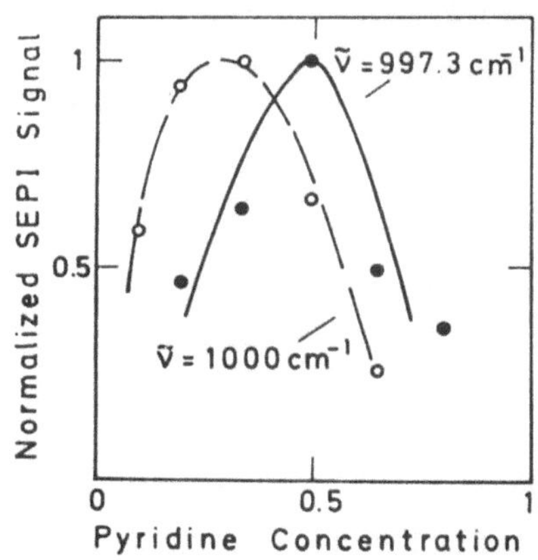

Fig. 4 Concentration dependences of the amplitudes of the two SEPI lines at 997.3 cm^{-1} and 1000 cm^{-1}

pyridine concentrations between x_{py} = 0.1 and 1.0 molar fraction. The concentration dependence of the amplitudes is shown in Fig.4 for the two SEPI lines at 997.3 cm^{-1} (full points) and 1000 cm^{-1} (circles). The band at 997.3 cm^{-1} rises to a peak value at x_{py} = 0.5 molar fraction, while the band at 1000 cm^{-1} shows a maximum at higher methanol concentrations with x_{py} = 0.25 molar fraction.

Our data suggest the following picture of the hydrogen bonded complexes: (i) There exist at least three distinct complexes of pyridine and methanol. (ii) The species at 997.3 cm^{-1} consists of one pyridine and one methanol molecule (PM). This complex has its highest concentration when pyridine and methanol are at equal numbers in the solution. (iii) The component at 1000 cm-1 appears to be made up of one pyridine and two or - more probably - four methanol molecules (PM_2 or PM_4). Indications for the latter complex are found in the literature; self-association of methanol is reported to occur at high methanol concentrations with a predominance of methanol tetramers /15/. (iv) The component at 1001 cm^{-1} is tentatively assigned to complexes PM_n with n > 4.

5. Fourier Transform Raman Spectroscopy of Transitions with Terahertz Frequency differences

High-resolution Fourier transform coherent Raman spectroscopy was first reported by Graener and Laubereau, who worked with light pulses of twenty picosecond duration. The authors studied vibrational-rotational transitions in CH_4 separated by less than 1 cm^{-1} with high resolution of 10^{-3} cm^{-1}. The data were taken in a time interval of 12 ns.

In this paper precise measurements are discussed of frequency differences between vibrational modes in liquids which are separated by tens to hundreds of wavenumbers. These large frequency differences may be studied in experiments which reveal the related beating phenomena. For example, a frequency difference of 200 cm^{-1} produces a beating pattern with peaks separated by 167 fs. In order to resolve this very short phenomenon, one has to work with femtosecond light pulses. Light pulses of approximately 50 fs duration became available only recently. We present two examples of time-resolved coherent Raman spectroscopy with femtosecond time resolution /16/. First we show data of pure liquid pyridine.

Liquid pyridine has Raman active vibrations at frequencies 991 cm^{-1} and 1030 cm^{-1}. They are assigned to two A_1 ring modes /17/. Both vibrations have similar spectral width (2.2 cm^{-1}) and similar Ramn cross sections; they are separated by 39 cm^{-1}. In the Raman excitation process a frequency difference of $(\omega_L-\omega_S)/2\pi c$ = 1010 cm^{-1} between the laser and the Stokes frequencies is applied. Due to the broad spectral width of the femtosecond exciting force of 200 cm^{-1} both pyridine modes are simultaneously excited in the experiment. Fig.5a shows the observed anti-Stokes signal plotted as a function of the time delay between

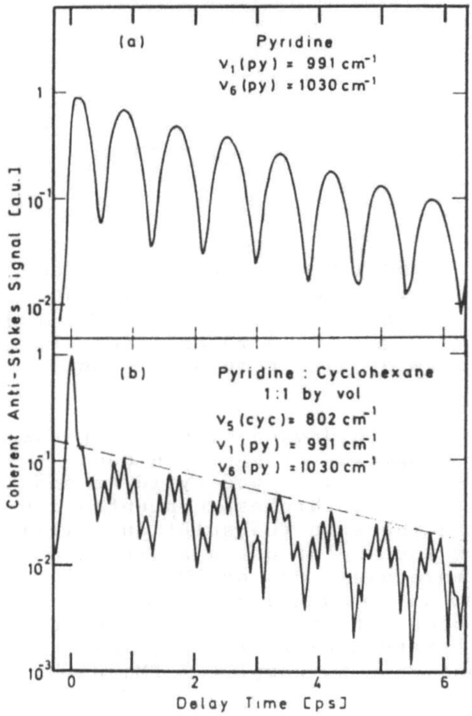

<u>Fig.5</u> Time-resolved coherent Raman scattering using femtosecond time resolution. (a) Beat pattern resulting from pure pyridine after excitation of the two pyridine modes at 991 cm^{-1} and 1030 cm^{-1}. (b) Beat pattern recorded from a mixture of pyridine and cyclohexane. Beating of three molecular modes is observed

exciting and probing pulses. During the excitation process, at time zero, the coherent signal rises to a pronounced peak. It subsequently decays quickly over more than an order of magnitude. Later, the signal recovers and shows strong oscillations. The modulation depth exceeds a factor of ten. Two features of the coherent signal are of special interest here.

(i) The period of oscillation is 0.85 ps. Consequently, the frequency difference between the two modes is 1.175±0.015 THz or $\Delta\tilde{\nu} = \Delta\nu/c \approx 39.2\pm0.5$ cm^{-1}. We emphasize that $\Delta\tilde{\nu}$ may be determined with high precision, mainly depending on the signal-to-noise ratio of the experiment and the accuracy of the optical delay line.

(ii) The envelope of the peaks of the oscillation decays exponentially with the decay time $T_2/2 \approx 2.55\pm0.15$ ps. The depths of the oscillations slightly decrease for long observation times, indicating a slight difference of 20% between the dephasing times of the two modes.

In Fig.5b we present an amazingly complicated but perfectly reproducible anti-Stokes pattern. A mixture of cyclohexane and pyridine (1:1 by volume) was investigated with femtosecond light pulses. (Note, that there are no hydrogen bonded aggregates in this mixture). In this case three vibrational modes with similar T_2 values beat together; they are: one cyclohexane mode at 802 cm^{-1} /18/ and the two pyridine modes at 991 cm^{-1} and 1030 cm^{-1}. We find a rich beating structure originating from the interference of the three modes.

The frequency differences of the investigated modes were determined by the following procedure. The exponential decay was removed by multiplying the signal with an exponential rising function. An appropriate window function was introduced to remove the influence of the boundaries of the time range. After these arithmetical manipulations a Fourier transformation of the time dependent data gave the results shown in Fig.6. Three sharp and pronounced peaks are found at 39 cm^{-1}, 189 cm^{-1}, and 228 cm^{-1}, which correspond to the frequency differences between the three modes excited in the mixture.

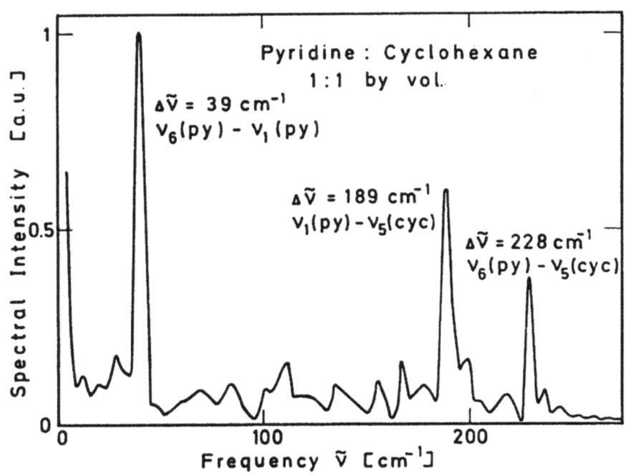

Fig. 6 Difference spectrum of pyridine and cyclohexane obtained by numerical Fourier transformation of the time resolved data from Fig.5b. Note the strong peaks at 39 cm^{-1}, 189 cm^{-1}, and 228 cm^{-1}, which correspond to the differences between the three excited modes. The highest difference frequency observed is $\Delta\nu = 6.84$ THz.

An important aspect remains to be discussed here: In order to improve spectral resolution in the time-domain experiment, the coherent signal has to be measured over delay times as long as possible. Taking into account the exponential decay of the signal, one immediately finds that the spectral resolution is directly related to the signal-to-noise ratio of the experiment. The following numbers should illustrate the situation. For dephasing times of approximately T_2 = 3 ps and with experimental data of the coherent Raman signal available over three orders of magnitude, one estimates an accuracy of the $\Delta\tilde{\nu}$ values of approximately 0.3 cm^{-1}.

6. Conclusions

We have shown that time resolved coherent Raman measurements provide spectral information hardly obtained from steady-state spontaneous spectra or from coherent steady-state spectroscopy. We have presented two different techniques which are of value in their own rights. The short excitation and prolonged interrogation (SEPI) technique is well adapted to resolve congested spectral regions; lines separated by a few wavenumbers are resolved. On the other hand, Fourier-transform coherent Raman spectroscopy using femtosecond light pulses is well suited to measure accurately frequency differences of modes separated by more than 200 cm^{-1}.

Acknowledgement
The authors acknowledge valuable contributions by M.C.Nuss, R.Leonhardt, W. Holzapfel, and R. Hackl at various stages of this work.

References

1. A. Laubereau, W. Kaiser: Rev. Mod. Phys. 50, 607 (1978); A. Penzkofer, A. Laubereau, W. Kaiser: Progr. Quant. Electron. 6 55 (1979)
2. D. von der Linde, A. Laubereau, W. Kaiser: Phys. Rev. Lett. 26 954 (1971)
3. S. Velsko, J. Trout, R.M. Hochstrasser: J. Chem. Phys. 79, 2114 (1983)
4. G.M. Gale, P. Guyot-Sionnest, W.Q. Zheng: Optics Commun. 58, 395 (1986)
5. W. Zinth, W. Kaiser: In Organic Molecular Aggregates, eds. R.Reinecker, H.Haken, H.C.Wolf, Solid State Science, Vol.49, (Springer, New York 1983) p.124
6. H. Graener, A. Laubereau: Optics Commun. 54, 141 (1985)
7. R.L. Carman, I. Shimizu, C.S. Wang, N. Bloembergen: Phys. Rev. A2, 60 (1970)
8. W. Zinth, A. Laubereau, W. Kaiser: Optics Commun. 57, 457 (1978)
9. W. Zinth: Optics Commun. 34, 479 (1980)
10. W. Zinth, M.C. Nuss, W. Kaiser: Chem. Phys. Lett. 88, 257 (1982)
11. M.A. Collins, P.A. Madden, A.D. Buckingham: Chem. Phys. 24, 291 (1985)
12. W. Zinth, M.C. Nuss, W. Kaiser: Phys. Rev. A30, 1139 (1984)
13. J. Dobler, H.H. Schulz, W. Zinth: Optics Commun. 57, 407 (1986)
14. H. Takahashi, K. Mamola, E.K. Plyler, J. Mol. Spectrosc. 21, 217 (1966) B.P. Asthana, H. Takahashi, W. Kiefer: Chem. Phys. Lett. 94, 41 (1983)
15. W.B. Dixon: J. Phys. Chem. 74, 1396 (1970); A.N. Fletcher, ibid. 75, 1808 (1971)
16. R. Leonhardt, W. Holzapfel, W. Zinth, W. Kaiser: Chem. Phys. Lett., to be published 1987
17. D.A. Long, F.S. Murfin, E.L. Thomas, Trans. Farad. Soc. 59, 12 (1963)
18. K.B. Wiberg, A. Shrake: Spectrochim. Acta 27A, 1139 (1971)

Coherence in the Transport of a Quantum Particle

P. Reineker[1] *and K. Kassner*[2]

[1]Abteilung Theoretische Physik, Universität Ulm,
D-7900 Ulm, Fed. Rep. of Germany
[2]Department of Chemistry, Massachusetts Institute of Technology,
Cambridge, MA 02139, USA

1. Introduction

If a molecular crystal, e.g. naphthalene or anthracene, is irradiated by light, energy is absorbed by electrons and they are excited from the highest occupied orbital of a molecule to its lowest unoccupied state. In this way energy is stored and a localized exciton, a quantum particle, is created. However, such a localized excitation is not a stationary state of the crystal. On account of the interaction between the molecules, it will be transferred to other crystal sites. One of the interesting questions in this connection is the following: is this particle motion incoherent or is it coherent, or in other words, does the excitation move via a hopping process [1], which may be described by a master equation [2] or does it propagate as a wave packet [3] via tunneling, whose time development is determined by the Schrödinger equation?

It is quite natural that H. Haken became involved into this problem. On the one hand, he had made fundamental contributions to the theory of excitons [4] which still today form the basis for a series of investigations on excitons in semiconductors. On the other hand, coherence is one of the characteristic features of lasers whose theory has been basically influenced by him [5]. The essential idea of Haken's approach to the exciton transport problem is as follows. At very low temperatures and in an ideal crystal a localized excited state is described by a superposition of excitonic eigenstates with definite phase relations, forming thus a wave packet which moves through the crystal with its group velocity. This coherent, wave-like motion is perturbed by the vibrations in the crystal, which scatter the exciton waves and thus destroy the phase relations in the course of time. For high enough temperatures phase relations decay so rapidly that the exciton motion can be described by a hopping process. In the mathematical formulation of this idea the phonons are simulated by a δ correlated Gaussian Markov process. This description of the exciton transport problem is now known as the HAKEN-STROBL [6-8] model for the coupled coherent and incoherent exciton motion.

Experimental methods and results, also taking into account to some extent the development of the field in the past, have been reviewed recently by WOLF [9] for molecular crystals. Some aspects of the energy transfer of photosynthetic systems have been discussed by PAILLOTIN [10]. Finally, it should be mentioned that the same questions, as discussed for the motion of excitons, have arisen recently in connection with the transport of hydrogen [11] and muons [12] in metals.

On the basis of the ideas of Haken the transport of a quantum particle (exciton) is described by the Hamiltonian

$$\begin{aligned} H = & \sum_n \varepsilon_n a_n^+ a_n + \sum_{n \neq n'} \sum J_{n-n'} a_n^+ a_{n'} \\ & + \sum_n V_n(\{Q_i\}) a_n^+ a_n + \sum_{n \neq n'} \sum V_{nn'}(\{Q_i\}) a_n^+ a_{n'} \\ & + \sum_i 1/2(P_i^2 + Q_i^2) \ . \end{aligned} \tag{1.1}$$

The first line describes the coherent transport of the particle, ε_n is the energy at site n, and $J_{n-n'}$ the transfer integral between sites n and n'. The operators a_n^+ and a_n create and annihilate, respectively, a quantum particle at site n. The third line is the Hamiltonian of the vibrational degrees of freedom and the second line describes the interaction between both. Replacing the phonon operators by classical fluctuating quantities we obtain the following stochastic Hamiltonian [13-15]

$$H = \sum_n \varepsilon_n a_n^+ a_n + \sum_{n \neq n'} \sum J_{n-n'} a_n^+ a_{n'}$$

$$+ \sum_n h_{nn}(t) a_n^+ a_n + \sum_{n \neq n'} \sum h_{nn'}(t) a_n^+ a_{n'} \ . \tag{1.2}$$

In the HAKEN-STROBL model [6-8] it is assumed that the $h_{n_1 n_2}(t)$ represent a δ correlated Gaussian Markov process with mean value and correlation function given by

$$< h_{n_1 n_2}(t) > = 0 \tag{1.3}$$

$$< h_{n_1 n_1'}(t_1)\, h_{n_2 n_2'}(t_2) = 2\Lambda(n_1, n_1', n_2, n_2')\ \delta(t_1 - t_2) \tag{1.4}$$

$$\Lambda(n_1, n_1', n_2, n_2') = \gamma_{|n_1 - n_n'|}\ \delta_{n_1 n_2'}\ \delta_{n_1' n_2} + \bar{\gamma}_{n_1 - n_1'}\ \delta_{n_1 n_2}\ \delta_{n_1' n_2'} (1 - \delta_{n_1 n_1'}) \ . \tag{1.5}$$

The equations of motion for the stochastically averaged density matrix elements read (for a derivation see [8,16])

$$\dot{\rho}_{nn'} = - i[H_0, \rho]_{nn'} - \delta_{nn'} \sum_{n_1} 2\gamma_{|n-n_1|} (\rho_{nn} - \rho_{n_1 n_1})$$

$$- (1 - \delta_{nn'})(2\Gamma\ \rho_{nn'} - 2\bar{\gamma}_{n-n'}\ \rho_{n'n}) \tag{1.6}$$

where

$$\Gamma = \sum_{n_1} \gamma_{|n-n_1|} \ . \tag{1.7}$$

In paragraph 2 of this contribution (1.6) is solved exactly for the motion of a quantum particle between two sites [16,17]. An exact solution for the mean square displacement of the particle on a linear chain [16,18,19] is given in paragraph 3. Paragraph 4 gives a recent generalization of the model [20] considering the influence of non-Markovian effects on the diffusion constant and presents a dynamical derivation of Anderson localization. Finally the last paragraph gives some hints on present and possible future developments in the field.

2. Two-site Model for the Motion of a Quantum Particle

For the two-site problem (n=1,2) the density matrix equation (1.6) reads explicitly

$$\dot{\rho}_{11} = - 2\gamma_1(\rho_{11} - \rho_{22}) - iJ(\rho_{21} - \rho_{12}) \tag{2.1a}$$

$$\dot{\rho}_{22} = - 2\gamma_1(\rho_{22} - \rho_{11}) - iJ(\rho_{12} - \rho_{21}) \tag{2.1b}$$

$$\dot{\rho}_{12} = - 2(\gamma_0 + \gamma_1)\rho_{12} + 2\bar{\gamma}_1 \rho_{21} - iJ(\rho_{22} - \rho_{11}) \tag{2.1c}$$

$$\dot{\rho}_{21} = -2(\gamma_0+\gamma_1)\rho_{21} + 2\bar{\gamma}_1\rho_{12} + iJ(\rho_{22}-\rho_{11}) \tag{2.1d}$$

where we have normalized the particle energy at site n to zero and assumed $J_1 = J_{-1} = J$, $\gamma_1 = \gamma_{-1}$. In this site representation the first two equations describe the change of the occupation numbers of the particle at the two sites, whereas from the two remaining equations we get the change of phase relations between them.

Let us first consider the case of a vanishing tunneling matrix element J. Then the equations for diagonal and non-diagonal elements of the density matrix are completely decoupled. The equations for the diagonal elements represent a set of rate equations for the particle occupation numbers. The transition probability for a particle between the two molecules is given by $2\gamma_1$, which means that it is determined by non-local fluctuations. From the equations for the non-diagonal elements we see that phase relations decay by local and non-local fluctuations.

In the case of a non-zero tunneling matrix element J the equations for diagonal and non-diagonal matrix elements are coupled and we have to solve a four-dimensional non-hermitian eigenvalue problem. Using the ansatz

$$\rho_{ij}(t) = e^{-\lambda t}\,\rho_{ij} \tag{2.2}$$

the eigenvalues are given by

$$\lambda_1 = 0 \tag{2.3a}$$

$$\lambda_2 = 2(\gamma_0+\gamma_1-\bar{\gamma}_1) \tag{2.3b}$$

$$\left.\begin{matrix}\lambda_3\\ \lambda_4\end{matrix}\right\} = \gamma_0+3\gamma_1+\bar{\gamma}_1 \pm [(\gamma_0-\gamma_1+\bar{\gamma}_1)^2 - 4J^2]^{1/2} \tag{2.3c, 2.3d}$$

and the (non-normalized) eigenvectors are represented in Table 2.1. (The square roots are defined in (2.3).)

Table 2.1: Eigenvectors in the localized basis system

	$\rho^{(1)}$	$\rho^{(2)}$	$\rho^{(3)}$	$\rho^{(4)}$
ρ_{11}	1	0	$i2J$	$i2J$
ρ_{22}	1	0	$-i2J$	$-i2J$
ρ_{12}	0	1	$-(\gamma_0-\gamma_1+\bar{\gamma}_1+\sqrt{\ })$	$-(\gamma_0-\gamma_1+\bar{\gamma}_1-\sqrt{\ })$
ρ_{21}	0	1	$(\gamma_0-\gamma_1+\bar{\gamma}_1+\sqrt{\ })$	$(\gamma_0-\gamma_1+\bar{\gamma}_1-\sqrt{\ })$

The general solution for this dimer problem is given by a superposition of the four eigensolution:

$$\rho(t) = \sum_{i=1}^{4} a_i\rho^{(i)}\,e^{-\lambda_i t}\ . \tag{2.4}$$

For a discussion of the coupled coherent and incoherent motion it is useful to consider the time development of a particle which initially sits at the origin, i.e. we assume that $\rho_{11}(0) = 1$ and all the other $\rho_{ij} = 0$. The time evolution of the density operator is then given by

$$\rho(t) = \frac{1}{2}\rho^{(1)} + i\,\frac{\gamma_0-\gamma_1+\bar{\gamma}_1-\sqrt{}}{8J\sqrt{}}\,\rho^{(3)}\,e^{-\lambda_3 t} - i\,\frac{\gamma_0-\gamma_1+\bar{\gamma}_1+\sqrt{}}{8J\sqrt{}}\,\rho^{(4)}\,e^{-\lambda_4 t}\,. \quad (2.5)$$

Especially the probability of finding the particle at time t at site 1, if it was there initially, becomes (we have assumed $\gamma_1 = \bar{\gamma}_1$):

$$\rho_{11}(t) = \frac{1}{2} + \frac{1}{4}\,\frac{\sqrt{\gamma_0^2-4J^2}-\gamma_0}{\sqrt{\gamma_0^2-4J^2}}\,e^{-[\gamma_0+4\gamma_1+\sqrt{\gamma_0^2-4J^2}]t}$$

$$+ \frac{1}{4}\,\frac{\sqrt{\gamma_0^2-4J^2}+\gamma_0}{\sqrt{\gamma_0^2-4J^2}}\,e^{-[\gamma_0+4\gamma_1-\sqrt{\gamma_0^2-4J^2}]t} \quad (2.6)$$

This solution shows that the probability of finding the particle at the origin decays to 1/2, i.e. for large times both sites are equally populated. However, the way in which this final distribution is obtained, strongly depends on the strength of the fluctuations. If $\gamma_0 < 2J$, the square roots in the exponent of (2.6) become imaginary and the time development of the occupation probability shows damped oscillations. On the other hand, for $\gamma_0 > 2J$ the square roots are real and the occupation numbers between the molecules reach their steady-state values without oscillations. Therefore it is reasonable to denote the exciton motion in the first case as coherent and in the second as incoherent. This criterion for the classification of the exciton motion is due to HAKEN and STROBL [6]. Equation (2.6) describes the escape time of a quantum particle from a localized state under the combined action of tunneling and fluctuations. This kind of question has been addressed recently in various papers [21-24] starting from a microscopic description of the vibrational degrees of freedom.

3. Mean Square Displacement of a Quantum Particle on a Linear Chain

Using the density matrix equation (1.6), the time dependence of the moments of the particle distribution function is obtained from the following exact set of equations:

$$\frac{d}{dt}\sum_n n^s\rho_{nn} = 2\sum_{t=1}^{s}\binom{s}{t}\sum_m m^t\gamma_m\sum_n n^{s-t}\rho_{nn}$$

$$+\sum_{t=1}^{s}\binom{s}{t}\sum_{m>0} m^t H_m\sum_n n^{s-t}\,i(\rho_{n+m,n}-\rho_{n,n+m}) \quad (3.1)$$

$$\frac{d}{dt}\sum_n n^s i(\rho_{n+m,n}-\rho_{n,n+m}) = -2(\Gamma+\bar{\gamma}_m)\sum_n n^s i(\rho_{n+m,n}-\rho_{n,n+m})$$

$$-2\sum_{t=1}^{s}\binom{s}{t}(-m)^t H_m\sum_n n^{s-t}\rho_{nn}$$

$$-\sum_{t=1}^{s}\binom{s}{t}\sum_{r(\neq m)} r^t H_r\sum_n n^{s-t}(\rho_{n+m+r,n}+\rho_{n,n+m+r}) \quad (3.2)$$

$$\frac{d}{dt}\sum_n n^s(\rho_{n+m,n}+\rho_{n,n+m}) = -2(\Gamma-\bar{\gamma}_m)\sum_n n^s(\rho_{n+m,n}+\rho_{n,n+m})$$

$$+\sum_{t=1}^{s}\binom{s}{t}\sum_r r^t H_r \sum_n n^{s-t}\, i(\rho_{n+m+r,n}-\rho_{n,n+m+r}) \ . \qquad (3.3)$$

The mean square displacement is given by the second moment

$$M_2 = \sum_n n^2 \rho_{nn} = \langle R^2(t)\rangle \qquad (3.4)$$

which, with the help of V_m and W_m

$$V_m = -\sum_n (2n+m)i(\rho_{n+m,n}-\rho_{n,n+m}) \qquad (3.5)$$

$$W_m = \sum_n (\rho_{n+m,n}+\rho_{n,n+m}) \qquad (3.6)$$

is calculated from the following set of equations [25]:

$$\dot{M}_2 = 2\sum_m m^2\gamma_m - \frac{1}{2}\sum_m m H_m V_m \qquad (3.7)$$

$$\dot{V}_m = -2(\Gamma+\bar{\gamma}_m)V_m - 4m H_m + \sum_{r(\neq -m)} 2rH_r W_{m+r} \qquad (3.8)$$

$$\dot{W}_m = -2(\Gamma-\bar{\gamma}_m)W_m \ . \qquad (3.9)$$

With the initial condition $\rho_{nn'}(0) = \delta_{n0}\delta_{n'0}$ i.e. the particle is initially at the origin and all non-diagonal elements of the density matrix disappear, we obtain for the time dependence of the mean square displacement [18,19,25] in the case of nearest neighbour interaction

$$M_2(t) = [4\gamma_1+2J^2(\Gamma+\bar{\gamma}_1)^{-1}]t + J^2(\Gamma+\bar{\gamma}_1)^{-2}\,(e^{-2(\Gamma+\bar{\gamma}_1)t}-1) \ . \qquad (3.10)$$

In the completely incoherent case $J = 0$, and we get the linear time dependence of a diffusion process:

$$M_2(t) = 4\gamma_1 t \ . \qquad (3.11)$$

On the other hand, the completely coherent case results in a quadratic time dependence [3,18,19,25]

$$M_2(t) = 2J^2t^2. \qquad (3.12)$$

In the long-time limit the mean square displacement is given by

$$M_2(t) \approx [4\gamma_1+2J^2(\Gamma+\bar{\gamma}_1)^{-1}]t - J^2(\Gamma+\bar{\gamma}_1)^{-2} \ . \qquad (3.13)$$

From this expression the diffusion constant is obtained via (3.14) with the result (3.15)

$$2D = \lim_{t\to\infty} t^{-1} M_2(t) = \lim_{t\to\infty} t^{-1}\langle R^2(t)\rangle \qquad (3.14)$$

$$D = 2\gamma_1 + J^2(\Gamma+\bar{\gamma}_1)^{-1} \ . \qquad (3.15)$$

This expression shows that the diffusion constant has two contributions. The first is determined by non-local fluctuations and increases with increasing strength of the fluctuations. The second contribution is proportional to the square of the tunneling matrix element and decreases with increasing strength of local and non-local fluctuations.

In the following paragraph we shall determine the diffusion constant for fluctuations which are not δ correlated. In that case it is more convenient to start from the Laplace transform of the mean-square displacement:

$$\langle \hat{R}^2(s)\rangle = \hat{M}_2(s) = \int_0^\infty dt\ e^{-st}\ M_2(t)\ . \tag{3.16}$$

Using again the initial condition $\rho_{nn'(0)} = \delta_{n0}\delta_{n'0}$ and neglecting non-local fluctuations, instead of (3.7-9) we get

$$s\hat{M}_2(s) + J\hat{V}(s) = 0 \tag{3.17a}$$

$$(s+2\gamma_0)\hat{V}(s) - 2J\hat{W}(s) = 0 \tag{3.17b}$$

$$(s+2\gamma_0)\hat{p}W(s) = 0\ . \tag{3.17c}$$

The Laplace transform of the mean-square displacement is given by

$$\hat{M}_2(s) = \frac{4J^2}{s^2(s+2\gamma_0)} \tag{3.18}$$

and the diffusion constant is obtained from

$$D = \lim_{s\to 0} \frac{1}{2}\ s^2\ \langle \hat{R}^2(s)\rangle = \lim_{s\to 0} \frac{1}{2}\ s^2\ \hat{M}_2(s) = \frac{J^2}{\gamma_0} \tag{3.19}$$

which is identical with (3.15) if non-local fluctuations are neglected.

4. Diffusion of a Quantum Particle in the Case of Coloured Noise and Anderson Localization

The previous calculations were based on the Hamiltonian (1.2) whose stochastic properties are defined in (1.3-5). In this section we shall again assume that the local excitation energies ε_n are site independent and can therefore be normalized to zero. Furthermore we shall neglect all non-local fluctuations. The model Hamiltonian is thus given by

$$H = H_0 + H_1(t) \tag{4.1}$$

$$H_0 = \sum_{n,a} J\ a^+_{n+a}\ a_n \tag{4.2}$$

$$H_1(t) = \sum_n h_n(t)\ a^+_n\ a_n\ . \tag{4.3}$$

To specify the stochastic properties of the Hamiltonian we shall assume that it is described by a dichotomic Markov process [26] with disappearing mean value and exponentially decaying correlation functions [27,28]. Mathematically its properties are defined by

$$\langle h_n(t)\rangle = 0 \tag{4.4}$$

$$\langle h_n(t_1)\, h_{n'}(t_2)\rangle = \delta_{nn'}\, \Delta^2\, e^{-\gamma(t_1-t_2)} \tag{4.5}$$

$$\langle h_n(t_1)\, h_n(t_2)\, h_n(t_3) \dots h_n(t_n)\rangle = \langle h_n(t_1)\, h_n(t_2)\rangle\, \langle h_n(t_3) \dots h_n(t_n)\rangle$$

$$t_1 \geq t_2 \geq t_3 \geq \dots \geq t_n \tag{4.6}$$

The procedure for the investigation of the mean-square displacement is analogous to that used in the case of the δ correlated stochastic process. We shall first derive an equation for the density matrix and in the second step use it to obtain an equation for the mean square displacement. In the case of coloured noise the calculations are rather involved and are only sketched in the following. A more elaborate presentation is given in [20].

Abbreviating the commutator by $H_0^\times \dots = [H_0, \dots]$, the density matrix equation and its formal solution are

$$\dot{\rho}(t) = -i(H_0^\times + H_1^\times(t))\, \rho(t) \tag{4.7}$$

$$\rho(t) = U(t)\, \rho(0) \tag{4.8}$$

$$U(t) = \overset{\leftarrow}{T} \exp[-i \int_0^t dt'(H_0^\times + H_1^\times(t'))] \; . \tag{4.9}$$

We shall now rewrite the time evolution operator using the disentanglement theorem. Disentangling the stochastic part $H_1^\times(t)$ of the Hamiltonian, we have

$$U(t) = V(t)\, U_0(t) \tag{4.10}$$

$$V(t,\tau) = \overset{\leftarrow}{T} \exp\{-i \int_\tau^t dt'\, H_1^\times(t')\} \tag{4.11}$$

$$U_0(t) = \overset{\leftarrow}{T} \exp\{-i \int_0^t dt'\, \bar{H}_0^\times(t')\} \tag{4.12}$$

$$\bar{H}_0^\times(t) = V^{-1}(t)\, H_0^\times\, V(t) \; . \tag{4.13}$$

$V(t)$ is an abbreviation for $V(t,0)$. The stochastic average is carried out with the help of a time-dependent projection operator

$$\bar{\mathcal{P}}(t)A = V^{-1}(t) \quad V(t)A \tag{4.14}$$

where $\mathcal{P}$ represents the stochastic average

$$\mathcal{P}B = \langle B\rangle \; . \tag{4.15}$$

Decomposing $\bar{H}_0^\times(t)$ in the usual manner [29] and applying the disentanglement theorem in (4.12), we get

$$\bar{H}_0^\times(t) = (1-\bar{\mathcal{P}}(t))\, \bar{H}_0^\times(t) + \bar{\mathcal{P}}(t)\, \bar{H}_0^\times(t) \tag{4.16}$$

$$U_0(t) = U_1(t,0) - i \int_0^t d\tau\, U_1(t,\tau)\, \bar{\mathcal{P}}(\tau)\, \bar{H}_0^\times(\tau)\, U_0(\tau) \tag{4.17}$$

$$U_1(t,\tau) = \overleftarrow{T} \exp\{-i \int_\tau^t dt' \, (1-\bar{\mathcal{P}}(t')) \, \bar{H}_0^\times(t')\} \, . \tag{4.18}$$

We use this result in (4.8,10) and apply $\bar{\mathcal{P}}(t)$ in (4.8) in order to obtain an integral equation for the stochastically averaged density matrix

$$\langle\rho(t)\rangle = K_1(t) \, \rho(0) - i \int_0^t d\tau \, K_1(t-\tau) \bar{H}_0^\times \, \langle\rho(\tau)\rangle \tag{4.19}$$

$$K_1(t-\tau) = \langle V(t) \, U_1(t,\tau) \, V^{-1}(\tau)\rangle \, . \tag{4.20}$$

If we assume that the density operator at the initial time is diagonal in the site representation, we have $[H_1(t),\rho(0)] = 0$, which implies $K_1(t)\rho(0) = \rho(0)$. The equation for the density operator, which is the basis for the following calculations, then reads

$$\langle\rho(t)\rangle = \rho(0) - i \int_0^t d\tau \, K_1(t-\tau) \, H_0^\times \, \langle\rho(\tau)\rangle \, . \tag{4.21}$$

Finally we wish to calculate the diffusion constant from (3.19). To that end we first determine the Laplace transform of the mean square displacement

$$\langle R^2(s)\rangle = \sum_n n^2 \langle\hat{\rho}_{nn}(s)\rangle \, . \tag{4.22}$$

We insert the Laplace transform of (4.21) and arrive at an equation where the power of n on the right-hand side has been reduced by one. This step can be repeated twice and we end up with a closed set of three coupled equations which is the generalization of (3.17) for the case of coloured noise:

$$\langle\hat{R}^2(s)\rangle = i\frac{J}{s} \sum_{n,a} (2na-a^2) \, \langle\hat{\rho}_{n,n-a}(s)\rangle \tag{4.23}$$

$$\sum_{n,a} (2na-a^2)\langle\hat{\rho}_{n,n-a}(s)\rangle = 2iJ \sum_{n,p,q,a,a'} aa' \, \hat{K}_{1n,n-a,p,q+a'}(s)\langle\rho_{p,q}(s)\rangle \tag{4.24}$$

$$\sum_{n,p,q,a,a'} aa' \, \hat{K}_{1n,n-a,p,q+a'}(s)\langle\hat{\rho}_{p,q}(s)\rangle = \frac{1}{s} \sum_{n,a,a'} aa' \, K_{1n,n-a,0,a'}(s) \, . \tag{4.25}$$

The solution of this set of equations is

$$\langle\hat{R}^2(s)\rangle = \frac{2J^2}{s^2} \sum_{aa'}\sum aa' \sum_n \hat{K}_{1\; n,n+a,0,a'}(s) \tag{4.26}$$

$$= \frac{2J^2}{s^2} \sum_{aa'}\sum aa' \sum_n \hat{K}_{n,n+a,0,a'}(s) \tag{4.27}$$

$$\langle\hat{R}^2(s)\rangle = \frac{2J^2}{s^2} \sum_{aa'}\sum aa' \, \hat{\kappa}_{aa'}(s) \, . \tag{4.28}$$

The definition of the contraction $\hat{\kappa}_{aa'}(s)$ is obvious by comparison of (4.27,28). The derivation of (4.27) from (4.26) is somewhat tedious and has been carried out in Appendix C of [20]. K(t), whose Laplace transform is $\hat{K}(s,)$ is defined by

$$K(t) = \langle U(t) \rangle \ . \tag{4.29}$$

The mean square displacement can be determined if we know the contraction $\hat{\kappa}_{aa'}(s)$. In the following step we therefore derive an equation for this quantity. Using the disentanglement theorem again, we can write

$$K(t) = e^{-iH_0^{\times}t} \langle \tilde{V}(t) \rangle \tag{4.30}$$

$$\tilde{V}(t) = \overset{\leftarrow}{T} \exp\{ -i \int_0^t dt' \ \tilde{H}_1^{\times}(t') \} \tag{4.31}$$

$$\tilde{H}_1^{\times}(t) = \exp\{iH_0^{\times}t\} \ H_1^{\times}(t) \ \exp\{-iH_0^{\times}t\} \ . \tag{4.32}$$

We differentiate (4.30) with respect to time, use the disentanglement theorem once more in order to rewrite the time derivative of $\langle \tilde{V}(t) \rangle$ and arrive at

$$\dot{K}(t) = - i \ H_0^{\times} \ K(t) - \int_0^t d\tau \ \tilde{M}(t-\tau) \ K(\tau) \ . \tag{4.33}$$

The kernel of this integrodifferential equation is given by the following equations

$$\tilde{M}(t-\tau) = \langle H_1^{\times}(t) \ W(t,\tau) \ U_0'(t,\tau) \ H_1^{\times}(\tau) \rangle \tag{4.34}$$

$$W(t,\tau) = \overset{\leftarrow}{T} \exp\{- i \int_\tau^t dt' \ (1-\mathcal{P}) \ H_1^{\times}(t')\} \tag{4.35}$$

$$U_0'(t,\tau) = \overset{\leftarrow}{T} \exp\{ - i \int_\tau^t dt' \ H_{0\tau}^{\times}(t')\} \tag{4.36}$$

$$H_{0\tau}^{\times}(t) = W^{-1}(t,\tau) \ H_0^{\times} \ W(t,\tau) \ . \tag{4.37}$$

Forming now the contraction of (4.33) we obtain

$$\dot{\kappa}(t) = - \int_0^t d\tau \ \tilde{\mu}(t-\tau) \ \kappa(\tau) \tag{4.38}$$

$$\{s+\hat{\tilde{\mu}}(s)\} \ \hat{\kappa}(s) = 1 \tag{4.39}$$

where the last line has been obtained by Laplace transformation. Equations (4.28) and (4.39) are still exact, and together with (3.19) form the basis for the calculation of the diffusion constant.

Obviously (4.39) has the form of equations for Green's functions, where $\hat{\tilde{\mu}}(s)$, the contraction of the Laplace transform of $\tilde{M}(t)$, plays the role of the self-energy. In the following we shall determine $\tilde{\mu}(s)$ up to second order in J, which gives the diffusion constant up to fourth order in the tunneling matrix element. For particle motion on a linear chain $\hat{\kappa}(s)$ can then be calculated from (4.39) [20]. In the general case one is still able to obtain a solution for

$\langle\hat{R}^2(s)\rangle$ without further approximations. For this purpose from (4.28,39) we derive recurrence relations for quantities $\hat{R}_n(s)$ defined in the following way

$$\hat{R}_n(s) = \frac{2J^2}{s^2} \sum_{a,a',a_1,\ldots,a_{2n}} aa' \hat{\kappa}_{a-a_1-\ldots-a_{2n},a'} . \qquad (4.40)$$

The comparison of this expression with (4.28) shows that

$$\langle\hat{R}^2(s)\rangle = \hat{R}_0(s) . \qquad (4.41)$$

The recurrence relations mentioned above are given by ($n = 0,1,2,\ldots$)

$$\frac{\phi(J,s)}{z} \hat{R}_0(s) + \frac{2n+1}{\zeta_{n+1}} \chi(J,s)\hat{R}_n(s) + \frac{2n+1}{\zeta_{n+1}}\psi(J,s)\hat{R}_{n+1}(s) = \frac{2}{s^2} J^2a^2 . \qquad (4.42)$$

Subtracting two consecutive equations, we get

$$\sigma_n\chi(J,s)\, \hat{R}_n(s) - (\chi(J,s) - \sigma_n\psi(J,s))\, \hat{R}_{n+1}(s) - \psi(J,s)\, \hat{R}_{n+2}(s) = 0 . \qquad (4.43)$$

This 3-step recurrence relation may be solved by continued fractions

$$y_n = \frac{\hat{R}_{n+1}(s)}{\hat{R}_n(s)} = \frac{\sigma_n\chi}{\chi-\sigma_n\psi + \psi y_{n+1}} . \qquad (4.44)$$

z is the number of nearest neighbours. The $\phi(J,s)$, $\chi(J,s)$ and $\psi(J,s)$ are known functions of J and s [20]. σ_n and ζ_n are given by

$$\sigma_n = \frac{2n+1}{2n+3}\,\frac{\zeta_{n+2}}{\zeta_{n+3}} \qquad (4.45)$$

$$\zeta_n = \sum_{a_1,\ldots,a_{2n}} \delta(0,a_1+a_2+\ldots+a_{2n}) \qquad (4.46)$$

where ζ_n is the number of possibilities to return to the origin after $2n$ steps each to nearest neighbours. With (4.41) the diffusion constant becomes

$$D = \lim_{s\to 0} \frac{s^2}{2} \langle\hat{R}^2(s)\rangle = \lim_{s\to 0} \frac{s^2}{2} \hat{R}_0(s) . \qquad (4.47)$$

The N-th approximation to R_0 is obtained by setting $y_{N-1} = 0$ in the denominator of (4.44) and in this way we obtain

$$D = za^2J^2 \frac{1\ |}{|\phi+\chi+} \frac{\sigma_0\chi\psi\ |}{|\chi-\sigma_0\psi+} \frac{\sigma_1\chi\psi\ |}{|\chi-\sigma_1\psi+} \cdots \frac{\sigma_{N-2}\chi\psi\ |}{|\chi-\sigma_{N-2}\psi} . \qquad (4.48)$$

It can be shown [20] that the continued fraction (4.48) converges in a neighbourhood of $s=0$, if

$$\frac{\Delta^2}{J^2} > z(z+1) \qquad (4.49)$$

i.e. for large enough values of Δ or small enough values of J. This is consistent with the approximation used in the evaluation (4.48) of the continued fraction because the term neglected in the denominator of (4.44) is proportional to J^2 for small values of s.

In the following we shall discuss our result for the diffusion constant (4.48) and compare it to several results obtained recently in the literature. Kitahara and Haus [30] study diffusion on a linear chain within a stochastic model which reduces to the present one if the stochastics are dichotomic Markov processes. Their perturbation theory approach is applicable to fluctuations with short correlation times, i.e. $\gamma \gg \Delta, J$. Furthermore, because they take the stochastic part of the Hamiltonian as a perturbation, their approximation will fail for $\Delta \gg J$, except for extremely short correlation times.

In our notation the result of Kitahara and Haus reads

$$D = a^2 J^2 \frac{\gamma}{\Delta^2} \left(1+3 \frac{J^2}{\gamma^2}\right) \tag{4.50}$$

Our treatment is valid, if $\Delta \gg J$. Therefore, one might conclude that it is not meaningful to compare our result to (4.50). However, as we claim that the second and fourth order (in J) terms of our result are exact, (4.50) must be included in it containing only J^2 and J^4 terms. For that reason (4.50) yields a useful contercheck of our theory.

Exactly the same model as the one presented here was investigated by Inaba [27]. He assumes that all density matrix elements apart from the diagonal and nearest neighbour off-diagonal ones vanish. This assumption is justified, if the correlation times are short, i.e. $\gamma \gg J$, and the amplitude of the stochastic fluctuations is large, i.e. $\Delta \gg J$. The assumptions become doubtful, if only one of these conditions is fulfilled. If the fluctuations are weak, the deterministic motion is not perturbed sufficiently to render the density matrix 'tridiagonal' in spite of the short correlation time. If, on the other hand, the correlation time is long, the coherence of the wave functions is not destroyed within a few lattice constants in spite of the magnitude of the fluctuation amplitudes. Inaba obtains for the diffusion constant

$$D = \frac{z}{2} a^2 J^2 \left(\frac{\gamma}{\Delta^2} + \frac{1}{\gamma} - \frac{J^2}{\gamma^3}\right) . \tag{4.51}$$

To compare our result with (4.50) and (4.51) we take the second approximant of (4.48) and neglect the ψ term in the second denominator. The resulting expression is still exact up to fourth order in J:

$$D = \frac{z}{2} a^2 J^2 \left\{\frac{\gamma}{\Delta^2} + \frac{1}{\gamma} + \frac{J^2}{\gamma^3} \left[- \frac{z+2}{4} + \left(z+\frac{\zeta_2}{3z}\right) \frac{\gamma^2}{\Delta^2} - \frac{27}{4} \frac{\gamma^2}{3\gamma^2+4\Delta^2}\right]\right\} . \tag{4.52}$$

Comparing the three results we find that (4.51) and (4.52) agree up to second order in J, whereas in (4.50) $1/\gamma$ is neglected against γ/Δ^2. A comparison of the fourth-order terms shows that for $z = 2$ (linear chain) the first term in formula (4.52) is identical to the J^4 term in Inaba's formula, the second to that of formula (4.50) ($\zeta_6 = 6$ for a linear chain); none of these formulae includes the third J^4 term of (4.52) stemming from four-time correlation functions of the primary process. For large Δ, Inaba's and our formulae agree, because the second and third J^4 terms may be neglected.

For $z > 2$, only (4.51) and (4.52) can be compared. The formulae do not even agree in the case of large Δ - in Inaba's formula the diffusion constant is just proportional to z whereas ours contains terms $\propto z^2$. To make clear the origin of such a difference, we remark that in Inaba's treatment all the memory of jumps of the diffusing particle beyond nearest neighbours is lost; our result, however, contains the memory effects of closed paths consisting of four jumps (to a nearest neighbour each) in the approximation (4.52) and of even more in the full continued fraction.

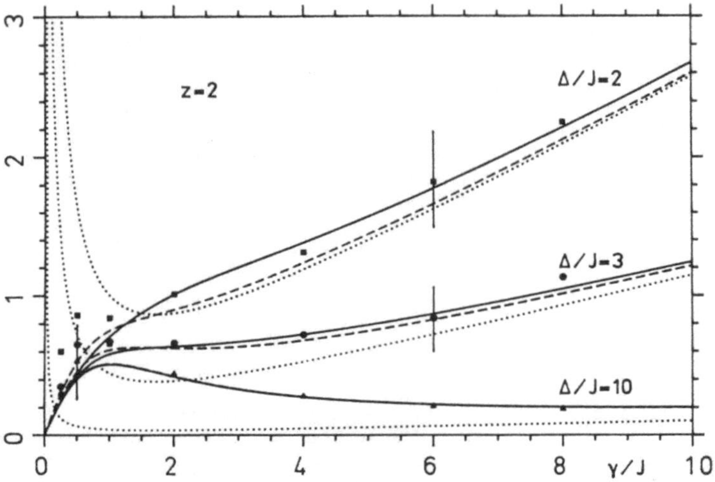

Fig.1 Dependence of the diffusion constant on the switching rate for several fluctuation strengths Δ/J. Linear chain case. Dotted lines represent the Kitahara-Haus result [30], dashed ones Inaba's result [27], full lines show approximation (4.52) of the present theory. Single points are numerical results (solution of the stochastic Schrödinger equation) obtained by Inaba [28] (the largest error bar is given for each case).

To illustrate the differences between the three results we plot the reduced diffusion constant D/a^2J corresponding to (4.50-52) as a function of the reduced switching rate γ/J for several values of Δ/J (Fig.1). The picture shows clearly that approximation (4.50) becomes poor for small values of γ/J and that the other two approximations (4.51,52) agree pretty well for large values of Δ/J (linear chain case). Results (4.51) and (4.52) are also in good agreement over most of the γ/J range with numerical results plotted in the same figure which were obtained as a solution to the stochastic Schrödinger equation [27].

A further treatment of the same diffusion problem has been given by Inaba [28]. In a CPA calculation he neglects diagrams describing backscattering of the diffusing particle. Therefore, one expects that the diffusion constant obtained from this approximation will be too large. From his density matrix approach [27] one should obtain a too small diffusion constant in the case of large γ/J, because backscattering effects are overemphasized; for small γ/J, there is no simple way of prediction, because backscattering between nearest neighbours is fully taken into account, but backscattering after going to a next to nearest (or farther) neighbour is neglected. These conclusions are confirmed by Fig.2 comparing the two solutions with our approximation (4.52). The CPA result is larger than ours, although at high switching rates the difference between the two curves is almost zero. The density matrix approximation yields a smaller diffusion constant than our theory for rapid fluctuations. In the case of slow fluctuations our result is smaller, indicating that more long-range coherence leading to an increased backtransfer is taken into account.

In Fig.3 we compare (4.52) and several approximants of the continued fraction solution (4.48) of the 1D case to the diffusion constant obtained from a direct solution of (4.39). For $\Delta/J = 1$ the latter solution does not exist over the whole range of γ/J, because the radicand of square roots contained in it becomes negative within a certain interval of γ/J values. In the same interval the continued fraction diverges. However, the figure also shows that for $\Delta/J \geq 1.5$ there is perfect convergence over the whole range of γ/J and that already the fifth approximant is a very good approximation of the total continued fraction.

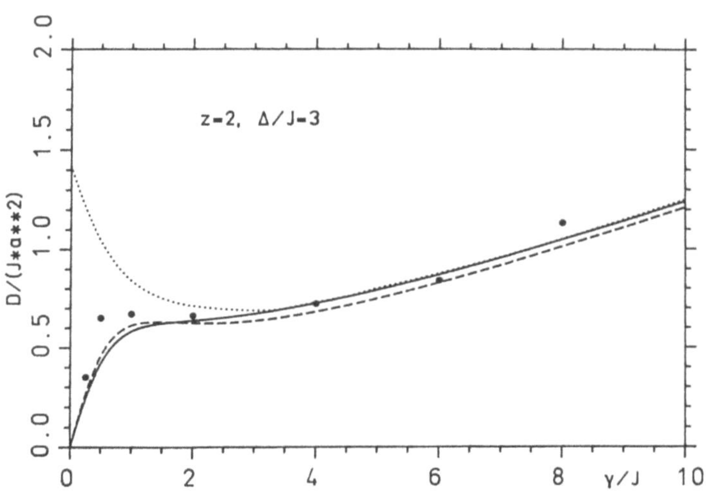

Fig.2 Comparison of approximation (4.52) (full line) to Inaba's density matrix approach [27] (dashed line), a numerical CPA calculation [28], and the results of a numerical solution of the Schrödinger equation [28].

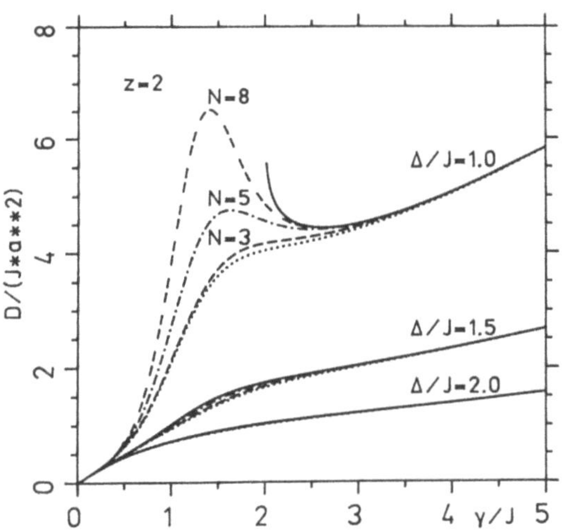

Fig.3 Approximants of the continued fraction representing the diffusion constant as compared to the solution derived from (4.39) (full line) in the linear chain case. The dotted lines show approximation (4.52). For $\Delta/J > 1$ only the third and fifth approximants are plotted. For $\Delta/J = 1$ the solution on the basis of (4.39) does not exist over the whole range of γ/J.

We have calculated some approximants of the continued fraction (4.48) for several values of Δ/J and different lattices as a function of γ/J. The following results are obtained. For sufficiently large Δ/J the continued fraction converges numerically within the whole range of switching rates. Examples for this behaviour are given in Fig.4 where Inaba's approximation [27] is plotted, too. The value of Δ/J, below which a region of non-convergence exists (in which our approximation fails) is increasing with the number of nearest neighbours; in the case of a linear chain we have a limiting value of $\Delta/J \approx 1.5$, for an fcc lattice $\Delta/J \approx 9$. These values are somewhat smaller than those estimated from (4.49).

We have seen that Inaba's formula is well behaved everywhere. One should not conclude, however, from this nice analytical property that (4.51) is superior in the range where our expression diverges. The density matrix approximation breaks down there as well, because the assumptions necessary for its derivation are no

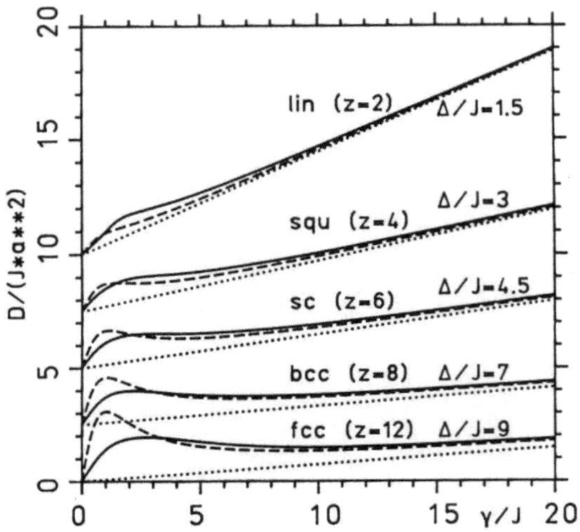

Fig.4 Continued fraction solution (N=10, full lines) for several lattices and fluctuation strengths. The origin of each curve is shifted 2.5 units upwards with respect to the preceding one to avoid intersections. For reasons of comparison, Inaba's result (dashed lines) and the total memory loss case (dotted lines) [27] are plotted as well.

longer valid; this is just not visible in the pictures. In fact, it is an advantage of our theory that the final result yields the possibility of estimating the range of its approximate validity. Basically, our theory is an unusual kind of perturbation scheme treating as perturbation the non-stochastic part of the Hamiltonian. For this reason, it should be a good approximation for sufficiently large Δ/J, no matter how small γ/J. On the other hand, for large γ/J taking into account only a few powers of J in the 'self-energy' is sufficient because of strong memory loss effects (the δ correlated limit is approached). As explained above, just these features show up in the convergence behaviour of the final solution.

The figures show that for a fixed value of Δ/J the continued fraction (4.48) goes to zero with $\gamma/J \to 0$. It may be meaningless in the case of small Δ/J, because there is a range without convergence between the regime of large γ/J where the theory is perfectly applicable and the limiting case we are interested in. But from the above considerations we conclude that in the case of large Δ/J it is allowed to interpret the behaviour of this limit as a prediction of **Anderson localization** [31]. To our knowledge this would be the first proof of Anderson localization starting from a time dependent stochastic model with coloured noise. In short, our prediction is that in the model investigated Anderson localization occurs, if $\Delta/J > \sqrt{z(z+1)}$. However, the figures also show that even a small time dependence in the fluctuations results in a non disappearing diffusion constant and thus in a delocalization of the particle.

Another important feature of (4.48) is that the diffusion constant is determined by **all** the numbers ζ_n of closed paths of given lengths (2na) a particle can take starting out at some lattice point. But this means that the whole structure of the lattice enters the mathematical expressions as soon as the stochastics is not δ correlated. This is a genuine memory effect; at a given time the particle 'remembers' all the loops it has already run through.

There is, however, no dramatic change in the absolute value of the diffusion constant. Figure 4 shows that it is enhanced in comparison to the approximation $D = (1/2)za^2J^2(\gamma/\Delta^2)$ describing the 'corresponding white noise case' [28] or case of total memory loss [27]. For large values of γ/J this enhancement is somewhat larger than expected from Inaba's density matrix approximation and well described by the CPA result whereas for slow fluctuations only the first of these approximations agrees qualitatively with our theory. This agreement gets worse with increasing number of nearest neighbors.

5. Concluding Remarks

In paragraphs 1 to 3 of this contribution we considered the Haken-Strobl model for the coupled coherent and incoherent motion of a particle. The model was originally devised to describe the energy transport by excitons in organic molecular crystals. Its essential ingredient is the description of the influence of the phonons by a δ correlated Gaussian Markov process. In this model it is possible to discuss the limiting cases of incoherent and coherent particle motion, but also the whole range inbetween. In the past, the model has been generalized to allow for a description of optical absorption [7,16,32] and spin resonance [16,33-35] experiments in molecular crystals.

Another generalization of the stochastic model, which includes coloured noise, was used in paragraph 4 for a discussion of the diffusion constant. The calculation shows that the effects on the diffusion constant are not too dramatic. This is to be expected because the time scale relevant for diffusion is infinite. Coloured noise effects should be more distinct if the correlation time of the noise is comparable to the time scale of the experiment. In the case of diffusion, for example, we have seen that for an infinite correlation time ($\gamma=0$) Anderson localization occurs. However, the transition from a delocalized to the localized state could not come out of our calculation because throughout our treatment we have assumed large amplitudes of the fluctuations. An approach to optical absorption within such a model is given in [36,37].

The quantum mechanical treatment of the influence of the phonons on the dynamical properties of a quantum particle has been approached in various ways. HAKEN and REINEKER [38] have considered the phonons a heatbath and used a projection technique to eliminate its variables from the density operator equation of the particle (exciton). This idea has been pursued later on [39] and applied to the discussion of the temperature dependence of ESR line shapes [40,41]. GROVER and SILBEY [42] treated the interaction between excitons and phonons via the small polaron transformation of HOLSTEIN [43,44] and used a Green's function technique to describe the transport properties of the exciton. A combination of this canonical transformation with the projection technique used in [39-41] has recently been applied to describe the temperature dependence of optical [45] and ESR [46] experiments on dimers in molecular crystals. KENKRE and RAMAN [47] derived a generalized master equation for the description of the coupled coherent and incoherent exciton transport, and the connection between the various methods used in this field have been investigated [48] and reviewed by SILBEY [49] and by KENKRE [50]. A series of recent papers [21-24] shows that this microscopic approach to the transport problem is at present vividly investigated also in connection with the problem of self-trapping of electrons and excitons and the formation of excimers [51-53].

References

1. T. Förster: Ann. Phys. (Leipzig) 2, 55 (1948)
2. M. Trlifaj: Czech. J. Phys. 8, 510 (1958)
3. R.E. Merrifield: J. Chem. Phys. 28, 647 (1958)
4. H. Haken: Fortschr. Phys. 6, 271 (1958)
5. H. Haken: Light and Matter, ed. by S. Flügge, Handbuch der Physik, Vol. 25/2c (Springer, Berlin, Heidelberg 1970)
6. H. Haken, G. Strobl: In The Triplet State, ed. by A. Zahlan (Cambridge Univ. Press, London 1967) p. 248
7. H. Haken, P. Reineker: Z. Phys. 249, 253 (1972)
8. H. Haken, G. Strobl: Z. Phys. 262, 135 (1973)
9. H.C. Wolf: In Organic Molecular Aggregates, ed. by P. Reineker, H. Haken, H.C. Wolf, Springer Ser. Solid-State Sci., Vol. 49 (Springer, Berlin, Heidelberg 1983) p. 2
10. G. Paillotin, J. Breton: In Organic Molecular Aggregates, ed. by P. Reineker, H. Haken, H.C. Wolf, Springer, Ser. Solid- State Sci., Vol. 49 (Springer, Berlin, Heidelberg 1983) p. 272

11. H. Wipf, A. Magerl, S.M. Shapiro, S.K. Satija, W. Thomlinson: Phys. Rev. Lett. 46, 947 (1981)
12. K.W. Kehr, D. Richter, J.-M. Welter, O. Hartmann, E. Karlsson, L.O. Norlin, T.O. Niinikoski, A. Yaouanc: Phys. Rev. B26, 567 (1982)
13. P.W. Anderson, J. Phys. Soc. Jpn.: 9, 888 (1954)
14. R. Kubo, J. Phys. Soc. Jpn.: 17, 1100 (1962)
15. G.L. Sewell: Phys. Rev. 129, 597 (1963)
16. P. Reineker: In Exciton Dynamics in Molecular Crystals and Aggregates, ed. by G. Höhler, Springer Tracts Mod. Phys., Vol. 94 (Springer, Berlin, Heidelberg 1982) p. 111
17. G. Strobl: Diplomarbeit, Universität Stuttgart (1966)
18. E. Schwarzer, H. Haken: Phys. Lett. 42A, 317 (1972)
19. P. Reineker: Phys. Lett. 42A, 389 (1973)
20. K. Kassner, P. Reineker: Z. Phys. B59, 357 (1985); B60, 87 (1985)
21. A.O. Caldeira, A.J. Leggett: Phys. Rev. Lett. 49, 211 (1981)
22. R.A. Harris, R. Silbey: J. Chem. Phys. 78, 7330 (1983)
23. H. Grabert, U. Weiss, P. Hänggi: Phys. Rev. Lett. 52, 2193 (1984)
24. P. Hänggi: J. Stat. Phys. 42, 105 (1986)
25. P. Reineker: Z. Phys. 261, 187 (1973)
26. R.C. Bourret, U. Frisch, A. Pouquet: Physica 65, 303 (1973)
27. Y. Inaba: J. Phys. Soc. Jpn. 50, 2473 (1981)
28. Y. Inaba: Thesis, University of Tokyo (1982)
29. R. Zwanzig: J. Chem. Phys. 33, 1338 (1960)
30. K. Kitahara, J.W. Haus: Z. Phys. B32, 419 (1979)
31. Y. Nagaoka, H. Fukuyama (eds.): Anderson Localization. In Springer Ser. Solid-State Sci., Vol. 39 (Springer, Berlin, Heidelberg 1982)
32. E. Schwarzer, H. Haken: Opt. Commun. 9, 64 (1973)
33. P. Reineker, H. Haken: Z. Phys. 250, 300 (1972)
34. H. Haken, E. Schwarzer: Chem. Phys. 27, 41 (1974)
35. P. Reineker: Phys. Rev. B19, 1999 (1979)
36. A. Blumen, R. Silbey: J. Chem. Phys. 69, 3589 (1978)
37. I.B. Rips, V. Capek: Phys. Status Solidi (b) 100, 451 (1980)
38. H. Haken, P. Reineker: In Excitons, Phonons and Magnons in Molecular Crystals, ed. by A. Zahlan (Cambridge Univ. Press, London 1968)
39. A. Scheuing: Diplomarbeit, Universität Ulm (1979)
40. U. Schmid, P. Reineker: Molecular Phys. 55, 77 (1985)
41. J. Köhler, P. Reineker: Chem. Phys. 93, 209 (1985)
42. M. Grover, R. Silbey: J. Chem. Phys. 54, 4843 (1971)
43. T. Holstein: Ann. Phys. (N.Y.) 8, 325 (1959)
44. D. Emin: Ann. Phys. (N.Y.) 64, 336 (1971)
45. P. Reineker, R. Winkler, H. Port: Phys. Stat. Sol. (b) 137, 137 (1986)
46. R.G. Winkler, P. Reineker: Mol. Phys. (in print)
47. V.M. Kenkre, T.S. Rahman: Phys. Lett. 50A, 170 (1974)
48. V.M. Kenkre: Phys. Rev. B11, 1741 (1975)
49. R. Silbey: Ann. Rev. Phys. Chem. 27, 203 (1976)
50. V.M. Kenkre: In Exciton Dynamics in Molecular Crystals and Aggregates, ed. by G. Höhler, Springer Tracts Mod. Phys., Vol. 94 (Springer, Berlin, Heidelberg 1982) p. 1
51. A.S. Ioselevich, E.I. Rashba: J. Lumin. 34, 223 (1986)
52. G. Venzl, S.F. Fischer: Phys. Rev. B32, 6437 (1985)
53. P. Reineker, A. Scheuing: to be published

Structure Formation in Polymer Systems

G. Strobl

Fakultät für Physik, Albert-Ludwigs-Universität,
D-7800 Freiburg, Fed. Rep. of Germany

1. Introduction

One of the major problems in polymer physics is the understanding of polymer crystallization. For chains with a regular chemical constitution formation of crystals is in principle possible. These macromolecular crystals are generally built up by chains in a helical low-energy conformation with a uniform orientation of the helix axes. Principally, in order to form the crystalline equilibrium state all chains have to be extended over their full length. Starting from a highly entangled melt of coiled macromolecules this ideal crystalline state is never reached. Cooling a melt below the equilibrium melting point generally leads to the formation of metastable structures which are only in parts crystalline. One observes layer-like crystallites with a thickness in the order of 10 nm which are separated by amorphous layers, thus setting up a lamellar two-phase structure /1/. Thicknesses of amorphous and crystalline layers change with the crystallization temperature; they are determined by the kinetics of transformation. Lowering the transformation rate by increasing the crystallization temperature generally results in thicker crystallites. Under normal conditions transformation starts from nuclei which expand into growing spherulites. There are however special cases where the transformation from the melt to the partially-crystalline solid state occurs continuously without a passage over a heterogeneous state. This second mode of polymer crystallization has been observed for some preoriented systems where chains were stretched in the melt /2/.

Two different modes of transformation are also found in another case: the phase separation initiated in binary polymer mixtures by a temperature jump from the one phase- into the two phase -region. Depending upon the temperature and composition unmixing proceeds either by nucleation and growth, or continuously by spinodal decomposition /3,4/. The two different paths of transformation lead to different characteristic structure patterns. Quasiperiodic phase fluctuations are typical for spinodal unmixing. The length scale of the fluctuation depends upon the temperature chosen for the unmixing, which shows that structure formation is again kinetically controlled.

The aim of this contribution is a discussion of the kinetic factors which govern structure formation in these two-phase polymer systems, crystalline-amorphous systems as well as binary mixtures. Some selected experimental results are presented for illustration.

2. Crystallization by Nucleation and Growth

Starting from the isotropic melt, a sufficient amount of supercooling below the equilibrium melting point $T_m(\infty)$ is necessary in order to initiate crystallization with reasonable crystallization rates. Spherulites are then observed which grow with constant radial speed. The spherulites have a complex internal structure and are composed of crystalline-amorphous stacks. Fig.1 shows as an example the layer structure in a surface which passes through the center of a sperulite /5/.

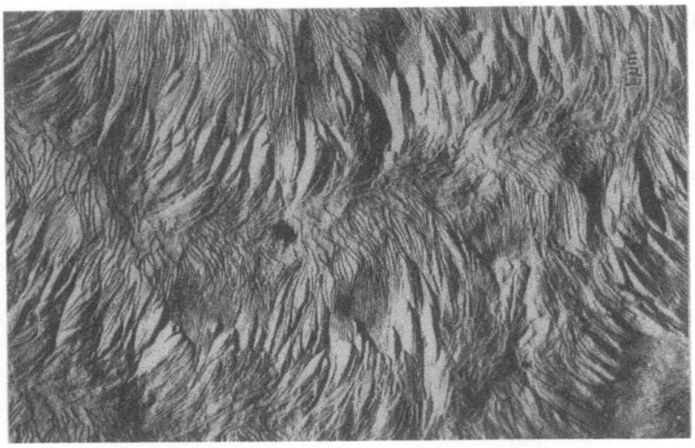

Fig.1 Electron micrograph of replica of a surface spherulite of polyethylene.

Spherulites expand through the lateral growth of the crystallites. Fig.2 sketches the situation given for an individual crystallite. Growth occurs only in lateral direction by an attachment of chain sequences straightened over a length corresponding to the crystallite thickness l. Growth in chain direction is largely suppressed by the presence of folds and entanglements at the crystallite surface. A thickening of crystals requires a replacement of whole chain sequences within the crystallites and therefore proceeds, if at all, only very slowly.

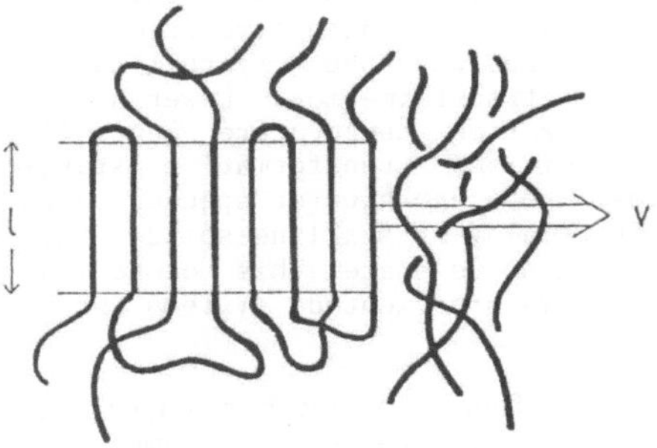

Fig.2 Structure and growth front of a polymer crystallite.

Crystallite thicknesses l and growth rates v change strongly with temperature. Studies were performed on numerous systems. The behaviour is essentially uniform and can be described by the following characteristic dependencies:

$$l \sim \frac{C_1}{T - T_m(\infty)} + C_2 \tag{1}$$

$$v \sim \exp - \frac{C_3}{T - T_m(\infty)} . \tag{2}$$

According to eq.(1) chains become more and more extended on approaching the equilibrium melting point. However, at the same time crystallization rates decrease drastically. Hence, growth of extended chain crystals at temperatures near $T_m(\infty)$ is practically impossible.

3. Kinetics of Lateral Growth

The microscopic steps associated with lateral growth are under discussion since the beginning of crystallization studies on polymers. General agreement has not been reached so far. However, a basic law, the selection rule which determines the thickness of the growing crystallites.has become clear: the lamellae which form at a given crystallization temperature are those with the largest growth rate. Analysis can be based upon this criterium.

Generally, crystal growth rates are determined by the thermodynamic driving force and relevant relaxation times. Dealing with crystallites of finite thickness one has first to inquire after their melting point. For Gibbs free energies g_a and g_c for a monomer in the melt and the crystal respectively and a surface free energy σ the equilibrium condition at the melting point of crystallites composed of sequences of n^* monomers, $T_m(n^*)$, becomes

$$g_c n^* - 2\sigma = g_a n^* . \tag{3}$$

Near to the equilibrium melting point $T_m(n^* \to \infty)$ the Gibbs free energy difference increases linearly with temperature

$$g_a - g_c = \frac{\Delta h}{T_m(\infty)} (T_m(\infty) - T) \tag{4}$$

(Δh denotes the melting enthalpy). Insertion of eq.(4) in eq.(5) gives the well-known result

$$T_m(n^*) = T_m(\infty) - \frac{2\sigma T_m(\infty)}{\Delta h} \frac{1}{n^*}. \tag{5}$$

Eq.(5) describes the melting point reduction resulting from the small thickness of the crystallites.

Clearly, crystals growing at a given crystallization temperature T_c have to be thicker than the crystals in phase equilibrium at this temperature (sequence length n^*):

$$n > \frac{2\sigma T_m(\infty)}{\Delta h} \frac{1}{T_m(\infty) - T_c} = n^* \qquad \text{or} \tag{6}$$

$$n = n^*(T_c) + \delta n . \tag{7}$$

The overlength δn determines the thermodynamic driving force and therefore the growth rate.

In linear approximation the velocity of the phase boundary can be generally written as

$$v = \frac{a}{\tau} \frac{f}{kT} , \tag{8}$$

where f denotes the driving force

$$f = (g_a - g_c)\, \delta n \tag{9}$$

and "a" the monomer diameter. τ^{-1} describes the equilibrium rate of sequence attachment which is equal to the rate of sequence detachment at a stationary boundary. The experimental observations indicate a strong exponential dependence of τ^{-1} on the sequence length n:

$$\tau^{-1}(n) = \tau_0^{-1} \exp - \alpha n . \tag{10}$$

The elementary monomer jump rate τ_0^{-1} and the parameter α are only weakly temperature dependent.

Insertion of eqs.(10) and (9) into eq.(8) gives the dependence of the growth rate v on the overlength δn:

$$v(\delta n) = \frac{a}{\tau} \frac{g_a - g_c}{kT} \exp - \alpha n^* \cdot \delta n \cdot \exp - \alpha \delta n \; . \tag{11}$$

It is schematically depicted in Fig.2. Crystals grow for $n > n^*$ and melt for $n < n^*$. The growth rate has its maximum at

$$n_{max} = n^* + 1/\alpha \tag{12}$$

where it becomes

$$v = \frac{a}{e\alpha\tau_0} \frac{\Delta g}{kT} \exp - \alpha n^* \; . \tag{13}$$

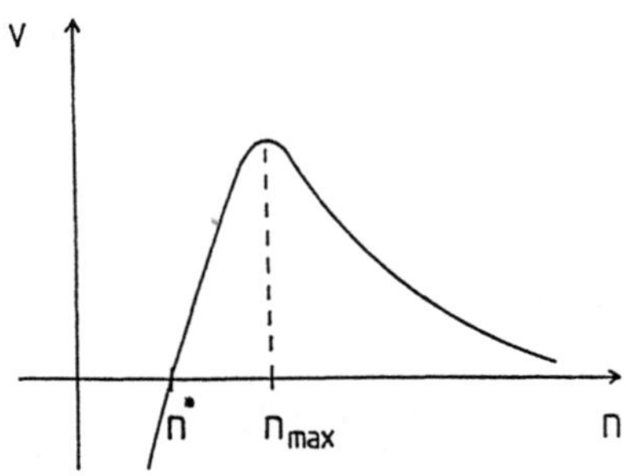

Fig.3 Dependence of crystallization and melting rate on the sequence length.

Choosing a fixed crystallization temperature, the large majority of crystallites will grow with the thickness and growth rate at the maximum as given by eqs.(12), (13). Insertion of eq.(6) yields for the dependence of the crystallite thickness $l_{max} = an_{max}$ and the growth rate v on the supercooling $\Delta T = T_m(\infty) - T_c$ the relations

$$l_{max} = \frac{a2\sigma T_m(\infty)}{\Delta h} \cdot \frac{1}{\Delta T} + \frac{1}{\alpha} \tag{14}$$

and

$$v \sim \frac{a}{\tau_0} \frac{\Delta T}{T_m(\infty)} \exp - \frac{C}{\Delta T} \tag{15}$$

in agreement with the experimental results (eqs.(1), (2)).

So far the situation seems clear. What remains, is an understanding of the exponential dependence of the equilibrium attachment rate on the sequence length (eq.10)

$$\tau^{-1} = \tau_0^{-1} \exp - \alpha n \; .$$

This factor mainly determines the temperature dependence of the structure. Its origin is still under discussion. There is an original suggestion by Hoffman and

Lauritzen /6/ which provided the basis for some detailed theories. These authors interpreted eq.(10) as being indicative for a secondary nucleation at the growth face. A smooth growth face is assumed onto which complete layers are added after nucleation steps. This secondary nucleation would require an activation energy proportional to the sequence length and therefore explain eq.(10). There exists, however, an alternative explanation. Experimental evidence for atomically smooth growth faces in crystallizing polymer melts is lacking. As pointed out recently by Sadler /7/ for polymer solutions, growth faces of polymer crystals can be rough. If this is also true for melts, "normal growth" rather than "layer growth" occurs /8/. For "normal growth" - conditions the dependence expressed by eq.(10) finds a straightforward simple answer: before a sequence, which lies coiled in the melt, can be incorporated into the growing crystal, it has to be stretched, and all intervening entanglements with other chains have to be removed. This requires an activation entropy proportional to the sequence length. Expressed in other words: only a fraction $\sim \exp - \alpha n$ in the melt shows the proper conformation required for an attachment. Alternatively the detachment rate rather than the attachment rate can be considered, since both are equal. The detachment rate is determined by the heat of melting as

$$\tau^{-1} \sim \exp - \frac{n\Delta h}{kT} . \qquad (16)$$

Hence, if "normal growth" onto rough growth faces occurs in polymer crystallization, growth kinetics and the resulting temperature-dependent structure formation finds a simple answer.

4. Continuous Crystallization Mode

If crystallization is conducted starting from an oriented melt or glass, a process different from nucleation and growth has been observed in a number of cases /2/. Studies in the electron microscope or by X-ray scattering suggest a continuous transition from the amorphous to the partially crystalline state. Density fluctuations appear all over the sample and grow continuously in amplitude up to the final state given by the crystalline-amorphous layer structure. Throughout the transformation the sample remains optically homogeneous. The observations indicate occurrence of a special mode of crystallization.

An example is given with Figs.4-7. The study /9/ deals with the crystallization process initiated by annealing a glassy oriented fibre of polyethyleneterephthalate (PET) at temperatures slightly above the glass transition. Fig.4 shows as a result of an X-ray scattering experiment the evolution of the structure factor I(s) with annealing time (s denotes the scattering vector $s = 4\pi \sin\theta/\lambda$). Two different ranges show up: for scattering vectors $s < s_c = 0.95\text{nm}^{-1}$ there is an increase of intensities with time, for $s > s_c$ intensities decrease. As shown by Fig.5, kinetics follow an exponential law, $I \sim \exp Rt$, with positive or negative rate constants R for $s < s_c$ or $s > s_c$ respectively.

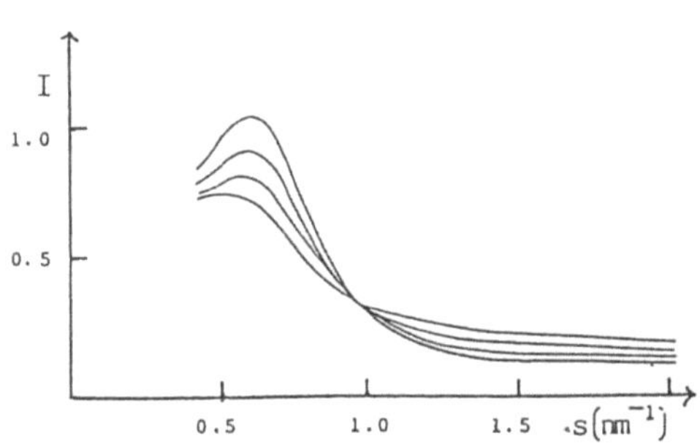

Fig.4 Crystallization of PET starting from oriented glassy state. Evolution of structure factor with time.

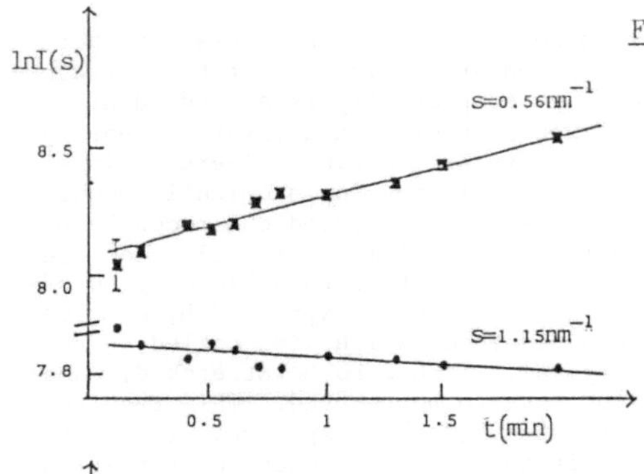

Fig.5 Exponential growth or decay of scattering intensities.

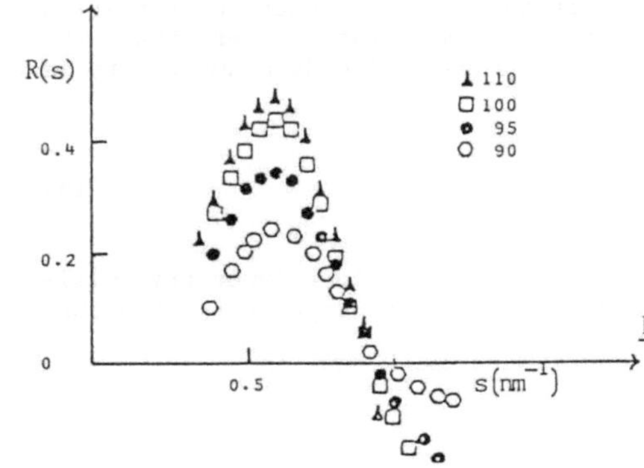

Fig.6 Rate constants in dependence on the scattering vector

The observed scattering curves reflect quite generally speaking the introduction of density fluctuations into the originally homogeneous sample. The experiment shows, that long wavelength fluctuations have the tendency to grow, and short wavelength fluctuations to decay. The structure which develops is primarily determined by those fluctuation waves which have the largest growth rates. Fig.6 indicates, that this maximum occurs at s ≈ 0.6 nm^{-1}. As a consequence the resulting structure will exhibit a quasi-periodicity L in the order of 10 nm.

The location of the growth rate maximum and therefore the long-spacing L change with the crystallization temperature. As demonstrated by Fig.7, L shifts to larger values with increasing temperature. The dependence can be described in good approximation by

$$L \sim (T_s - T)^{-1/2} \tag{17}$$

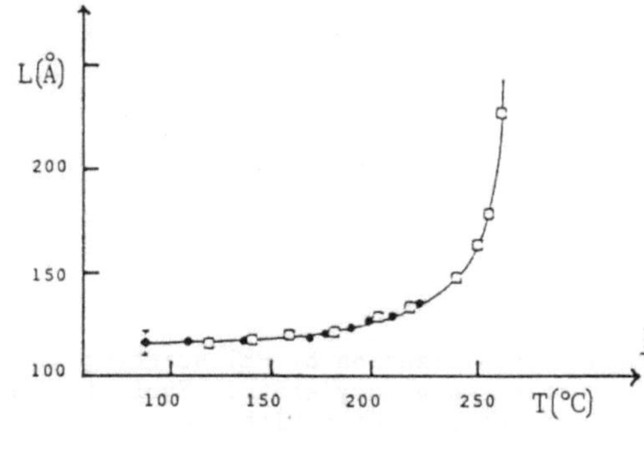

Fig.7 Crystallization of oriented PET. Long spacing in dependence on the crystallization temperature.

where T_s is located near to the equilibrium melting point $T_m(\infty)$ of polyethyleneterephthalate.

The observed kinetics are reminiscent of a well-known process occurring in a different system: the spinodal decomposition of binary mixtures. Here the phase separation initiated by a temperature jump from the homogeneous phase into the two phase-region below the spinodal can proceed via a specific unmixing process. A theoretical analysis was given by Cahn and Hilliard /10/. Theory shows that concentration fluctuations with wavelengths above a critical value become intrinsically unstable. As a consequence phase separation sets in without any activation. The structure pattern is dominated in its initial stages by the concentration wave with the largest growth rate. Its wavelength shifts to higher values on approaching the spinodal curve. All these characteristics are also found for the continuous mode of polymer crystallization. It looks therefore appealing to try to transfer the mechanism of spinodal decomposition to the case of polymer crystallization.

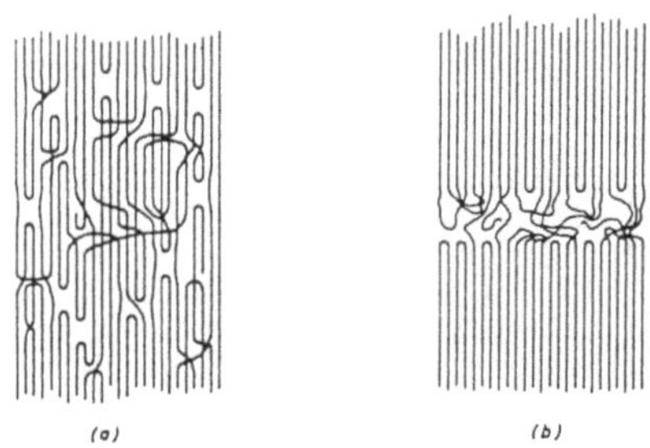

Fig.8 Clustering of conformational defects annealing an amorphous polymer fiber

Of course, from the chemical point of view, one is dealing with a one-component system. However, the monomers which constitute the polymer chain can greatly differ in their mobilities and flexibilities and hence, their potential to crystallize within the given limited time. For example, chain parts contributing to an entanglement essentially act as crosslinks which cannot be incorporated into a crystal. Other unfavorable conformations with long times of relaxation are sharp folds or chain torsions. These "conformational defects" are not fixed at certain monomers, but can move along the chain. Considering these properties a process of "defect clustering" can be envisaged which continuously builds up two phases, crystalline regions free of defects, and amorphous regions where all the defects become concentrated. Fig.8, taken from a work of Fischer and Goddar /11/, illustrates this idea. Clustering processes can only take place if the defects move along the chains. Preorientation of chains in the amorphous state apparently leads to a high axial mobility and thus promotes the continuous crystallization mode.

5. Phase separation in polymer mixtures

Depending upon temperature and composition, two polymers can either be miscible or will separate into two phases. It is possible to control the building up of two phase structures and prepare specific patterns. Usually one starts from the homogeneous state and induces phase separation by a temperature jump into the two phase-region. Since diffusion constants of polymers are always small, unmixing proceeds in general slowly. The process can always be stopped by quenching samples to temperatures below the glass transition. By this kind of treatment samples can be kept under continuous control. Structures can be studied in detail through the different states of development.

Figs.9 and 10 show two characteristic structure patterns. They were obtained for mixtures of polystyrene (PS) and partially brominated polystyrene (PBr_xS)

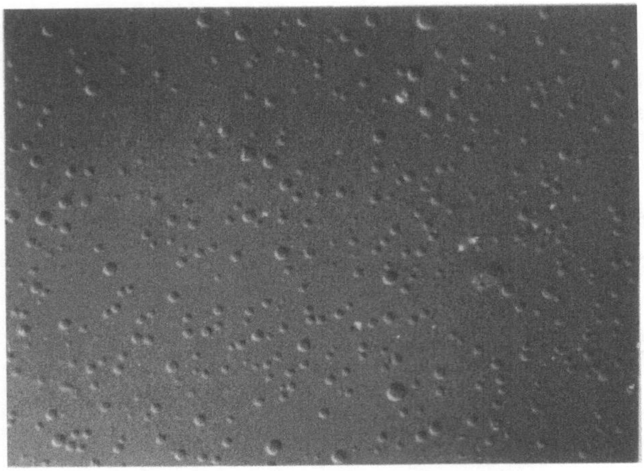

Fig.9 PS/PBr$_x$S(0.8/0.2)-mixture. Structure pattern indicating phase separation by nucleation and growth

/12/. The two components are fully miscible for temperatures above 220°C. Below this critical temperature a miscibility gap exists. Phase separation was induced by a temperature jump from 230°C into the miscibility gap at 200°C. Two different volume fractions were chosen, Φ (PS) = 0.8 and Φ (PS) = 0.5 in the cases of Fig.9 and Fig.10 respectively. The two patterns are qualitatively different and suggest occurrences of different mechanisms of phase separation. Fig.9 shows spherical precipitates in a continuous matrix which indicates nucleation and growth. The pattern Fig.10 exhibits strongly interconnected domains. The structure is symmetric with regard to the two phases. A pattern like this can be considered as indicative for a spinodal mechanism of unmixing.

The early stages of spinodal unmixing can be analysed by theory. The original treatment of Cahn and Hilliard /10/ which was devised for metals and anorganic glasses has been transferred with appropriate modifications to the polymer case /13,14,15/. We give here only a short summary of relevant results.

The concentration fluctuations in the homogeneous phase can be described in a molecular field treatment by the random phase approximation /16/. Measurements of the corresponding structure factor are possible using neutron scattering or X-ray scattering, provided that there exists sufficient contrast between the blend

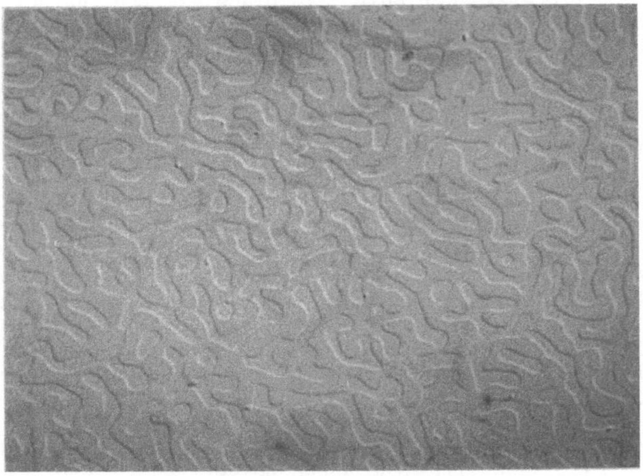

Fig.10 PS/PBr$_x$S(0.5/0.5)-mixture. Structure pattern suggesting phase separation by spinodal decomposition.

components. For symmetric mixtures with identical degrees of polymerization of the two components, $N_A = N_B = N$, the structure factor $I_\chi(s)$ becomes

$$I_\chi^{-1}(s) = \frac{1}{N\Phi(1-\Phi)} \quad f_D(s^2R_G^2) - 2\chi , \tag{18}$$

where Φ gives the volume fraction of component A, χ denotes the Flory-Huggins interaction parameter and R_G is the coil radius. f_D describes the Debye structure factor of noninteracting ideal chains:

$$f_D = \frac{2}{x} (1 - (1-\exp(-x))/x), \; x = R_G^2 s^2 . \tag{19}$$

Experiments are in full agreement with eq.(18) /17,18/ thus proving the validity of the mean field treatment for dense polymer systems. The reason for this accordance seems clear. Each macromolecule contacts a large number of other macromolecules and therefore interacts in fact with the mean field. It is also impossible to build up a shell structure with compositional short-range order as is usually found in blends of low molecular weight compounds. Due to this property theoretical treatment of polymer mixtures can sometimes be simpler than that of alloys of atoms or small molecules.

The interaction parameter χ is temperature dependent. As a consequence the reciprocal structure factor $I_\chi^{-1}(s)$ shows a parallel shift when temperature changes. Fig.11 depicts the general behaviour. Curve I corresponds to a homogeneous mixture. Curve II is that found exactly at the spinodal line, where the intensity diverges, $I(s = 0) \to \infty$. The spinodal point is reached for $\chi = 2/N$. For values $\chi > 2/N$ phase separation occurs. Formally the structure factor eq.(18) can be extended into the two phase-region. It then shows negative values for scattering vectors s below a critical value s_c (curve III).

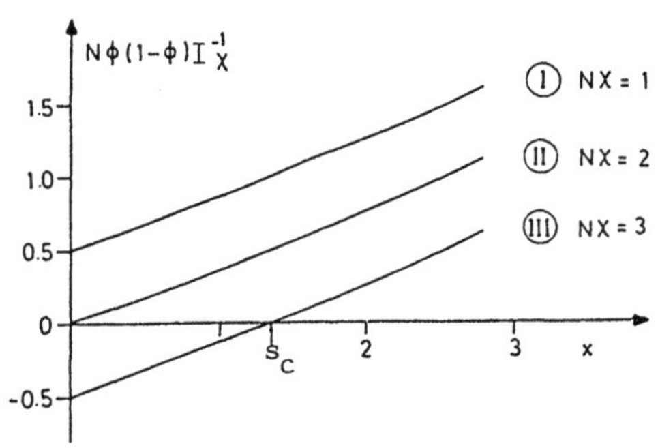

Fig.11 Structure factors of symmetric polymer blend as given by the random phase approximation (eq.(18)): (I) one-phase region (II) spinodal, (III) virtual structure factor in two - phase region.

After a temperature jump the sample relaxes into the new equilibrium structure. T-jumps can be performed either within the one phase-region or into the two phase-region crossing the spinodal. It can be shown theoretically that the equation of motion is the same for both cases, relaxation to a new one-phase structure and spinodal unmixing /15/. Both kinetics can be described by a first order relaxation equation:

$$\frac{dI(s,t)}{dt} = -\tau(s)^{-1}(I(s,t) - I_\chi(s)) \tag{20}$$

For T-jumps within the one phase-region $I_\chi(s)$ denotes the new equilibrium structure factor; in the case of spinodal unmixing it corresponds to a "virtual" structure factor, both being given by eq.(18). $\tau(s)$ describes the relaxation time in dependence on the scattering vector s. Binder /13/ has given the following expression:

$$\tau^{-1}(s) = \frac{6}{\tau_R} R^2 s^2 (1 - 2N\Phi(1-\Phi) \cdot \chi \cdot f_D(s)) \ . \tag{21}$$

Here τ_R denotes the time required by a polymer molecule to diffuse over a length comparable to its own diameter. Eq.(21) shows that for temperatures in the one phase-region ($\chi < 1/(2N\Phi(1-\Phi))$) relaxation rates $\tau^{-1}(s)$ increase continuously with s. For temperatures in the two phase-region ($\chi > 1/(2N\Phi(1-\Phi))$) $\tau^{-1}(s)$ takes on negative values for scattering vectors below a critical value s_c. This means that fluctuations grow rather than decay for wavelengths $\lambda > 1/s_c$, which is indicative of an intrinsically unstable situation. A sample quenched to this temperature immediately begins to unmix. During the initial stages structure formation is governed by that concentration wave which has the largest growth rate. The corresponding value s_{max} can be directly derived from eq.(21). Upon approaching the spinodal line s_{max} decreases, i.e. the length scale of the two - phase pattern increases.

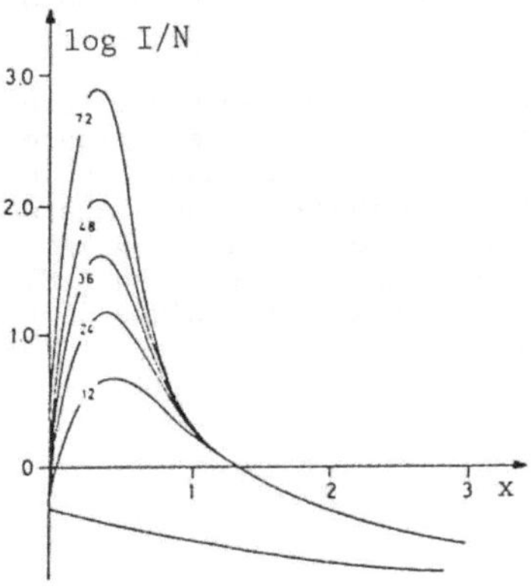

Fig.12 Time dependence of structure factor after a quench from the one - phase region into the two - phase region. Model calculation according to eqs.(20),(21). Times are given in units $6t/\tau_R$.

Fig.12 presents the result of a model calculation based on eqs.(18) - (21). It shows the evolution of the structure factor after a quench from the one - phase region ($N\chi = 1.0$) into the two-phase region ($N\chi = 2.5$). The peak which appears grows in amplitude exponentially with time. Curves simulate the spinodal unmixing during its initial stages. For a sample this is followed by coarsening processes (Ostwald-ripening driven by surface tensions).

Experimental results are in qualitative agreement with the calculations. Studies have been performed either with time - dependent light scattering experiments at temperatures near to the critical temperature T_c /18/ or by small angle X-ray scattering experiments /19/ conducted near to the glass transition T_g. Light scattering and X-ray scattering probe the relaxation kinetics for different s-ranges. Near to T_c and T_g structure evolution becomes sufficiently slow to enable a resolution in a time-dependent experiment. Generally measurements yield the relaxation time as a function of s and the temperature. Fig.13 gives an example. It deals again with the properties of blends of polystyrene and partially brominated polystyrene. Phase separation was induced at temperatures near to the glass transition. The wavelength of the dominating fluctuations for this sample is located at about 50 nm. This microscopic fluctuation structure provides the starting point for a structure development which leads in a later stage to the kind of pattern shown in Fig.10.

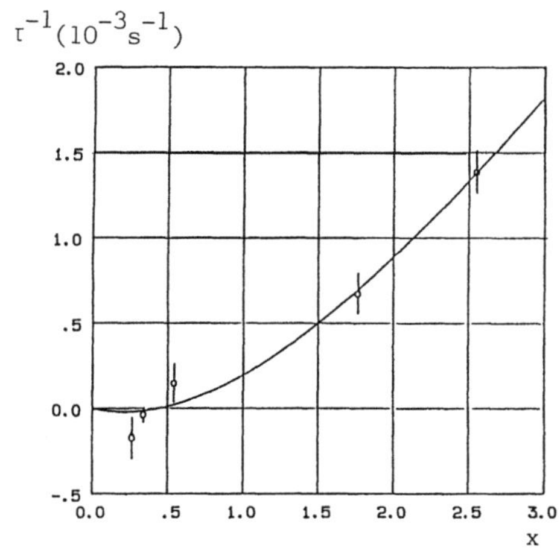

Fig.13 PS/PBr$_x$S(0.5/0.5)-mixture. Relaxation rates for spinodal unmixing induced by quenching a homogeneous mixture to 132°C.

Surveying the presented results the feature in common is evident. In all cases discussed here structure formation is controlled by the same principle: the structure which develops is dominated by the fastest growing mode. This basic selection rule is operative in both modes of polymer crystallization, and also for the spinodal mode of unmixing. It was the intention of this contribution to show that structure formation in polymer systems in many cases can be understood and analysed on the basis of this fundamental criterium.

Acknowledgement

The author thanks Professor E.W. Fischer, Max-Planck-Institut für Polymerforschung (Mainz), for the permission to include in this contribution results of a forthcoming paper (Figs.4-7, Ref.9).

1. B. Wunderlich: Macromolecular Physics, Vol.1, 2 (Academic Press, New York 1973, 1976)
2. J. Petermann, J.M. Schultz, R.M. Gohil, R.W. Hendricks, J.S. Lin: J. Polymer Sci., Physics Ed. 20, 523 (1982)
   R. Günther: Dissertation (Mainz 1981)
3. L.P. McMaster: Polym. Prepr. (Am. Chem. Soc., Div. Polym. Chem.) 15, 254 (1974)
4. T. Nishi, T. Wang, T. Kwai: Macromolecules 8, 227 (1975)
5. R. Eppe, E.W. Fischer, H.A. Stuart: J. Polym. Sci. 34, 721 (1959)
6. J.D. Hoffmann, G.T. Davis, J.I. Lauritzen: In Treaties on Solid State Chemistry, ed. by N.B. Hannay, Vol.3 (Plenum Press, New York, 1976)
7. D.M. Sadler: Polymer 24, 1401 (1983)
8. A.A. Chernov: Modern Crystallography III. Crystal growth. (Springer Verlag, Berlin, 1984)
9. R. Günther, E.W. Fischer: to be submitted (1987)
10. J.W. Cahn: Trans. Metall. Soc. AIME 242, 166 (1968)
    J.E. Hilliard: In Phase Transformations, ed. by H.L. Anderson (American Society for Metals, Metals Park, OH, 1970)
11. E.W. Fischer, H. Goddar: J. Polymer Sci. 16, 4405 (1969)
12. G.R. Strobl, J.T. Bendler, R.P. Kambour, A.R. Shultz: Macromolecules 19, 2683 (1986)
13. P.G. de Gennes: J. Chem. Phys. 72, 4756 (1980)
14. K. Binder: J. Chem. Phys. 79, 6387 (1983)
15. G.R. Strobl, Macromolecules 18, 563 (1985)
16. P.G. de Gennes: Scaling Concepts in Polymer Physics, Ch. IV (Cornell University Press, Ithaka, N.Y. 1979)
17. M. Shibayama, H. Yang, R.S. Stein, C. Han: Macromolecules 18, 2179 (1985)
18. T. Izumitani, T. Hashimoto: J. Chem. Phys. 33, 3694 (1985)
19. H. Meier, G.R. Strobl: Macromolecules, to be published (1987)

Part VII

Selforganization in Biological Systems

A General Approach to Complex Systems in Bioholonics

H. Shimizu

Faculty of Pharmaceutical Sciences, University of Tokyo,
Hongo, Bunkyo-ku, Tokyo, 113 Japan

1. Introduction

Thanks to the slaving principle postulated by Professor Haken and his collaborators, we are able to understand how order is self-organized in a uniform system, such as a laser, composed of synergetic elements. In fact, the slaving principle gives us a generalized way of studying self-organization in physical, chemical and biological systems. It is a great contribution to the progress of science [1], [2].

Now, synergetics has found one of its major applications in the study of complex systems, which provides us with complexity beyond our cognitive ability [3]. Haken proposed an operational approach based on information compression relevant to a system's character. However, we have not found a generalized scheme that can be applied to understand a complex system. Haken told me that a global point of view is necessary to understand the essence of a complex system and that, for this approach, the Western way of thinking had limitations. He pointed out the necessity of amalgamating the Western and Eastern ways of thinking to establish a new science. The amalgamation of Western and Eastern civilizations is an important topic, and according to Haken the study of complex systems gives a concrete objective to this important subject. The following is a proposition from our Eastern side in answer to the problems discussed in this dialogue; or, more exactly, it is only the beginning of our proposition. It is my pleasure and happiness to dedicate it to him for his 60th birthday.

The following discussion will explore a new scientific approach to complex systems. Our efforts will be centered on finding a description of complexity, in particular of biological complexity. This problem is, of course, related to scientific observation of complex systems. We will approach this problem by studying the general properties of information processing in a "cognitive computer", a kind of autonomic parallel processor, which is capable of autonomically compressing the irrelevant parts of an input signal out of a complex system to discover order, *cognitive order*, in the signal. It also shows us how a general rule is found to describe the characteristic features of cognitive order. In the following discussion we will name such a rule *semantic order*.

2. Biological Complexity

A biological system is a complex system composed of many different sorts of elements. Animals which are visible to the naked eye are generally composed of a number of cells with different characters. The diversity in the character of elements plays a crucial role in the self-organization of biological functions. The complexity of the biological system arises from the fact that not only do such elements have different characters but also their characters are modified to produce a new (but often instantaneous) function allowing them to survive in a changing environment. This beccmes clear if we think of a human organization like a company under conditions of competition. The members change their "characters" to create a new (global) function in the company for survival in severe competition. It is said that the human brain has fifteen billion neurons, each of which is connected with 10^3 - 10^4 other neurons in a particular manner. About 200 types of neurons are present in the cerebrum, and the total length of axons connecting neurons in 1 mm^3 is about 15 km (see e.g. [4]). We are not able to discuss the function of the brain while ignoring such particularities in the neurons. Our problem is to study the general principle of self-organization in such a complex system. How is the presence of elements with different and particular properties compatible with the self-creation of order in the system? We may call such a problem an "order-through-diversity" (OTD) problem.

We soon realize the limitations of our cognitive ability when we try to tackle this kind of problem. Unfortunately, the human brain may not be able to process such a large amount of information at the same time. Probably, the limit of a tractable amount of information is not so large. For instance, in the case of statistical physics, we are able to solve the equations of motion for hundreds of interacting particles with a computer. However, it is almost impossible to obtain meaningful information simply by tracing the particles one at a time. Even for ten particles there is cognitive confusion or disorder. Hence, for the study of such a system we must reduce the amount of information. In the case of statistical physics, an enormous reduction in the amount of information is obtained by taking statistical averages of microscopic physical quantities, which enable us to describe the system in terms of thermodynamic quantities, and to find the dynamical laws of the system using such quantities. The same spirit is found in the derivation of the slaving principle by Haken and his collaborators [1]. It should be stressed that the above situation is caused by a limitation of our cognitive ability to find order in the solutions found with a computer, and not from an inability to obtain the solutions.

The most important aspect of the study of complex system is therefore the method of reducing the amount of information. Clearly, statistical averaging is not applicable to many-particle systems composed of elements with different particularities and functions, as it is impossible to take the average of different kinds of functions. Moreover, a new method must satisfy the following two requirements: first, cognitive order must be obtained by information compression; and second, the information compression must be objectively defined. Now, we are up against a new kind of difficult problem, i.e., to greatly reduce the amount of information without utilizing the method of statistical averages.

3. Hidden Order

A biological system self-organizes various kinds of order: structural orders, temporal orders and functional orders. These orders may be briefly named biological orders. There is an essential difference in self-organization for nonbiological and biological systems. To remain alive it is necessary for a biological system not only to describe itself but also to describe the external world, which may change from time to time. However, this is not the case for a laser, for instance. To have coherent dynamics of laser atoms, it is required only that the laser describes itself. In other words, a laser has a closed information system while a biological system has an open information system. A biological system has the property of having a number of different orders, cognitive orders, depending on input signals. (A self-organized order is transmitted to the "self" of the system where it is "interpreted" in terms of stored semantic information.) [5]

As an open information system, a biological system has many unstable internal states which are substantially degenerate but become mutually distinguishable in the final stage expressing cognitive orders. The presence of many different and particular kinds of elements in a biological system affords many distinguishable internal states according to the particular connections of elements. Generally, these internal states are selectable, or accessible. An input signal to a biological system selects an initial state that normally leads the system to a particular final state, a stable or quasi-stable steady state. Thus, the biological system has a definite number of selectable orders which rather sharply depend on the particularity of relations among the elements. Generally, the macroscopic properties of a biological system are not well separated from microscopic properties.

Here, we have a problem: if we fix our attention at the macroscopic level, we are not able to see microscopic details of the particular relations among elements. This is not a suitable way of studying a biologically complex system. On the other hand, if we fix our eyes on the microscopic level, macroscopic order will be embedded in microscopic details. We have to find the macroscopic order hidden in microscopic details by selectively suppressing microscopic details which are irrelevant to the order to be studied. (The method of information compression is rather subjectively determined.) Such a process will be called semantic compression of information. In any case, in contrast to uniform systems composed of identical elements, we do not generally find any "objective methods" for information compression, such as calculating the statistical average of the behaviors of elements. We are therefore interested in seeing how the semantic compression which is particular to a system is found. This is one of our major tasks.

4. Cognition of Order in "Chaos"

We are able to see examples of such semantic compression in the study of chaos. The important characteristics of chaos are not obtained simply by taking a statistical average of orbits in a phase space. The Lorenz plot is one method of semantic compression which clearly picks up the characteristic order from a chaos with orbital instability [6]. There is no proof that the Lorenz plot is the only available method and that all the characteristics of the chaos are obtained by this method. Similarly, a large amount of information on the strange attractor is embedded in the Poincare map, where no clear description of time sequence is given [7].

It should be noted that such an order is found with partial or complete compression of information on time. This reflects the fact that because of the complex behavior of the system caused by orbital instability, we are not able to trace the motion of the system based on mechanical causality and that the description of a system based on the concept of "orbital motion" becomes irrelevant if we want to continuously observe the system for a span of time which is much longer than the mean period of the orbital motion. Therefore, we are not able to find order in the behavior of the system if we want to describe it in terms of Newtonian time. If we use such a description, we are not able to compress irrelevant information caused by the complexity of orbital motion. Hence, to have a cognitive order, the information on time has to be more or less compressed. The system will be described by means of a coordinate system where time is not explicitly given as one of the coordinate axes. These examples of "chaos" indicate that cognitive order will be found in complex system if we discard a causal description and introduce a new kind of time, which will be named *semantic time*. What is important are events which can be related to each other.

Biological complexity will result in a similar but more complicated problem on the cognition of order. Although efforts have been made in the history of molecular biology to describe biological systems in terms of mechanical causality, biological complexity will prevent it. We have to find a relevant description of biological systems in which information due to irrelevant complexity is compressed. In contrast to the case of "chaos" with few degrees of freedom, biological complexity is caused by the presence of a huge number of variables which lead to many kinds of frustrations. Alternation of our approach from a causal to a relational description inevitably introduces compression of information on time and understanding of the object by means of a kind of "spatial pattern". Therefore, our cognitive computer self-organizes an ordered pattern from an input signal. Only microscopic relations contributing to the formation of the pattern are left as relevant relations; the others are suppressed as irrelevant. A general prescription for suppression of irrelevant information has not been explored. We do not even know how relevance and irrelevance are defined in biological science. The present paper aims to sketch a prescription for semantic compression based on the amalgamation of the Eastern and Western ways of thinking.

5. Science as a Universal Language

Scientific research may be regarded as the creation of a universal language for describing nature. As was postulated by SAUSSURE [8], any language is a way of describing something in terms of categorized meanings of the words. And the categorization of meanings is determined by the culture of the area where the language has been spoken. One of the important characteristics of Western civilization is undoubtedly its universal character. Clearly, such a universal character of culture will be required for the creation of a science which "objectively" describes nature.

Modern science is based on the way of thinking in Western countries, where languages are spoken in a much more logical and reductive way than in Oriental countries. When I was in California many years ago, I was impressed to recognize that the way of thinking in daily conversations such as, for example, at a party can be extended naturally to modern science and technology, at least some important part of them. However, most Japanese, for example, cannot keep up such logical and reductive thinking in conversations taking place in a relaxed atmosphere, because effort is needed for them to follow the logical stream of conversation. Recently I heard from Professor Feigenbaum, of Stanford University, who is famous as the creator of expert systems of artificial intelligence, that the information processing used in the expert systems was quite natural for him because it was just his way of thinking in daily life (private communication).

It is not easy to be creative if one must think about something new which needs imagination and deep meditation using an unfamiliar way of thinking. This will also be true in the case of technology. Japanese technology made great progress when Japanese workers discovered how to apply their way of thinking to the modern technology which had been the product of Western modern culture.

6. Eastern and Western Way of Thinking

The greatest difference between Eastern and Western countries in ways of thinking might not be the difference between global and analytical approaches, which seems to the author superficial, since we are able to find a number of examples against such a classification. Perhaps a more fundamental cause lies behind the general tendencies. The Western way of thinking is probably based on serial information processing, while the Eastern way is based on parallel information processing. This was also pointed out by Haken (private communication). As a result of this, logical inference and reduction may become important characteristics in Western culture and it would be natural that events are analyzed on the basis of causality. Such a way of thinking would create mechanical descriptions of events. In other words, the Western countries have compressed information in order to clarify the causal relations in events. On the other hand, the main object of parallel information processing is to find relationships between different pieces of information. Therefore, the Eastern countries have been interested in finding "hidden" relationship between different events. (The author personally thinks that Japanese is a language which is suitable for expressing relations between different events.) In parallel processing the trace of an event based on causality soon becomes fuzzy, and it is therefore useless as an approach to finding cognitive order. As a result, synchronicity is a more useful concept than causality to describe a group of events. It is interesting to think that a Poincare map of a strange attractor of a type of "chaos" provides a description of a complex event with a kind of synchronicity.

Both kinds of information processing, the serial one and the parallel one, possess universality to describe events. But they are by no means complete and are therefore complementary in their descriptions. The serial approach will lead to much clearer results than the parallel one if the object can be decomposed into elements which are not so deeply involved with each other. On the other hand, the parallel approach will be more suitable for describing a complex system than the serial one. Therefore, it will be worthwhile to explore a new way of describing complex systems based on the Eastern way of thinking.

7. Cognitive Computer

We hope to develop a generalized approach to discover the order existing in signals transmitted out of a complex system. As was mentioned previously, this problem is equivalent to discovering the mechanism of an autonomic cognitive computer which is able to compress irrelevant information to generate cognitive order. Generally speaking, such a cognitive computer has at least three hierarchical levels of information. These three levels differ in the order of abstraction (compression). As schematically shown in Fig. 1, in the lowest level of the hierarchy, a signal transmitted from a complex system is given in microscopic detail.

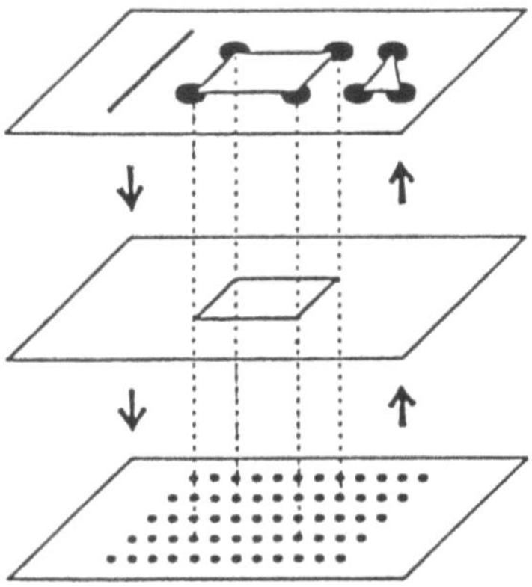

Fig. 1 A schematic representation of information hierarchy for the input of a square lattice pattern; it denotes elementary, organized and semantic information from the bottom.

(The input signal is divided into elementary signals, elementary information, at the microscopic level.) At the middle level self-organization of relevant elementary signals into *organized information* occurs accompanied by compression of the irrelevant part of the elementary signals to make the organized information consistent with one of the items of semantic information, i.e. semantic order, stored at the highest level. The organized information describes the cognitive order of the input signal. Generally the organized information is further compressed into more abstract information to extract the characteristic features of the cognitive order, as if to create a law (a semantic order) determining a local rule (a cognitive order). Then such abstract information is solidified in conjunction with the semantic order to which it refers.

Exactly speaking, the decomposition of the input signal into elementary signals may be influenced by the semantic information in such a way as to easily produce organized information which is relevant to the semantic information. (The so-called "cocktail-party effect", referring to our ability to have a conversation amid the noisy sounds of a cocktail party, would be responsible for choosing elementary signals that are convenient for forming organized information in a way relevant to a particular semantic order). In the self-organization of organized information, a kind of semantic separation occurs dividing the elementary signals into relevant and irrelevant parts, which are organized and compressed into "figure" and "background". If all the elementary signals are connected, we will not be able to find any order in the input signal. The creation of order therefore requires the suppression of irrelevant signals. When the complex system varies in time, the input information will depend on time. Then we have a time series of input information, which may also be decomposed into an assembly of elementary signals along the time axis. (We have still left an important question unanswered: what is cognitive order?)

Let us imagine that we have a cognitive computer which spontaneously creates organized information giving cognitive order according to the above scheme. We assume that organized and semantic information is created by means of parallel information processing. The necessary conditions for such a cognitive computer will be as follows. First, it must have many unit receivers which accept elementary signals one by one. Then the above process occurs spontaneously. We may assume that input information from a complex system is also complex in the sense that complexity is present in the distribution of elementary signals in the computer's information space. Clearly, such a cognitive computer must itself be a complex system. If the observed system is a uniform system like a gas laser, we are able to greatly reduce the number of elementary signals by taking a local average of the elementary signals, and are able to arrive at a cognitive order based on such averaged elementary signals. However, in the case of a complex system composed of elements with particular properties and functions, we are not able to take an average in the above sense, meaning that the number of elementary signals cannot be reduced as in the uniform system. Thus, if we want to "see" a complex system, a cognitive computer must have sufficient complexity that the elementary signals reflecting the complexity of the observed system can be encoded. Generally speaking, we need a complex system whose complexity exceeds that of the observed system.

A well-known difficulty, the so-called hermeneutic circle, prevents a cognitive computer from relating elementary signals to produce cognitive order based on reductive information processing. To understand this difficulty, let us tentatively assume that our cognitive computer has no semantic information, and that each elementary signal is surrounded by many other elementary signals. Furthermore, we assume that our computer has the ability to connect relevant neighbors and to suppress irrelevant ones. A relevant relation is defined as a relation that is consistent with cognitive order that is an organized assembly of such relevant neighbors. Now, relevancy in the relations between elementary signals is determined only when the cognitive order is given. However, the cognitive order can be given only when all the relevant relations between the elementary signals are determined. Evidently, this difficulty cannot be solved just by logical inference. Parallel information processing at least is needed.

The cognitive computer should have the ability to avoid such a hermeneutic circle. Parallel processing is necessary, but it is not enough. Owing to its self-organizing ability, the cognitive computer might find relevant relations based on the slaving principle. However, the slaving principle, alone does not enable us to separate figure from background because no guiding principle is given for compressing details of the input information. As a result, organized information is not free from irrelevant details in the input information, and such organized information is still not tractable because of its complexity. It might be generally difficult to obtain a law or principle describing the essential features of similar complex systems. In addition, because of its complexity, the system may self-organize a number of local orders in which the slaving principle is only partially applicable. However, no unique relations will be found among such local orders. Moreover, the introduction of relevant connections among such local orders might change local relevance by modifying the local slaving principle.

In any case, a self-organizing complex system might escape from a hermeneutic circle based on competitive growth of orders, but it is still not possible to compress some of the elementary signals to get cognitive order in a simplified form. If, for any reason, a semantic order, i.e. an abstract and characteristic description of cognitive order (to be discovered), has already been present in the computer, it will be utilized as a guiding principle for semantic separation with information compression, and the cognitive computer will escape being trapped in a hermeneutic circle generating cognitive order.

Goedel's incompleteness theorem [9] indicates that such an abstract and general description cannot be obtained only from an input signal by means of logical inference based on reductive thinking. The crucial problem is how such a description can be present before the formation of the organized information. The cognitive system should have the ability of abduction to generate such semantic information according to the principle to be discussed below, where information is simultaneously obtained: the creation of organized information and semantic information being coupled with a feedback loop. What is to be studied is information processing which makes such processes possible. The cognitive computer must be a self-ascending system.

We would like to note that the concepts of relevance and irrelevance can be defined in elementary signals as well as in organized information only when semantic information has been specified. Relevance between the organized information and semantic information indicates that they are consistent in (physical or biological) meaning. Therefore, the meaning of the semantic information can be transferred to the organized information. Or, more exactly, the meaning of the organized information is included in the meaning of the semantic information. The relevancy between a group of elementary signals and the organized information means that the latter can be organized by integrating the former.

8. Holons as Semantic Correlators of Elementary Signals

In a cognitive computer elementary signals are related to each other to generate organized information. In this process, relevant correlations between elementary signals are discovered also with the neglect of some correlations. In other words, semantic correlations are found in the assembly of elementary signals. We shall call autonomic unit processors for elementary signals, or semantic correlators of elementary signals, "holons". The holons are local-rule generators. As will be discussed later, holons are a kind of synergetic element with selectable internal states, internal modes, with internal dynamics. The elementary information is processed by the internal dynamics, as will be discussed below. (As a typical example of a holon, one may imagine a columnar assembly of identical self-sustained oscillators oscillating with phases particular to the internal modes.)

In a cognitive computer creating cognitive order as a spatial pattern, there is an internal information space in which holons are placed. For the sake of simplicity we may assume that the holons are fixed in the information space with a certain regularity. A holon has positive (excitatory) interactions or connections with some of the neighboring holons and negative (inhibitory) ones with other neighbors. The positive or negative nature of the interactions is determined by the following factors. The direction in which a holon has positive interactions specifically depends on its internal mode, while it has negative interactions in the other directions. Holons interacting with positive interactions have the same internal mode; their dynamics are coherent, and define them as having (dynamical) relations or connections. Holons interacting in such a coherent state mutually excite the partner's internal dynamics. When a holon alters its internal mode, it changes the directions of positive and negative interaction.

We have not yet decided whether or not all the internal modes are equally accessible. We may introduce the particularity of a holon in the rule of the selection of an internal mode, in the number of internal modes, or in its position in the information space. The elementary meaning of an elementary signal is encoded in the internal dynamics (e.g. in the phase and amplitude of a self-sustained oscillator) of a holon. Therefore, it depends on the selection of an internal mode and is determined only when dynamic relations are completely formed in the assembly of the holons.

From the above definition, it seems clear that the organized information is expressed in the cognitive computer as a kind of dissipative structure or dynamic order with coherent dynamics self-organized in the internal states of the holons. On the other hand, those holons which are inhibited in their internal dynamics compress coded elementary signals. When figure and background are determined by holons, the meaning -- the elementary meaning -- of an elementary signal encoded to a holon is determined. Thus, we may consider each different internal mode of a holon to encode the different elementary meanings to be given to an elementary signal.

9. Holons as Generators of Elementary Semantic Information

In thinking about the theory of the semantic dynamics of information, three important objectives are:

1) to define the properties of a holon and its dynamic laws;

2) to discover dynamic methods for expressing relevant relations and irrelevant relations between holons;

3) to discover ways of expressing semantic consistency between different pieces of information.

How can we achieve these? Li Po, a famous Chinese poet, wrote

Heaven and Earth are wayside inns for all things
Day and Night, an everpassing wayfarer.

The late Dr. Hideki Yukawa tried to establish a unified understanding of an elementary particle by advocating his nonlocal field theory [10]. Citing the above poem, he explained his theory as follows:

> "Recently, I have come to think that the role of physical space is to provide a seating place for matter. Space has many different seats to offer, and the character of an elementary particle is determined by the seat it takes. Particles do not have names or characteristics that are intrinsic to them, by which they might be identified. Time therefore could be understood as something that represents the changing process in the seating chart. *Heaven and Earth* may be understood as the physical space, and all things as the elementary particles. *Wayfarers* who stay at the inn are usually called not by their own names, but by their room numbers. In this sense, the rooms occupied by the particles do not determine their nature, and their state of motion. If we read *Day and Night* as time, the *wayfarer* may be read as the particle which arrives at and departs from the inn."

Dr. Yukawa's idea resembles very closely the idea of the holon that we have been advocating. The elementary signals to be put in each holon have no meaning and no name in themselves. Elementary meaning is determined by which inner modes the elementary signal will excite -- that is to say by which seats the elementary signal will occupy inside the holon. Changes caused by the way the seats are determined add temporal changes in cognized information. Generally speaking, the chronological passage that animals feel appears to be made by temporal changes of cognitive order -- that is, it is self-organized within themselves. Every elementary signal and piece of information the animal creates are in the hierarchical relationship of "elements" and the "whole", not indicating an absolute level.

It is interesting to note that cortical sensory fields, various areas in the visual field [11] and auditory field [12], are composed of a rather regular assembly of elements, so-called hypercolumns composed of a number of columns, neural network units. According to our interpretation, the columns correspond to the internal modes of a holon, or a hypercolumn. In other words, in each column of a hypercolumn elementary meaning to be given to an input elementary signal is coded. And the meaning is determined owing to interactions of hypercolumns under a strategy of semantic information. Input elementary signals which find appropriate positions in hypercolumns determine their meanings.

10. Holonic Loop, a Key Mechanism of Semantic Information Dynamics

If words were regarded as elementary signals, then sentences made by words could be organized information. Generally speaking, therefore, an elementary signal needs to have its own individuality. A holon, too, which receives an elementary signal needs to have its own individuality. Accordingly, what we face is the dynamic law of order that is self-organized among these distinctly individual elementary signals. The following fundamental thought of Buddhism is a good guide to the laws of self-organization under such a system:

When the dynamics of all living things
are connected with each relevant relation
there will be a circular relation
and the great order (Buddha) will be
self-organized and incorporate all.
And for all living creatures
within this relevant relation
a great order will be immanent.

In our view, the presence of the relevant relations between the dynamics of holons bearing elementary signals determines the existence of an object, a complex system, and one that is immanent in that these holons become holonic information, the holistic information shared by the holons.

Relevant relations are defined as relations that give a "whole" to the "elements" due to a steady circulation of information named a **holonic loop.** Semantic consistency between the information that is self-organized out of elementary signals and the information that exists forming its own inner world means that a relationship with semantic relevance is established between holons that bear both of these, and therefore that the dynamics of both relations are connected by a holonic loop.

When this condition is viewed from the level of elementary signals, correspondingly, there are two kinds of local holonic loops that are connected by a still larger holonic loop (or hyperloop). Based on such an observation, the self-organization of information in a cognitive computer can be dealt with strictly by scientific methods, which is regarded as a problem where a kind of temporal order is autonomically tied to the collective dynamics of holons coding elementary signals. We call this mechanism of self-organizing dynamical coherence a "synchronization" or "entrainment" in the broadest sense of the word.

11. Holons, Units of Autonomic Parallel Processors

There are several necessary conditions for a holon as a dynamical unit of a parallel processor [13]. First, it must possess coding ability; i.e., it must have many internal modes, which can be utilized as the semantic states coding the elementary meanings -- a loaded elementary signal. All the semantic states must be accessible. Second, excitability is crucial; as a switching mechanism, these semantic states must be weakly and uniformly excited by a suitable stimulator which can be utilized as a carrier of an input elementary signal. At this stage of uniform excitation the meaning of the elementary signal is fuzzy, and we refer to this as the initial fuzzy state. Third, the system must possess the processing ability to select the relevant internal modes: it can select the relevant one(s) out of the uniformly excited internal modes encoding elementary meanings as a selective increase in excitation according to the operational information generated in interactions between neighboring holons. Tentative connections of elementary signals introduced to the group of the holons are given by means of related excitations of some of holons. However, this mechanism leads to still-fuzzy local connections with many possibilities in the formation of images. Fourth, there is the controllability by operational information brought by feedback from evoked memory holons coding semantic information: the feedback of semantic information from the memory holons as holonic information gives holons operational information for selection of their elementary meanings in order to self-organize organized information consistently with the semantic information. The local relations due to the direct interactions among the holons are more or less modified by this feedback of semantic information, being brought about by excitation, suppression and induction of the excitation of the relevant internal modes of the holons. Finally, there must be coding stability: holons in relevant relations maintain selected internal modes stably against noises and fluctuations, at least during the time necessary for perception.

In the primary visual cortex, a hypercolumn might be regarded as a holon, and the simple cells -- more exactly orientation columns -- belong to the hypercolumn as the internal modes of the holon. Therefore, each of the elementary meanings expresses the meaning of the received elementary signal as a line element with a specific orientation. A holon selects its connected partners in a manner depending on its internal mode: when the relative activation of the internal modes of a holon changes, the holon alters its partners. The selection of an internal mode is influenced by operational information by two different routes, i.e., by local patterns self-organized in the primary visual cortex, an autonomic parallel processor, and by feedback from the semantic information, a general and abstract description of organized information, encoded in the memory holons. There are two kinds of operational information, positive and negative. Positive operational information selectively activates a specific internal mode while the negative one suppresses others.

12. Transformation to Semantic Information

Based on the preceding discussion, we will generalize the creative processing of semantic information as follows [13].

(1) Any input signal to a cognitive computer is received by a number of unit receivers, and the signal is decomposed into small elementary signal fragments. The spatial distribution of the elementary signals expresses the characteristic feature of the input signal in a syntactic configuration space. The decomposition of the input signal results in *syntactic fuzziness* due to an increase in syntactic information entropy, which means that there are many ways to relate the elementary signals.

(2) To obtain organized information, specifically related to the semantic information as a figure, from the ensemble of the elementary signals which are noth-

ing but syntactic signs, as much syntactic fuzziness has to be removed as possible, and the result must be related to the semantic information with the holonic loop. This step involves the transformation of input information into a kind of semantic information, i.e., a figure, pattern. The information organizer, a parallel processor composed of holons, has not only a syntactic space for receiving elementary signals at appropriate positions, but also a semantic space to give a relevant elementary meaning to the elementary signals. These two kinds of information space are mutually orthogonalized, as will be shown below. The organized information is expressed with a self-organized pattern in the whole information space composed of the syntactic and semantic parts.

(3) Holons have many internal modes and also have sufficient autonomy to control their internal dynamic activities according to operational information generated in their ensemble as holonic information. Owing to this mechanism internal modes become selectable. The operational information is information brought to each holon through a holonic loop as the feedback of a self-organizing pattern. The spatial positions of holons in the information organizer provide a syntactic space, while the internal modes of the holons provide a semantic space. The holons are capable of cooperatively processing the elementary information carried by the elementary signals. The elementary meaning of an elementary signal has already been prepared in the holon, having been encoded one by one in its internal modes. (A "hypercolumn" is a holon composed of many orientational columns corresponding to the internal modes.)

(4) Soon after the introduction of elementary signals to the syntactic space, holons at relevant positions in the information organizer receive elementary signals one by one. An elementary signal changes a resting holon to a weakly active one: the resting internal modes of the holons receiving the elementary signals are brought to a fuzzy state in which all the internal modes are weakly and equally activated. Such fuzziness in the semantic space will be called *semantic fuzziness*. To obtain organized information, the information entropy that arises because of semantic fuzziness has to be suppressed by introducing competition in forming a holonic loop with semantic information. Selective activations of some of the internal modes correspond to the selection of elementary meanings encoded in those modes. The chief factor influencing the competition is operational information. There are two kinds of operational information, local and global. Corresponding to these two are two kinds of rule for the selection of relevant elementary meanings.

(5) Local selections work to compress the semantic fuzziness as much as possible according to local syntactic rules among the holons. This compression is caused by selections of internal modes, i.e., by making dynamic connections of internal modes among relevant holons according to the local syntactic rules. To give the local syntactic rules, anisotropic interactions of holons that depend on their internal modes are necessary. The underlying principle is as follows: first, a holon selecting a particular elementary meaning connects with other holons selecting the same kind of elementary meaning in the neighborhood. The connected holons mutually activate the internal modes encoding the common elementary meanings and suppress the other modes. As the compression proceeds, the activities of the internal modes become stronger. When a holon alters its internal activity for any reason, it changes partners. In this sense, the holons are local-rule generators. Thus, local connections between holons compress semantic fuzziness to a certain extent. Note that, in order to have configurational consistency with the input signal, local compression leaves the original configuration of the elementary information almost unchanged. On the other hand, the connections between active holons are not always uniquely determined.

(6) In addition to this local compression, the system has compression due to feedback, i.e., semantic feedback, from semantic orders. (For the sake of understanding, we will tentatively assume that relevant semantic orders have already been stored in memory holons.) From highly excited internal modes of connected holons, elementary meanings are transmitted to the memory holons where semantic orders are stored. Relevant memory holons are evoked according to upward information flow from the information organizer. There might be competition among memory holons when they are aroused. Operational information is transmitted from the aroused memory holons as downward information flow to the organizer. The operational information acts as positive feedback for the internal modes coding the elementary meanings that are consistent with the meaning of the semantic items, and as negative feedback to the other kinds of internal modes. Some of the active holons that are irrelevant to the semantic order are more or less inactivated by the feedback of semantic information. Thus, the original configuration of elementary information is transformed into a kind of semantic information by providing relevant connections to a group of elementary meanings encoded by the connected holons. Elementary meanings on holons in disconnected states are neglected. Therefore, the organized information is consistent with the evoked semantic items as well as with the input signal self-organized in the information organizer. As the holonic loop composed of both upward and downward information flow is recursively applied, semantic compression proceeds successively in both the organizer and memory toward some limit of convergence, until finally those elementary meanings that are consistent with the meaning of the referred semantic item remain owing to successive activation.

(7) As a result of the semantic compression associated with the formation of connections and compression of input elementary signals in the syntactic space, organized information is created. The figure indicating the order of the organized information is determined by the way elementary signals are connected to produce semantic compression. The figure is given by the "view" of the self-organized pattern projected onto the syntactic subspace, while the holistic meaning of the figure is given by the pattern onto the semantic subspace. Not only is the holistic meaning of the figure determined by the elementary meanings of related elementary signals; the elementary meanings of the elementary signals are also specifically determined by the holistic meaning of the organized information. Thus, a holonic loop relates the meanings of elementary signals to the holistic meaning of the figure.

(8) When no relevant semantic order is found in memory, one has the following situation. Some items are weakly and simultaneously aroused but none of them become predominant. If one of them is selected by means of external constraints, a figure is self-organized from only a small part of the whole input elementary signals, and the major part has to be compressed. Better orders could be tentatively generated as "working hypotheses" by suitably amalgamating the weakly aroused semantic orders. Generally, new parts would be incorporated into the orders to form better hypotheses. We may assume such hypotheses are created by a trial-and-error method. What is important is the simplicity of the semantic order to be generated -- the simpler the better, as a general rule. If "order" is too complicated, it is no longer an order: it loses general applicability. On the other hand, if the order is too simple, it cannot explain the complexity of the input signal.

13. The Creative Dynamics of Cognitive Order

In the following discussion, a cognitive system, *holovision* [13] [14], will be utilized as a concrete example. Space does not permit us to explain it in detail, however.

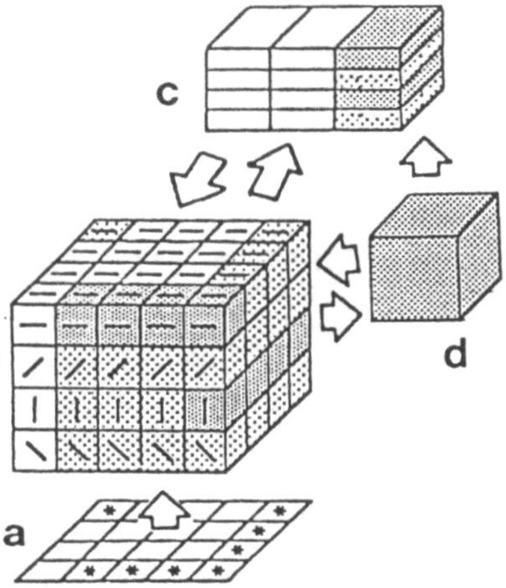

Fig. 2 The structure of Holovision. a: R-plane, b: information organizer, c: memory, d: reference processor. Arrows indicate the flow of information. Each cube in the organizer denotes the line processor and they compose vertical arrays called hypercolumns, i.e. holons. When elementary signals are transmitted from the R-plane to the organizer, mutual interactions between holons selectively activate densely shadowed line processors accompanied by the evocation of a semantic order in memory.

As illustrated in Fig. 2, holovision has three hierarchical levels of information: an R-plane, an information organizer where cognitive order is produced and whose memory stores many items of semantic information, and an auxiliary part, the reference oscillator. The R-plane corresponds to the LGN of the brain and is composed of a regular array of square meshes, R-cells, representing cells in the LGN (or in the retinal plane). The input signal is given on the R-plane as a number of elementary signals indicated by dots. The information organizer corresponds to the primary visual cortex. Each of the vertical columns in the information organizer represents a "hypercolumn", and cubic units, "simple cells". The horizontal planes are called "hyperplanes". A "hypercolumn" in the "visual cortex" is regarded as a set of autonomous oscillators whose phase depends on the internal modes. Different values in the phase correspond to different values in the orientation of a line element, and the amplitudes of the oscillators correspond to the length of the line elements. We will call the autonomous oscillators *line processors*; each oscillator functionally corresponds to each "simple cell" in the "hypercolumn" discovered by HUBEL and WIESEL [11]. The grid-like structure of the hyperplanes is topographically identical to that of the R-plane. The line processors in the generator cooperatively process encoded elementary signals according to the following excitatory and inhibitory interactions leading to the self-organization of an image: the excitatory interactions enhance the amplitudes of rhythmic activities by mutual entrainment with the same phase, while an inverse phase relation that leads to suppression is given in the inhibitory interactions.

As illustrative and instructive examples for understanding the creation of cognitive order, let us examine the simple problems of Holovision. (Details of calculations can be found in the references.) Elementary signals are given on an R-plane of 8×8 meshes. They are then transmitted to the information organizer, where each group of 4 line processors in each of 8×8 hypercolumns, holons, receives the same elementary signal. Three kinds of semantic order have been

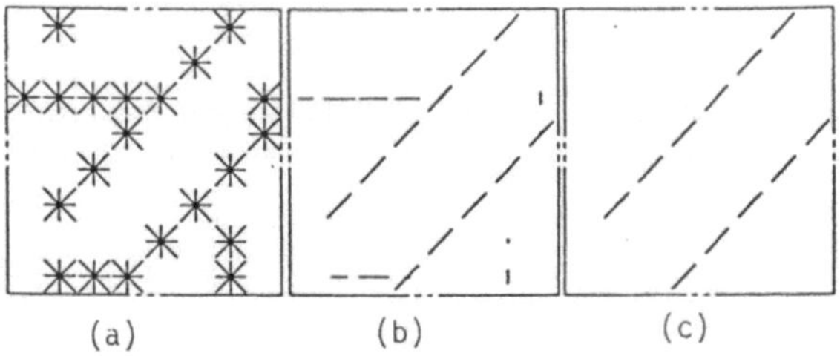

Fig. 3 (a) An input signal on the R-plane; (b) a primary pattern organized in the information organizer; (c) a perceptible pattern which is relevantly connected by a semantic order. The length of a bar in (b) and (c) shows the magnitude of the activity of the processor.

prepared in the memory: horizontal, right-up oblique, and vertical lines.

The input signal for the first example is shown in Fig. 3a. In Fig. 3b the pattern self-organized in the information organizer is shown, it is spontaneously connected with a right-up oblique line in memory. Fig 3b shows the primary pattern obtained by decoding from the internal dynamics of each holon. However, this primary pattern is distinguished from a cognitive order, because it has some parts, irrelevant parts, uncoupled with the semantic order. The cognitive order is the relevant part of the primary pattern related with the semantic order by a holonic loop. The "perceptible pattern" corresponding to the cognitive order is shown in Fig. 3c. In this case only oblique lines are left in the perceptible pattern.

Let us compare the above example with those when one of the others of the above semantic orders are selectively evoked. The initial condition is modified to select each one of the semantic orders. (Holovision reached a steady state within our calculation time.) The virtual evocation of a particular semantic order in memory due to this technique will be called forced selection, in contrast to the spontaneous selection illustrated in the above example. The results of forced selection of a specific semantic order are shown in Figs. 4a, b, which illustrate the primary and perceptible pattern for the same input signal as in the above example. Fig. 4c shows the results in the absence of semantic orders. It is now clear how the cognitive orders are relevant to the semantic orders, and that the cognitive orders in Figs. 4a, c only poorly describe the input signal.

Now, let us imagine that no semantic order in memory gives a satisfactory description of an input signal. In such a case, a new semantic order must be created in the memory as a kind of working hypothesis. Let us "create" an order composed of two lines in different directions, i.e., right-up oblique and

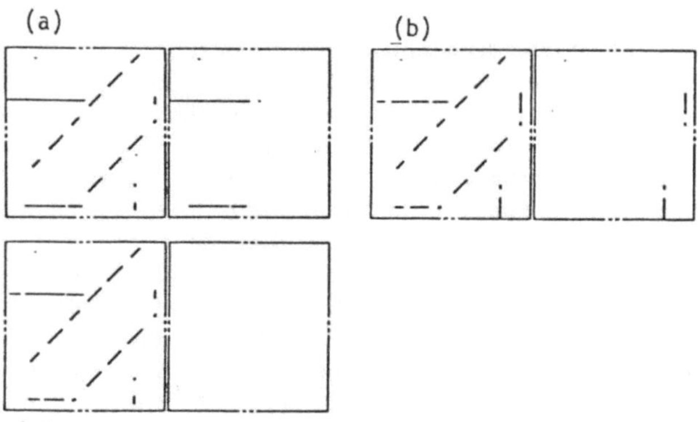

Fig. 4 A resultant pattern in the information organizer under forced selection of a semantic order: (a) horizontal line and (b) vertical line. Semantic order is absent in (c). The left and right figures show a primary pattern and a perceptible pattern, respectibely.

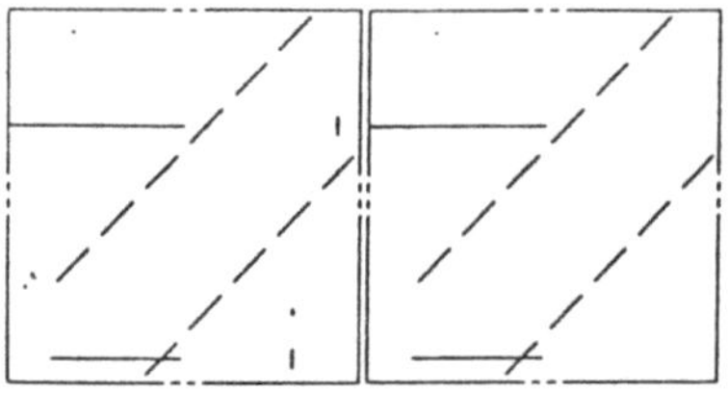

(a) (b)

Fig. 5 (a) The effect of the addition of a new semantic order composed of horizontal and oblique lines : (a) a primary pattern and (b) a perceptible pattern.

horizontal directions, and add it to the semantic orders in memory. Let Holovision start with the four semantic orders. As shown in Figs. 5a, b, among these four orders the new one is spontaneously selected, resulting in a perceptible pattern shown in Fig. 5b. Evidently, the largest amount of information is created in Fig. 5b as a cognitive order, indicating that the added semantic order may be the best one among the four semantic orders. However, on the other hand, the new semantic order is more complex than the others. Generally, the simpler the better, as a rule.

The final values of information entropy in the holons in the organizer when cognitive orders are created in the above cases are calculated from Eq.(A5) in the absence and presence of a holonic loop between the semantic and cognitive orders. Figure 6 shows the results. It is obvious that the evocation of semantic orders always diminishes the amount of information entropy, indicating that the amount of semantic information created by connection with semantic information in memory, Eq.(A6), is positive. It should be noted that the amount has the largest value in the case of combination of horizontal and oblique lines, which evoked spontaneously. In our computer experiments a general tendency is observed for the semantic order: the semantic order coupled with the cognitive order with the largest creation of information is spontaneously selected.

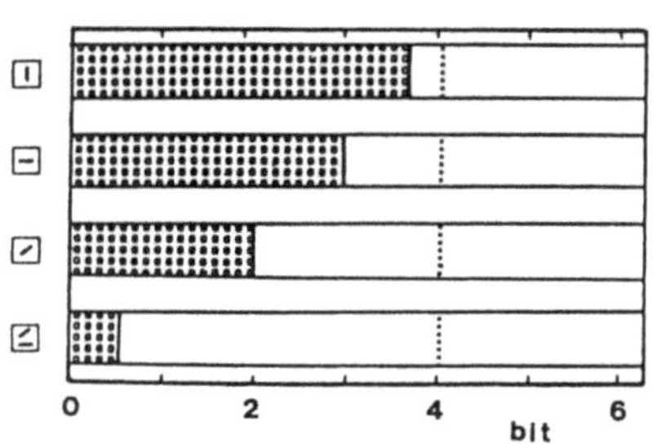

Fig. 6 The comparison of information entropy. Inserted figures on the left respectively denote the types of evoked semantic order. The amount of information entropy at the initial time is indicated by the full length of the band, and a dotted line denotes the value in the absence of semantic order.

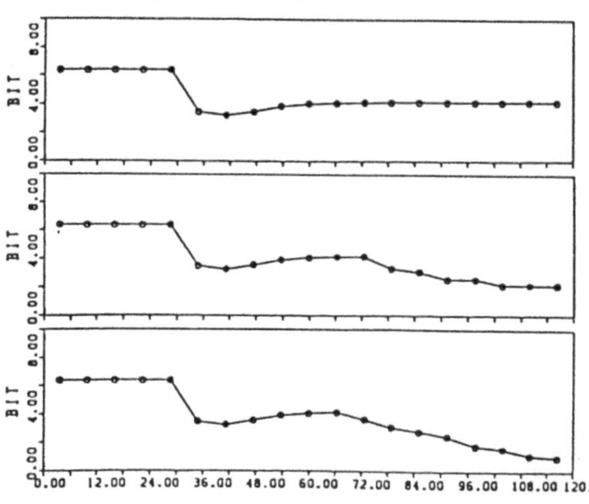

Fig. 7 The time-evolution of information entropy in the absence of semantic order (a), under spontaneous selection in the presence of three types of semantic order (b), and in the presence of four types (c); (a) - (c) correspond to Figs. 4(c), 3 and 5, respectively.

The time evolution of information entropy during the self-organization of the cognitive order with spontaneous selection is shown in Figs. 7b, c in the case of the three types of semantic order and of the four types of semantic order, respectively. Fig. 7a illustrates the time evolution of information entropy in the absence of a holonic loop. These results support the view that the largest amount of information is created in the group of holons by competitions among the semantic orders in forming a holonic loop.

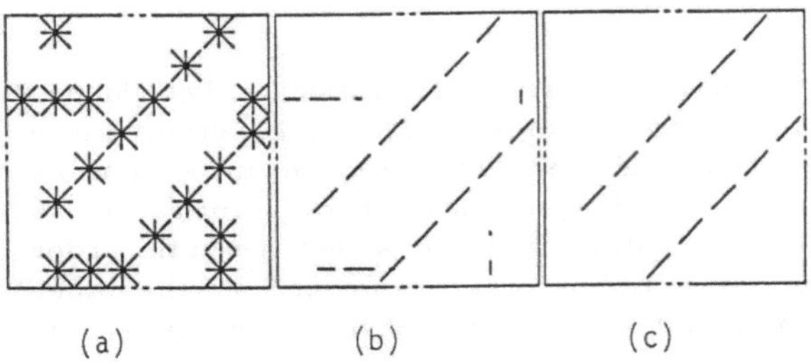

Fig. 8 (a) Input data; (b) primary pattern; (c) perceptible pattern.

A slight change, subtraction of only one dot from Fig. 3a, is made in the input signal shown in Fig. 8a. In this case, as shown in Figs. 8b, c, a cognitive order composed of oblique lines is spontaneously selected in spite of the presence of the four types of semantic order. In this case, forced selection with the amalgamated semantic order reduces the amount of information entropy more than in the case of spontaneous selection, indicating that the simplicity of semantic order is a factor in the selection of the cognitive order, because the simpler the semantic order, the faster the evocation.

14. Semantic Hierarchy

If the complexity of a complex system is too great, information has to be compressed many times before we get a semantic order. Thus, we have information carried by the dynamics of holons in a set of hierarchical levels which are linked step by step by means of a holonic loop. The semantic order is placed in the highest level. We will name such a hierarchy of information a semantic hierarchy. All the holons in a semantic hierarchy are directly or indirectly linked by relevant relations. We may compare the structure of such a semantic hierarchy with the structure of a mandala. The local rules for the information dynamics of holons are generally different for different hierarchical levels, since a new sort of information appears successively at every step of information compression. At each step an order is self-organized in the dynamics of holons.

A sort of parallel information processing with several layers is utilized in perceptron [15]. A group of distributed signals is given to the receptive elements in the first layer. The input signals are then transmitted through converging connections between the layers and are classified into suitable categories at the final layer. A similar principle is used in spin-glass models [16], [17]. However, our approach differs from these approaches in the following points. (1) Elementary signals are encoded in the dynamics of holons in such a way that the elementary information is spontaneously organized when self-organization occurs in the collective dynamics of the holons. As a result, the hierarchical presence of the dynamical orders leads to semantic hierarchy with informational orders with various degrees of information compression. (A new sort of information appears when dynamical order is self-organized in the behavior of holons coding a signal and when the dynamics has relevant connections with those of the holons in higher levels.) (2) Our cognitive computer utilizes a self-organizing phenomenon for the self-organization of information with compression. This idea was also proposed by HAKEN [18] independently under the name of the synergetic computer. The synergetic computer is not a classification machine as perceptron but an organization machine. The dynamical coding of information results in flexibility in the self-organization of information: there are a number of different ways of controlling the self-organization of dynamical orders in a cognitive computer which is a complex system. It should be noted that, strictly speaking, in a cognitive computer such as Holovision *a cognitive order is encoded in dynamical carriers of information (such as pulse waves from elementary oscillators in Holovision) circulating in a holonic loop* and not in the dynamics of holons; the latter code a primary pattern. This coding is necessary for connections of cognitive order with semantic order.

The author wishes to acknowledge Professor Hermann Haken for valuable conversation and suggestions. He is indebted to Dr. Y. Yamaguchi for kindly helping him in the preparation of the present paper, including the calculations and discussions on Holovision. He thanks Dr. I. Tsuda for his comments on the improvement of the manuscript.

REFERENCES

1. H. Haken: Synergetics -- An Introduction, 3rd ed. Springer Ser. Syn., Vol.1(Springer, Berlin, Heidelberg 1983)
2. H. Haken: Advanced Synergetics Springer Ser. Syn., Vol. 20 (Springer, Berlin, Heidelberg 1983)
3. H. Haken: in Complex Systems -- Operational Approaches in Neurobiology, Physics, and Computers, ed. by H. Haken, Springer Ser. Syn., Vol. 31 (Springer, Berlin, Heidelberg 1985)
4. S. W, Kuffler, J. G. Nicholls, A. R. Martin: From Neuron to Brain, 2nd ed. (Sinauer Associates Inc., Sunderland, Mass. 1984)
5. L. Wittgenstein : Philosophical Investigation. Anscombe, G.E.M, Phees, R., ed. with an English Translation by Anscombe, G.E.M. (Basil Blackwell 1953, 1958, 1967)
6. E. N. Lorenz: J. Atmospheric Sci. 20, 130 (1963)
7. H. Poincaré: Les méthodes nouvelles de la mécanique céleste. (Gauthier-Villars, Paris 1892, 1899), Reprint (Dover Publ., New York 1960)
8. Saussure: Cours de linguistique générale, publié par Charles Bally et Albert Sechehaye avec la collaboration de Albert Riedlinger, Lausanne et Paris, Payot.
9. K. Goedel: On Formally Undecidable Propositions. (Basic Books, New York 1962)
10. H. Yukawa: Kodansya Library, in Japanese (1971)
11. D. H. Hubel, T. N. Wiesel: Proc. R. Soc. Lond. B 198, 1-59 (1977)
12. M. M. Merzenich, J. H. Kaas, G. L. Roth : J. Comp. Neurol. 166, 387 (1976)
13. H. Shimizu, Y. Yamaguchi: Physica Scripta, in press
14. H. Shimizu, Y. Yamaguchi, I. Tsuda, M. Yano: in Complex Systems -- Opera-

tional Approaches in Neurobiology, Physics and Computers, ed. by H. Haken, Springer Ser. Syn., Vol. 31 (Springer, Berlin, Heidelberg 1985) p.225
15. F. Rosenblatt: Psychol. Rev. 65, 386 (1958)
16. J. J. Hopfield: Proc. Natl. Acad. Sci. USA 79, 2554 (1982)
17. W. Kinzel: in Complex Systems -- Operational Approaches, ed. by H. Haken Springer Ser. Syn., Vol. 31 (Springer, Berlin, Heidelberg 1985) p.107
18. H. Haken: "Pattern Formation and Pattern Recognition -- An Attempt at a Synthesis" in: Pattern Formation by Dynamic Systems and Pattern Recognition, ed. by H. Haken, Springer Ser. Syn. Vol. 5 (Springer, Berlin, Heidelberg 1979) p.2

Appendix
Information Theory for the Creation of Cognitive Order

When an ensemble of elementary signals is introduced to the information generator in which holons are regularly arranged in space, the syntactic fuzziness in the generator may be given by the following increase in the information entropy [13]:

$$H_{syn} = -\Sigma_n \Sigma_r P_{nr} \log P_{nr} = N\log 2, \tag{A1}$$

where N is the number of the holons; P_{nr} stands for the probability that the n-th holon is present in the r-th state with $r = 0$ or 1. In other words, we need this amount of information to specify the decomposed input signal. Then, in addition to (A1), we are able to define information entropy due to the semantic fuzziness

$$H_{sem} = -\Sigma_n \Sigma_k p_n{}^k \log p_n{}^k \tag{A2}$$

with probability $p_n{}^k$ for the activation of the k-th internal mode of the n-th holon. After compression H_{sem} is reduced to

$$H_{sem} = -\Sigma_n \Sigma_k P_n{}^k(\{P^k{}_n{}^{[1]}\},\{P^k{}_n{}^{[2]}\}) \times \log P_n{}^k(\{P^k{}_n{}^{[1]}\},\{P^k{}_n{}^{[2]}\}), \tag{A3}$$

where $P_n{}^k$ is the activity of the k-th internal mode of the n-th holon normalized over all the internal modes of the whole active holons. This quantity may be tentatively interpreted as the probability per unit time that the internal mode is activated by randomly impinging "photons". In our calculation for Holovision, we may assume $P_n{}^k$ as the pulse power of the k-th line processor of the n-th hypercolumn. In equation (A3) the symbols $P^k{}_n{}^{[1]}$ and $P^k{}_n{}^{[2]}$ indicate the activities of the k-th internal mode in the nearest and next nearest-neighbor hypercolumns. These terms represent, rather symbolically, the fact that the activity $P_n{}^k$ depends on the averaged activity of a local image, i.e. a local order parameter, to which the line processor belongs. It is worth noting that the syntactic entropy defined by (A1) is unchanged during information processing even when some of the elementary signals are compressed in an inappreciable state.

When the generator is influenced by the m-th semantic item with activity a_m, the normalized activity $P_n{}^k(a_m)$ is specifically enhanced to amplify the internal modes $\{k\}$ which participate in the self-organization of an image that is consistent with this semantic item. The other kinds of internal modes are more or less compressed. We may conveniently classify the line processors belonging to the holons that receive the elementary signals into two categories. One is composed of those in a primary image maintaining a time-coherent relation with the memory processors that encode the m-th semantic item. The other is composed of the other line processors which are not directly related to the perceptible image. It is not permissible for the former processors to be regarded as a whole or "macroscopic" processor for a collective mode encoding the image. The other processors belonging to the latter category may be regarded as independent from each other. Thus, $H(m)_{sem}$ is composed of two parts corresponding to the former and the latter,

$$H(m)_{sem} = -P_m(a_m)\log P_m(a_m) - \Sigma_{n'}\Sigma_{k'}P_{n'k'}(a_m)\log P_{n'k'}(a_m), \tag{A4}$$

where for simplicity's sake we do not designate explicitly the dependence of the probabilities $P_m(a_m)$ and $P_{n'k'}(a_m)$ on the local order parameter. The first and the second terms in the right hand side of (A4) represent respectively the contributions from the collective and independent processors. The creation of information occurs only in the semantic space. The amount of created semantic information in the generator is given by

$$I^{(0)}{}_{sem} = \Sigma_n\Sigma_k P_n{}^k\log P_n{}^k - \Sigma_n\Sigma_k p_n{}^k\log p_n{}^k \tag{A5}$$

in the absence of connection with semantic items. The connection with the m-th semantic item leads to further creation of information,

$$\Delta I_{sem}(m) = P_m(a_m)\log P_m(a_m) + \Sigma_{n'}\Sigma_{k'}P_{n'k'}(a_m)\log P_{n'k'}(a_m) - \Sigma_n\Sigma_k P_n{}^k\log P_n{}^k. \tag{A6}$$

The semantic information of

$$I_{sem}(m) = I^{(0)}{}_{sem} + \Delta I_{sem}(m) \tag{A7}$$

is *created* as a whole when the assembly of input elementary signals are related to the m-th semantic item. Generally, the input elementary signals will be most easily connected with the semantic item that results in the maximum creation of semantic information.

Toward a Physical (Synergetic) Theory of Biological Coordination

J.A.S. Kelso and G. Schöner

Center for Complex Systems, Florida Atlantic University, Boca Raton, FL33431, USA

"I do not see any way to avoid the problem of coordination and still understand the physical basis of life" (PATTEE [1])

1. Introduction

In spite of, or perhaps because of, the successes of modern molecular biology, the great unresolved problem of all biology remains: the manner by which complex biological systems are coordinated to produce functionally-specific ordered behavior or spatiotemporal patterns. Faced with complexity (in terms of the number of potential degrees of freedom), a conventional approach is to unravel the material substrate, component by component, to find the cause of pattern generation at an ever smaller scale. Witness, for example, neurobiology's search for the command neuron in "simple" invertebrate preparations, the role of which is to evoke or trigger a pattern stored in the neural structure (e.g., [2]). Or the apparently self-evident existence of central pattern generators--patterned neuronal activities that can persist independent of afferent feedback influences--whose neurophysiological mechanisms are currently the subject of much investigation. The noun form "generator" expresses the material reductionist's hope that the circuitry can be completely identified and isolated. One is reminded of POINCARE [3] in this regard, that "...the aim of science is not things themselves, as the dogmatists in their simplicity imagine, but the relations among things; outside these relations there is no reality knowable". It is the quest for laws and principles of pattern generation, rather than particular neural mechanisms or instantiations of those laws, that spurs the present approach.

Why should an understanding of complex, biological systems and the behavioral complexity they exhibit be sought in terms of organizational principles rather than specific mechanisms? One reason is that often well-defined behavioral patterns can be produced by many different material substrates. Examples are manifold. Consider the numerous propulsive structures used in locomotion--from echinoderms to arthropods to anthropods (see [4], for review). The rhythmic behavior of all such anatomically distinct species is characterized by a constant phase relationship among adjacent component structures (e.g., tube feet, cilia, swimmerets, parapodia etc.). Or consider findings that show that a single, optically derived variable, time-to-contact, governs the perception of obstacles by moving animals--from gannets to houseflies to humans (see [5], for review). Such information is available to, and capable of, guiding all seeing creatures, regardless of their anatomically diverse visual systems. Whether the eye is simple or compound does not matter. Finally, at a neural level, the very concept of a "neural code" arises because similar patterns of electrophysiological activity have been observed in anatomically distinct species (see e.g., [6], for review).

On the other side of the coin, biological structures are clearly multifunctional. That is, the same anatomical components may support a variety of functions. Speaking and chewing for example, use overlapping sets of articulators to accomplish different functional requirements. Other well-studied examples are: 1) flight and walking of the locust [7] in which pairs of muscles in the thorax act as antagonists during flight and as synergists during walking; 2) forward and backward walking of the lobster [8] in which the same motoneurons switch their respective firing patterns for one activity or the other. Such observations indicate that the same neuronal network supports multiple functions. They also suggest strongly that function is not rigidly coded into neural structure. Rather, these findings are consistent with mounting evidence that flexibility (ability to adjust to changing environmental demands) and stability (ability to persist under varying environmental conditions) are characteristic features of softly assembled cooperative states, not hard-wired mechanisms.

How might functionally-specific patterns be (self) assembled? What are the materially-independent principles underlying pattern formation in complex, biological systems? This is where we turned several years ago to Haken's Synergetics for inspiration. In a theory of the spontaneous formation of structure, HAKEN [9, 10] was able to show--for a large class of (open, far from equilibrium) systems--how very many microscopic degrees of freedom can self-organize to form spatio-temporal patterns that are in turned governed by low-dimensional order parameter dynamics. This compression of degrees of freedom (the so-called slaving principle) allows for a micro to macro derivation of order that depends only on the coupled dynamics of the subsystems and not on their material properties. Structure and change of structure occur spontaneously in the nonequilibrium phase transitions studied by synergetics, as "control" parameters (often reflecting quite unspecific environmental conditions) change beyond critical points.

As HAKEN [11] himself has noted, however, a general, across the board application of synergetics to complex, biological systems is hazardous, and an operational approach--linking measurement and modeling--is to be preferred. In contrast to certain physical systems like the laser, for example, in biology the path from the microscopic dynamics to collective order parameters is not readily accessible to theoretical analysis. On the other hand, it may still be possible to develop an understanding of biological order using a synergetic strategy that involves:

- identifying the essential, collective variables (order parameters) and their low-dimensional dynamics
- determining the relevant "control" parameters that move the system through its collective states
- relating different levels of description through a study of component dynamics and their coupling

Such a program of research can be implemented most successfully near points of pattern change, i.e., where differential characterization of pattern is possible. In the following, we first illustrate the operational approach of synergetics to biological coordination, through our studies of movement patterns. Then we summarize the insights derived from such studies in a set of seven theoretical propositions that aim at a more general characterization of pattern stability and change in biological systems. A variety of examples (by no means exhaustive) is provided in support. But more important, insight into organizational principles leads us to ask (and hopefully to answer) new, biologically relevant questions, some of which we raise in the final part of the paper. Thus synergetics continues to--and must--evolve to form the language for understanding biological coordination.

2. Biological coordination: From phenomena to models to predictions to understanding

A good experimental test field for the synergetic strategy in biology is multidegree of freedom movements. These are rich enough in their dynamics, yet sufficiently accessible to observation--in part because of recent technological developments--that they are open to analysis. Indeed, it was the discovery of a phenomenon--a nonequilibrium phase transition in coordinated biological motion--that led us to adopt the present approach in the first place. Let's briefly describe the finding (for details see [12], [13], [14]).

The task for human subjects is to move their index fingers or hands rhythmically. The driving frequency, F (in Hz), corresponding to the control parameter is then increased in a stepwise fashion. In such a situation subjects can stably and reproducibly perform in only two phase-locked modes, either in-phase (homologous muscle groups contracting simultaneously) or anti-phase (homologous muscle groups contracting in an alternating fashion). The finding that only two stable phase-lockings exist was demonstrated quantitatively in experiments in which the relative phase between the two fingers was manipulated and the variability of the produced relative phase was measured. Much less variable performance was observed at relative phase $\phi = 0$ deg. (in-phase) and $\phi = \pm 180$ deg. (anti-phase) than any other specified relative phase [15], [16]. In our experiments, when subjects, initially moving in the anti-phase mode, were instructed to increase cycling frequency, an involuntary abrupt shift to the in-phase mode was observed at a critical frequency. Beyond this critical frequency value only the in-phase mode was performed stably. Similar findings have been obtained by several other groups of investigators ([17], [18]).

It seemed at least intuitive that the foregoing experimental observations were consistent with other bifurcation phenomena in nature. Indeed, their essential features were successfully modelled by HAKEN, KELSO and BUNZ [19] introducing relaxational dynamics

$$\dot{\phi} = -\frac{dV(\phi)}{d\phi} \tag{1}$$

for the relative phase ϕ and mapping the observed modes onto point attractors. The potential

$$V(\phi) = -a\cos(\phi) - b\cos(2\phi) \tag{2}$$

complying with periodicity and symmetry requirements, captures the bifurcation diagram in that it has minima at $\phi = 0$ and $\phi = \pm 180$ deg. for $a/b<4$ with the latter minimum turning into a maximum for $a/b>4$ (see fig. 1). The order parameter dynamics (1), (2) for ϕ were then derived from nonlinear oscillator equations for the two hands with a nonlinear coupling between them (see also [13] and [20].

The foregoing modelling work inspired further attempts to explicitly test the phase transition character of the phenomenon. For example, a stochastic generalization of the above model by SCHONER, HAKEN and KELSO [21] led to additional predictions regarding certain characteristic features, including critical fluctuations of the hypothesized order parameter and critical slowing down. These predictions reflect the centrality of the concept of <u>stability</u> in the understanding of biological coordination,not just as a characterization of the two attractor states, but also because <u>loss of stability</u> is hypothesized to be the generic mechanism effecting coordinative change.

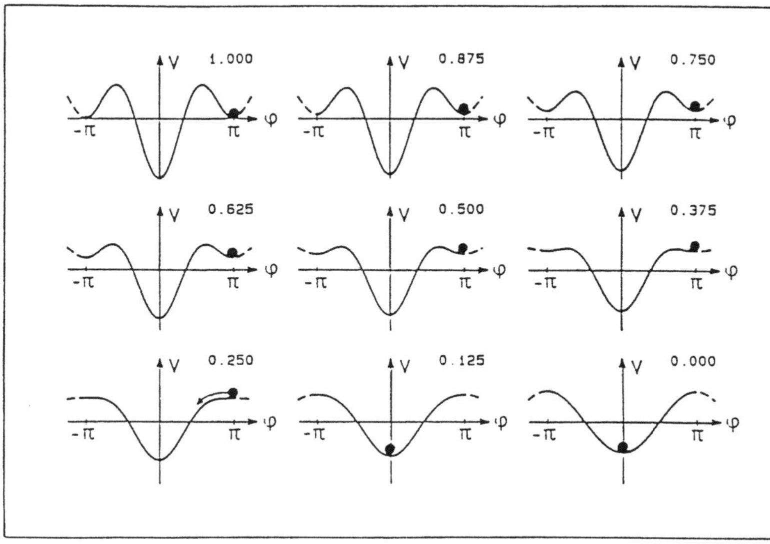

Fig. 1. The potential (2) as the ratio b/a is changed. The little ball illustrates the behavior of the system if it is initially (upper left corner) prepared in the anti-phase state [19].

In [21], stochastic forces act as continuously applied perturbations, thereby producing deviations away from the attractor state. The size of these fluctuations, as measured, for instance, by the SD of ϕ is thus a metric for the state's stability. A recent paper [14] examined the mean relative phase and its SD as frequency was scaled systematically in each of the two coordinative modes. When parameter scaling began in the in-phase mode, both variables remained roughly constant (see [14], figs. 2 and 4). In contrast, a clear enhancement of fluctuations was observed for the anti-phase mode both before and during the transition. Thus the first major prediction of the stochastic dynamic model--the presence of critical fluctuations--received striking experimental confirmation.

The second main prediction from stochastic dynamic theory concerns critical slowing down. If a small perturbation is applied to the system driving it away from its stationary state, the time it takes for the system to return to that state--the so-called local relaxation time, τ_{rel} --is a measure of the stability of the attractor. The smaller τ_{rel}, the more stable the attractor. In the present context, as the system loses its stability, the time it takes to reach the stationary anti-phase state following a slight perturbation should increase as the frequency of oscillation is scaled toward the transition. No strong increase in τ_{rel} is predicted for the in-phase coordinative state.

A test of this prediction [22] required considerable modification from the experimental set-up used to study critical fluctuations. In particular, special allowance was made to incorporate the injection of perturbations. Subjects inserted their index fingers into the metal sleeves of manipulanda whose axes of rotation coincided with those of the first metacarpophalangeal joints. The manipulanda restricted finger motion to flexion and extension at this joint in the horizontal plane. Finger movement trajectories were transduced by 0.1% linear potentiometers mounted over the respective axes of rotation. These data were recorded on FM tape and later digitized with 12-bit resolution at 250 samples/s for subsequent computer analysis. The latter involved: (1) Numerical differentiation of finger position (with a symmetric five sample window to avoid phase shifts) in order to obtain velocity; (2) calculation of each finger's oscillation phase in the phase plane (position, x and velocity, $\dot{x}$ normalized for each cycle to the square [-1,1] x [-1,1]) defined as ϕ = arctan $(\dot{x}_i/x_i)$ where i is left or right; (3) calculation of a continuous estimate (i.e. at the sampling rate) of relative phase ϕ defined as $\phi = \phi_{left} - \phi_{right}$. The subject's task was to oscillate both fingers rhythmically at the same frequency in one of the two modes of coordination. The movement was paced by an auditory metronome pulse with the instruction to perform one complete movement cycle per pulse. The initial pacing frequency was determined individually for each subject based on a series of 10 scaling test trials that determined the frequency of movement at which the subject switched spontaneously from the anti-phase to the in-phase mode of coordination with highest incidence. The initial pacing frequency was then chosen to be five 0.2 Hz steps below that most frequently occurring transition frequency. In the actual experimental runs the pacing frequency was increased every 10s in nine 0.2 Hz steps (frequency plateaus). Trials began either in the in-phase or in the anti-phase modes of bimanual coordination.

In order to estimate the system's relaxation time, the pattern of bimanual coordination was disturbed by perturbing the right index finger with a 50 ms. torque pulse that was generated by a dc torque motor mounted above the finger's axis of rotation. The torque pulse was timed to the peak velocity of finger flexion movement using a servo-control circuit. The size of the torque pulse was adjusted individually to produce approximately comparable displacements of the right index finger. The range of applied torque across subjects was 0.028 to 0.039 Nm. In each trial, movement perturbation occurred on up to four non-adjacent frequency plateaus. Perturbations were randomly distributed over a block of trials such that each of the nine frequency plateaus was perturbed a total of ten times.

Using interactive computer displays, an estimate of the relaxation time was obtained from the time of torque pulse offset until the relative phase time series stabilized at its pre-perturbation mean value. The results of these experiments for the relaxation time estimate are shown in figure 2 for all five subjects. The following features are notable: (1) Except for the lowest frequencies, the relaxation time in the anti-phase mode is consistently higher than in the in-phase mode. (2) As the frequency approaches the transition frequency, the relaxation time in the anti-phase mode increases yet remains constant or decreases in the in-phase mode. A mode by pacing frequency analysis of variance performed individually for each subject's data showed that this difference was statistically significant in all but one case. Even for this subject (BK), who showed an overall decrease of relaxation time in both modes, a sharp increase occurs in the anti-phase mode immediately prior (2.2 Hz) to the transition.

Overall pre-transitional increases in relaxation time thus prove the presence of critical slowing down in this biological coordination problem, entirely consistent with earlier theoretical predictions and experimental studies of relative phase fluctuations showing that: (1) The anti-phase mode is dynamically less stable than the in-phase mode; and (2) the transition from anti-phase to in-phase mode is connected with a loss of stability.

Specifically, in the theoretical model for the stochastic dynamics of relative phase ϕ [21]:

$$\dot{\phi} = -a\sin(\phi) - 2b\sin(2\phi) + \sqrt{Q}\,\xi_t \tag{3}$$

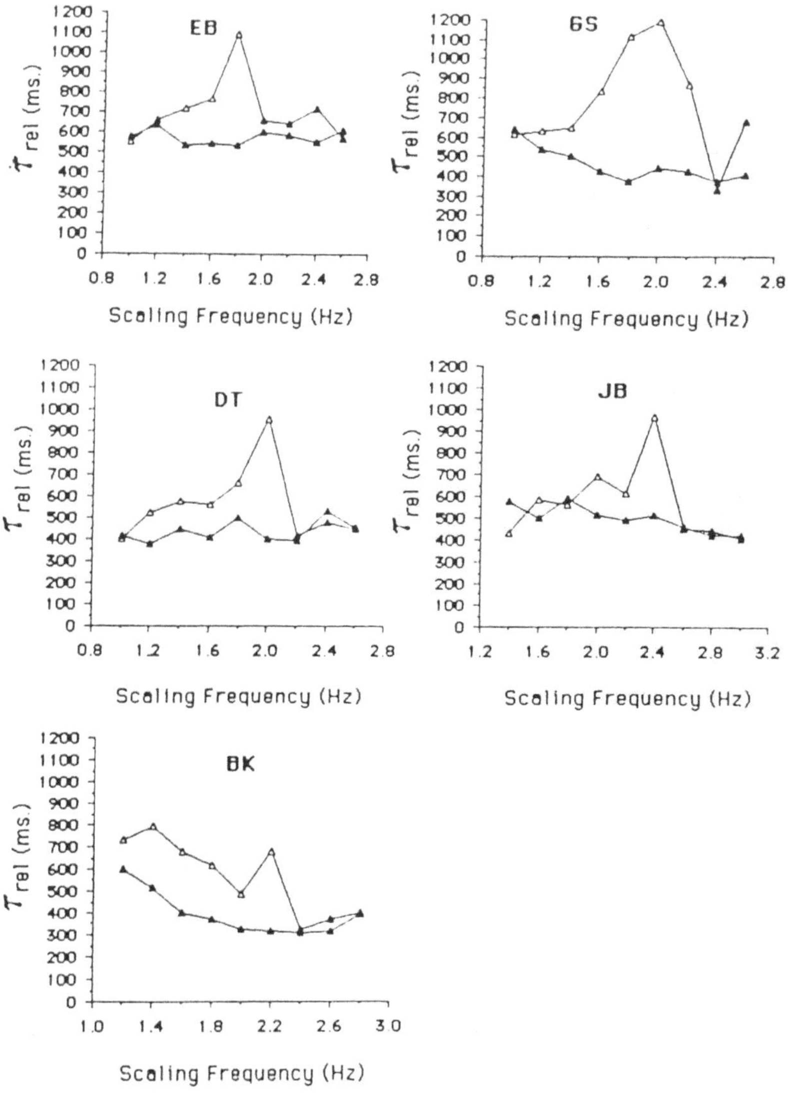

Fig. 2. The mean relaxation times as a function of pacing frequency for the five subjects (EB, GS, DT, JB and BK). The open triangles refer to the anti-phase mode, the closed triangles to the in-phase mode. The mean transition frequencies for the five subjects were: EB: 2.02 ±0.15 Hz (N=34), GS: 2.27 ±0.15 Hz (N=33), DT: 2.21 ± 0.17 Hz (N=30), JB: 2.56 ± 0.19 Hz (N=26) and BK: 2.35 ±0.19 Hz (N=28). In these means, transitions that were induced by a mechanical perturbation were discarded. Note that beyond the transition frequency trials started in either mode are in-phase [22].

with ξ_t as Gaussian white noise of unit variance, and model parameters a, b and Q the relaxation times, τ_{rel}, were predicted as:

$$\tau_{rel,0} = \frac{1}{4b+a} \qquad ; \qquad \tau_{rel,\pi} = \frac{1}{4b-a} \tag{4}$$

where zero refers to the in-phase mode and π to the anti-phase mode. When we determined the parameters a and b from the measured relaxation times in the two modes, we found that a/4b ≃ .39 on the last pre-transition frequency plateau for all subjects. This is much further from the critical point (a/4b = 1.0) than found in earlier studies of relative phase fluctuations (in [14] a/4b ≃0.64). Consequently the critical fluctuations predicted

$$\frac{SD_\pi}{SD_0} = \left(\frac{\tau_{rel,\pi}}{\tau_{rel,0}}\right)^{\frac{1}{2}} \tag{5}$$

to be comparatively small (cf. [14], where this ratio is 2.13). Indeed, in [22], the fluctuation enhancement on the pretransition plateau in the anti-phase mode was not statistically significant. The SDs in both modes were at a level of 20 deg., except for a transient enhancement on the transition plateau in the anti-phase mode. The parameter Q was estimated from these data as 0.47 Hz.

We used these parameter estimates to test the consistency of our stochastic-dynamic modelling (3). As discussed in [21] and [23], the system is predicted to switch, as soon as the time scales relation:

$$\tau_{rel,\pi} << \tau_p << \tau_{equ} \tag{6}$$

is violated. Here τ_{equ} is the equilibration (or global relaxation) time of (3) and τ_p is the time scale of parameter change which is identical here to the observed time scale, and is given by τ_p = 10 s. An estimate of τ_{equ} can be obtained from the mean first passage time (MFPT) for the passage from ϕ = ±180 deg. to ϕ = 0 deg. Using model parameter estimates for the frequency plateau immediately before the transition we calculated the MFPT numerically from a standard formula (cf. [21], equ. (4.26)) and found MFPT = 13.0 s. For critical parameters (a = 1.6 Hz, b = 0.40 Hz, Q = 0.47 Hz) we found MFPT = 5.35 s. Thus switching indeed occurs as τ_{equ} becomes shorter than τ_p, entirely consistent with the stochastic theory. In earlier experiments [13], [14] which found critical fluctuations, τ_p was smaller (4 s) allowing the system to come closer to the critical point before (6) was violated (MFPT = 9.62 s pre-transitional and 5.28 s critical there). As we have emphasized before ([21], [23]), and as the present observations clearly indicate, time scales relations are crucial to the interpretation of pattern stability and change in biological systems.

Finally, stochastic dynamic theory contains a further, quite novel feature, the so-called <u>switching time</u>, i.e., the duration of the transient from the anti-phase to the in-phase state. The basic idea is that during the transition, the probability density of ϕ--initially concentrated at ϕ = ±180 deg. flows to ϕ = 0 deg, and accumulates there until the new peak at ϕ = 0 deg. becomes dominant and stationary. In [21] predictions for the distribution and the mean duration of this process are provided that we have recently evaluated experimentally [22].

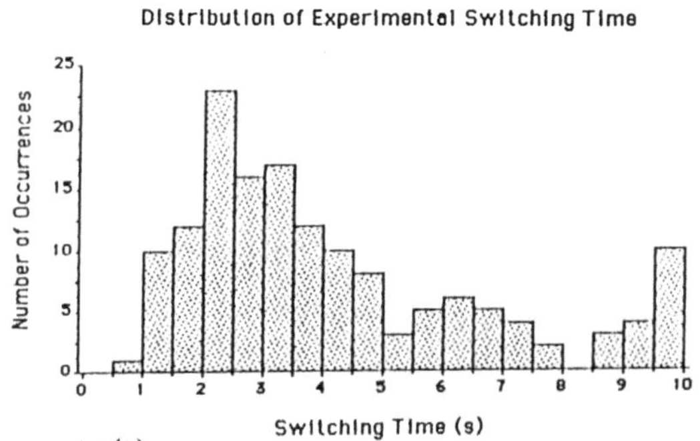

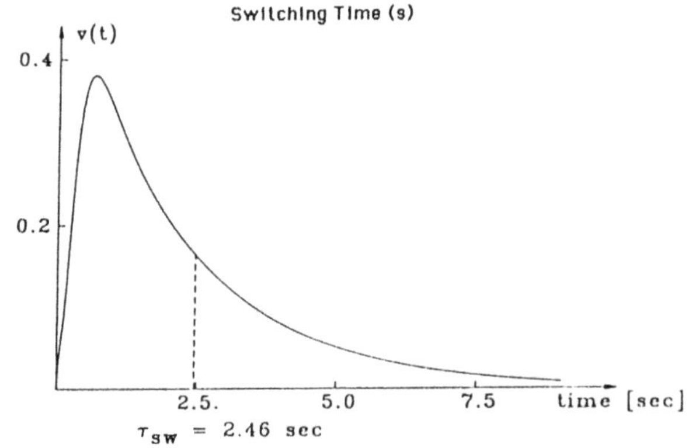

Fig. 3 Top: The distribution of switching times (time from last change of control parameter frequency to the completion of switching) from all subjects and trials. Again we discarded trials in which the transition was induced by one of the torque pulse perturbations [22].

Bottom: The distribution of switching times as calculated theoretically from the transient probability distribution of relative phase at the critical point [21].

Interactive computer displays were again used to measure the switching time on frequency plateaus in which a transition occurred. Here the estimate was determined as the time from the beginning of the frequency plateau to the point where the relative phase time series stabilized at a zero (or 360 deg.) mean value corresponding to the completion of the transition.

The histogram shown in figure 3 represents the experimentally determined switching times for the present experiment. A quite astounding agreement with the theoretical curve (also shown) is observed, both in terms of the mean switching time and the shape of the distribution. Hence yet another feature of the dynamics of biological coordination is shown to follow in detail simple stochastic dynamics for the collective variable ϕ. In particular, we find that noise is qualitatively and quantitatively relevant for the switching behavior studied here, closely paralleling similar features in physical switching processes [24]. In short, data and theory converge on a nonequilibrium phase transition interpretation (in the sense of synergetics) of the present biological coordination problem. Confirmation of theoretical predictions regarding critical fluctuations, critical slowing down and switching times reveals that the emergence of coordinated movement patterns may be understood--in considerable detail--in terms of the physics of nonequilibrium processes. We see also how the language and tools of synergetics, when closely allied to experimental data, offer a window into processes underlying pattern stability and change in a complex, biological system. It may therefore be useful to summarize the insights gained from these studies in a set of propositions that may facilitate an understanding of order and pattern formation in many natural systems at multiple levels of organization. The term "behavioral pattern" below should be conceived quite broadly. The "truth" of the following propositions must, of course, be individually evaluated in the various experimental systems observed.

3. Propositions for behavioral pattern stability and change

1. Behavioral patterns can be characterized by low-dimensional collective variables (or order parameters), whose nature and dynamics are specific to biological functions and tasks.

2. Experimentally well-defined behavioral patterns (reproducible, stationary over certain observation times) correspond to stable collective states (or attractors) of the order parameter dynamics. Behavioral patterns can change only from one collective state to another.

3. Environmental context, task variables and certain biological boundary conditions act as parameters on the collective dynamics in the sense that they can modify the behavioral patterns but are themselves not dependent on these patterns. A parameter that moves the system through different collective states is a control parameter in the sense of synergetics. Such control parameters may be quite unspecific in nature, rather than detailed prescriptions for behavior.

4. Fluctuations in the dynamics of the order parameters are conceptually important to reconcile the stability of behavioral pattern with the ability to change behavioral patterns. Fluctuations establish time scales relations that govern the switching dynamics among collective states, e.g., whether stationary or transient behavior is observed, whether multistability or averaging over multiple states is observed, and so forth.

5. Loss of stability leads to behavioral change. Fluctuations allow the system to feel its stability and loss of stability by exploring the environment of a collective state. Fluctuations thus govern the process of behavioral change. How a system changes depends on the availability and relative stability of the system's collective states.

6. The level of description used in experimental observations of behavioral patterns (e.g., neuronal, muscular or kinematic) is a question of convenience. Patterns at all levels are governed by the dynamics of collective variables. In this sense, no single level has ontological priority over any other.

7. In a number of specific cases (which may nevertheless, be quite common) collective states (e.g., multiple phase and frequency-lockings) may arise from a coupling of self-excited nonlinear (limit cycle) oscillatory processes. The dynamics of such collective states are always low dimensional (cf. Proposition 1). In such

cases, the relation between levels can be defined precisely, and the manner by which the degrees of freedom are compressed may be understood.

4. Some additional patterns in biology open to synergetic study

How general might the application of the foregoing propositions be to understanding other observed changes in behavioral pattern? Surveying the literature on biological coordination, we are reminded of the impression Hermann Haken had ten years ago with regard to the ubiquity of nonequilibrium phase transitions ([10]) using Kohelet's words: "There is nothing new under this sun". Let us just illustrate in a few examples (which are not theoretically worked out in any comparable detail to Section 2, but rather take the form of sketches) how the synergetic strategy can throw new light on old (and not so old) findings.

As long as 50 years ago, VON HOLST observed the patterns of limb coordination during locomotion of the centipede Lithobius as a function of the number of legs that were amputated. He found that the relative phase among neighboring limbs (about 1/7 of a cycle in the intact animal) increases as more limbs are removed, until the Lithobius walks as an insect if only six legs remain, as a quadruped when four legs are left over and finally, as a biped (anti-phase relation) when all but two legs are left ([25]). Here the control parameter is the number of legs and the order parameter is again relative phase. This example is particularly striking in that it makes the explanatory strategy of a "central pattern generator" with fixed, stored patterns very uneconomical, but is easily understood in terms of self-organizing coupled oscillators.

Another striking example of higher order behavioral patterns comes again from older work by VON HOLST and VON SAINT PAUL [26] on the domestic fowl. These researchers were able to elicit certain well-classified behaviors by electrical stimulation of regions in the hen's brainstem. For a fixed electrode position, an increase in voltage led the animal through a reproducible ordered series of different behaviors (e.g., looking about, standing up, walking, crouching, flying away, see figure 4). Similarly, a series of different behaviors resulted if the depth of injection of the electrode was varied.

Fig. 4. a: Behavior sequence of fleeing from a ground enemy with slowly increasing stimulation of the field (or constant stimulation of moderate intensity); b: The reaction to a sudden, strong stimulus (adapted from [26], fig. 8.29 there).

Experiments of this type are often interpreted as evidence for localization of functions in the brain. It turns out however, that the stimulus-behavior relation is very sensitive to the general "mood" of the animal, the environmental context, the sequence of stimulus variations, the speed of stimulus variations etc. In view of these effects, and adhering strictly to the behavior observed, an interpretation in terms of phase transitions may seem more general. Here stimulus strength or electrode depth may act as control parameters (see also [27]). The collective variables for the behavioral pattern are more difficult to quantify. Still, a measure of stability (i.e., switching back and forth between two patterns as the stimulus strength approaches a critical point) could be used to investigate the consistency of the phase transition description. The aforementioned plasticity is an expression of the dependence of the collective behavior on boundary conditions and other parameters.

An example from the recent neurobiology literature is the so-called scratch reflex of the turtle studied by STEIN and colleagues ([28], [29]). Here three different forms of periodic scratching movement can be elicited as a function of where the stimulation (in this case, mechanical) is applied. Kinematically, the three forms differ again by the relative phasing of hip and knee joint, and consequently by the anatomical component used to scratch. These patterns can also be characterized on the basis of electromyographical and electroneurographical observations where they differ by the relative timing of different components. Again, transition areas of the receptive field have been noted where it is possible to observe 1) either of two adjacent scratching patterns; 2) a switching from one to the other during several cycles of scratching; or 3) a 'blend' of two adjacent patterns. Work remains to be done here in precisely characterizing the collective states so that 'blending' (perhaps a superposition of two patterns?) may be better understood. This and similar experiments have led neurobiologists to introduce more and more complicated (so-called multipartite) pattern "generators" [30]. However, from a synergetic perspective, we may view such findings as a welcome chance to study different patterns and transitions among patterns as a control parameter (stimulation site) is varied, in a system that is accessible on various levels of observation.

In our own experimental work we have begun to apply these ideas to the multiple articulator patterns of speech production, i.e. how a moving vocal tract (with many potential degrees of freedom) structures sound in a lawful fashion for a listener [31]. The collective variables for speech production and their dynamics are just now being explored. For example, in recent experiments lip and jaw movements were monitored along with the opening and closing of the glottis as subjects produced simple syllables (e.g., /ip/ and /pi/) repetitively. As expected, the phase relation between lips and glottis was quite different for the two utterances. For the syllable /pi/, this phase relation remained stable as subjects increased speaking rate (up to approx. 7 syllables/sec.). Interestingly, however, for the syllable /ip/, when speaking rate was increased, an abrupt shift in the phasing pattern was observed (at approx. 4 syllables/sec) to the <u>same</u> phase relation as in /pi/. Although both forms of coordination were quite stable below the critical point, only the coordinative pattern characteristic of the consonant-vowel form existed beyond the transition. Linguists have long assumed <u>a priori</u> that the consonant-vowel (CV) unit is fundamental, on no other grounds than that it is universally observed in all known languages of the world. Our results suggest an intriguing articulatory reason based on relative stability, as to why this may be the case. Finally, the <u>development</u> of behavior--exemplified by observations of kinematic patterns during infant kicking, crawling and walking--is now open to the synergetic strategy. Here, the emergence of patterned forms and qualitative changes in behavior are being examined in detail, (e.g., [32]) affording, in principle, a much closer link between experiment and theory (see [33]).

5. Perspectives and Future Directions

Obviously one can find many phase transition-like phenomena in the behaviors produced by biological systems. Here we have only scratched the surface, selecting a few examples in which the theory-experiment relation (an operational synergetics) might be fruitfully explored. It may be more important, however, to use the present insights gained into organizational principles to motivate new problems i.e., to characterize them in a way that allows them to be attacked (and solved). We raise these issues here (in no particular order) as a kind of placemarker or torchlight guiding our particular synergetics program on the coordination of movement. A sample of some of the questions we would like to answer in the future are the following:

. Are changes in behavior always associated with instabilities? Up to now, only spontaneous (so-called involuntary) changes in behavior have been shown to be governed by stochastic, nonlinear dynamics. Are environmentally induced or voluntary changes from one collective state to another also governed by relative stability and fluctuations (e.g., easier in the direction of more stability than vice-versa)? If so, the same processes would be shown to play a major role in intentional switching (typically defined in non-dynamical language) as in spontaneous switching.

. Movement, or more generally, action is often intimately related to perception. For example, in the experiments reported in Section 2, a metronome serves to impose a certain frequency on the movements. More precisely, then, it is the coupled environment- movement system from which an ordered state, or behavioral pattern is formed. The quite generally important question thus arises: what are the essential collective variables of this collective perception-action system? What are the dynamics of this relation? How can this relation be changed? In our work such perspectives have already arisen ([23]).

. In our experiments, we found a collective variable and its dynamics in synchronized, rhythmic movement. Obviously there are many functional movements that are not synchronized or not even rhythmic. Another task of generalization is thus to identify the variables and dynamics for these cases. Thus a study of entrainment and the relative stabilities of the different entrained states is crucial. For discrete movements, earlier work [34] showed already that interlimb coordination is characterized by fixed temporal relationships (e.g., a trend toward simultaneity). Another example is trajectory formation, where invariants found in the geometry of the trajectories hint at the existence of fixed temporal relationships (relative phasing) among different joints of a limb ([35]).

. Another direction of biologically relevant generalization poses new theoretical questions: How can one characterize patterns in multilimb movements? If many patterns coexist, what is their relation? Does a potential exist that governs the dynamics of several collective variables? Are there preferred transition paths from one pattern to another?

. Finally the rather detailed dynamical understanding of biological coordination in the language of synergetics, offers a unique opportunity to study such long-standing problems as learning, memory and adaptation. How does the stability of a collective state evolve during learning? Does learning a new state affect the stability of other states? Does the system increase its dimensionality during learning? Can one measure the different time scales involved as parameters defining the collective dynamics are themselves made dynamic?

6. Epilogue

A global aim of the research program discussed here is to establish the relevance of the synergetic approach--originally formulated by H. Haken in a physical context--to complex biological systems. The extent to which the synergetic strategy must be tailored to the needs of biology and psychology is an open question. In these disciplines various phenomena such as multiplicity of patterns, perception-action coupling, behavioral change, learning and so on demand to be understood. In view of the fact that very intricate patterns (including deterministic chaos) may be generated by low dimensional dynamics, it seems not unreasonable to suggest that such behaviorally complex phenomena may be encompassed by the present strategy. It is important to emphasize that the meaning of "understanding" depends on the language used. In using synergetics, understanding is sought within the abstract level of the essential (collective) variables and their dynamics, regardless of scale or material substrate. The approach, nevertheless, must be operational, that is, it must formulate dynamical laws for observable variables only and it must make experimentally testable predictions. By showing that behavioral patterns on several levels of description can be understood using the concepts of synergetics, our hope is that a common language may emerge. Thus the language barrier that currently exists among scientists who observe complex systems at very different scales of analysis may be overcome. Synergetics' recognition that the cooperation among many subsystems is governed by the same principles irrespective of the subsystems was a major step in this regard. A theory of biological coordination ultimately rests on a reductionism to a minimal set of laws governing dynamic patterns, and not the materialistic reductionism that currently predominates.

Acknowledgments

The present work was supported by a Joint Contract of the U.S. Office of Naval Research and the U.S. Air Force Office of Scientific Research. G. Schöner was supported by a Forschungsstipendium of the Deutsche Forschungsgemeinschaft, Bonn, FRG.

References

1. H. H. Pattee: in Topics in the Philosophy of Biology, ed. by M. Grene, and E. Mendelsohn, Boston Studies in Philosophy of Science, Vol. 27 (Reidel, Boston, Dordrecht Holland 1976) p. 171
2. I. Kupfermann, K.R. Weiss: Behav. Brain Sci. 1, 3 (1978)
3. H. Poincaré: Science and Hypothesis (First English translation, Walter Scott Pub. Co., 1905, Reprinted Dover, New York, 1952) p. xxiv
4. M. A. Sleigh, D. I. Barlow: in Aspects of Animal Locomotion, ed. by H. Y. Elder, E.T. Truemann (Cambridge Univ. Press, Cambridge, 1980)
5. D. N. Lee, D.S. Young: in Generation and Modulation of Action Patterns, ed. by H. Heuer, C. Fromm, Experimental Brain Research Series, Vol. 15 (Springer, Heidelberg, Berlin, New York, 1986)
6. D. H. Perkel, T. H. Bullock: Neurosciences Res. Prog. Bull. 6, 223 (1968)
7. D. M. Wilson: J. Exp. Biol. 39, 669 (1962)
8. J. L. Ayers, W. J. Davis: J. Comp. Physiol. 115, 1 (1977)
9. H. Haken: Rev. Mod. Phys. 47, 67 (1975)
10. H. Haken: Synergetics - An Introduction (Springer, Heidelberg, Berlin, New York, 3rd ed., 1983)
11. H. Haken: in Complex Systems-Operational Approaches in Neurobiology, Physics and Computers, ed. by H. Haken, Springer Ser. in Synergetics, Vol. 31 (Springer, Heidelberg, Berlin, New York, 1985).
12. J.A.S. Kelso: Am. J. Physiol.: Reg. Int. Comp. Physiol. 15, R1000 (1984)

13. J.A.S. Kelso, J.P. Scholz: in Complex Systems-Operational Approaches in Neurobiology, Physics and Computers, ed. by H. Haken, Springer Series in Synergetics, Vol. 31 (Springer, Heidelberg, Berlin, 1985)
14. J.A.S. Kelso, J.P. Scholz, G. Schöner: Phys. Lett. A118, 279 (1986)
15. J. Yamanishi, M. Kawato, R. Suzuki: Biol. Cybern. 33, 199 (1979)
16. B. Tuller, J.A.S. Kelso: paper presented at the Int. Neuropsych. Soc. meeting, Houston, TX, (Feb. 1984)
17. F. Baldissera, P. Cavallari, P. Civaschi: Neuroscience Lett. 34, 95 (1982)
18. C. L. McKenzie, A.E. Patla: Abstracts Soc. Neuroscience (1983)
19. H. Haken, J.A.S. Kelso, B. Bunz: Biol. Cybern. 51, 347 (1985)
20. B. A. Kay, J.A.S. Kelso, E.L. Saltzman, G. Schöner: J. Exp. Psych.: Hum. Perc. Perf. (in press)
21. G. Schöner, H. Haken, J.A.S. Kelso: Biol. Cybern. 53, 247 (1986)
22. J.P. Scholz, J.A.S. Kelso, G. Schöner: submitted to Phys. Lett. A
23. J.A.S. Kelso, G. Schöner, J.P. Scholz, H. Haken: Physica Scripta (in press)
24. R. Landauer: J. Appl. Phys. 33, 2209 (1962)
25. E. von Holst: in The Behavioral Physiology of Animals and Men, collected works of E. von Holst (U. of Miami Press, Coral Gables, FL, 1973) (original article published 1937) p. 3
26. E. von Holst, U. von Saint Paul: in The Behavioral Physiology of Animals and Men, Collected works of E. von Holst (U. of Miami Press, Coral Gables, FL, 1973) (original article published 1960) p. 220
27. M. L. Shik, F. V. Severin, G. N. Orlovskii: Biophysics 11, 756 (1966)
28. L. I. Mortin, J. Keifer, P.S.G. Stein: J. Neurophysiol. 53, 1501 (1985)
29. G. A. Robertson, L.I. Mortin, J. Keifer, P.S.G. Stein: J. Neurophysiol. 53, 1517 (1985)
30. S. Grillner, Science 228, 143 (1985)
31. J.A.S. Kelso, E.L. Saltzman, B. Tuller: J. Phonetics 14, 29 (1986)
32. E. Thelen, K. Skala, J.A.S. Kelso: Developmental Psychology (in press)
33. E. Thelen, J.A.S. Kelso, A. Fogel: Developmental Review 7 (in press)
34. J.A.S. Kelso, D.L. Southard, D. Goodman: Science 203, 1029 (1979)
35. J.F. Soechting, C.A. Terzuolo: Neuroscience (in press)

Synergetics and Social Science

W. Weidlich

2nd Institute of Theoretical Physics, University of Stuttgart,
Pfaffenwaldring 57, D-7000 Stuttgart 80, Fed. Rep. of Germany

1. INTRODUCTION

In the last 15 years synergetics has developed towards becoming the general theoretical frame for the description of nonequilibrium phase transitions and for the explanation of selforganization of order in space and time within open systems.

Hermann Haken, to whom this volume is dedicated, has contributed the main building stones and construction plans of this new wing of the house of science. Due to the wisdom of the architect, who used universal building principles, the wing of synergetics is situated in a very fortunate position and connects many parts of this house, for instance physics, chemistry and biology, which lay separate before. The author of this article, knowing the architect of the synergetics building as friend and colleague over the years, tried to find out whether there also exists a passage between this new wing and a building lying somewhat farther apart, the house of social science. Indeed I hope to demonstrate that this pathway exists and that the dwelling of social scientists is also constructed making use of the architectonic rules of synergetics, although the builders perhaps prefer to paint their house in a more colourful style than natural scientists do.

Let us now wander along the passage connecting the social sciences with the natural sciences, taking the overview from the central place, from synergetics!

We start with one central problem, which can be found in all sciences: The choice of the level of description. The level can be chosen to be more fundamental and microscopic, or to be more phenomenological and macroscopic. Synergetics, the theory of cooperative phenomena in multicomponent systems, evidently has to do with this problem. Physics, in particular statistical physics, but also chemistry and biology provide many examples for the adequate choice of micro- and macrolevels: In the physics of the atomic nucleus the proton and neutron are taken as the elements, but in statistical mechanics or chemical kinetics atoms or molecules are considered as the elementary units, and galaxies are seen as composed of stars and interstellar dust as their elements. Similarly, in biology the molecules composing the cell, the cells composing the organ, and the organs composing the individual are seen as the elementary units on the respective level of description. In any case it must be stated that "microlevel" and "macrolevel" are not absolute, but relative concepts.

But one main question must now be raised: Which property predestines certain "elements" to serve as the elementary units composing the next higher macrolevel in the hierarchy of levels? Looking at the examples taken from physics, chemistry and biology, we see that the "elementary units" need not necessarily be void of internal structure; but their internal structure including their dynamics must by strong inner interaction be more or less selfcontained so that the internal degrees of freedom of the units only weakly interact with the variables of the macrolevel composed by these units. (In statistical physics for instance the inner degrees of freedom of the nucleus can be considered as "frozen", if the macrolevel of the thermodynamics of a gas is considered. And only a few vibrational or rotational states of a molecule interact with the macrolevel, for instance by contributing to the specific heat of a molecular gas.)

Simultaneously and vice versa this means that the dynamics of the macrolevel is relatively independent of the detailed structure of its subunits. Even more: The macrolevel is governed by a few order parameters obeying a selfcontained dynamics. The influence of the microlevel here only shows up in some control parameters and in fluctuation terms, so that the macroequations generally take the form of nonlinear Langevin equations. The fact that the properties of such macrolevel equations can be studied independently of their formation from whatsoever microlevels is of extreme importance: It establishes the interdisciplinary relevance of synergetics, which by definition studies the space-time structure of open multicomponent systems on the macrolevel.

Let us now come to the social sciences. Here we find an intuitively clear separation into microlevel and macrolevel. The unit of the microlevel is the individual, and the macrolevel is the society consisting of collectives of individuals acting and behaving in more or less definite ways. It is evident that here the unit is itself a highly complex system. But its inner structure from metabolism to braindynamics is selfcontained in first order and clearly delimits the individual from its environment. On the other hand the individual acts on, interacts with, and reacts to the society and the modes and intensity of its interaction may even vary with time.

As we have already noted, the dynamics of the macrolevel is governed by a few order parameters only. But this very fact poses a problem: Which are the properties distinguishing the order parameters from other macrovariables which also could be formed out of the microvariables? And more generally: Can order parameters change the global mode of their dynamics and can there arise "new" and/or decay "old" order parameters under special circumstances? In other words: Do there exist phase transitions between different modes of the macrodynamics and is the space of order parameters an open one, so that its dimension can vary?

The answer is "yes" and we shall come back to the explanation of macro-order provided by synergetics. But first let us compare the system of natural science and social science with respect to the dynamics and phase transitions of their respective order parameter.

Although many equilibrium- and nonequilibrium phase transitions in systems of natural science have been studied in detail now, and also in this book, it can be said that a phase transition that means a global change of macroscopic order, still remains a rare and dramatic event in such systems, appearing under very special constallations of the control parameters. The same holds for the appearance or disappearance of order parameters.

In contrast, phase transitions of different order of magnitude are a frequent phenomenon in the human society! There do not only exist fundamental phase transitions like revolutions and the transition from peace to war or vice versa, but also smaller ones like political shifts, structural changes of the economy etc. Also the appearance of new order parameters can frequently be observed in the society: New political parties, new ideologies, new products, new ways of communication or of traffic refer to new collective behavioral modes and hence must be considered as phase transitions by arisal of new order parameters.

The fact, that phase transitions are rare events in natural science, but frequent events in social science, can at least partially be traced back to the simplicity or complexity of the respective units of the microlevel.

Whereas the interactions between the units (i.e. particles, atoms, molecules) of physical or chemical macrosystems are defined once for all by fundamental laws of nature and are rigid in this sense, the range of possible interactions between human individuals is more rich and flexible: Although the fundamental interactions between individuals as well as their internal structure remain selfdetermined and independent of the society in a first approximation, it may happen, that new states of the individuals and new modes of their interaction are activated (excited), and others suppressed, if the global state of the society changes. Evidently this interrelation between the global state and the interaction modes between microunits can increase the diversity of possible macrophases and of transitions between them.

It is wellknown that critical fluctuations of microvariables play a fundamental role in phase transitions. Since microfluctuations are not predictable, the result of a phase transition with a multiple of macro-alternatives is also not predictable. Hence it is clear that systems with frequently and permanently arising phase transitions completely lose their predictability. In some sense phase transitions transport the microchance to the macrolevel! Taking the frequency of phase transitions in the respective systems as a characteristic measure, one could define a sequence of sciences from "hard" to "soft" science, which simultaneously explains the gradual loss of predictability along this sequence.

After studying the levels of description we come to the problem of time scales. It was the ingenious idea of Hermann Haken to combine and connect both problems by showing that the fast or slow time scale on which interacting variables move decides the question which of the variables is an appropriate candidate for becoming a macrovariable and an order parameter, and which is doomed to remain a microvariable. It became clear that a variable grows towards a macrovariable because its dynamics is slow and persistent (but becoming unstable), whereas a variable remains a microvariable because its dynamics is fast and transient. The reason is, that fast modes hastily adapt to the ruling modes, whereas the slow but permanently ruling order parameters develop their eigendynamics fed and supported by the manifold of adaptive and "slaved" micromodes.

This is Haken's slaving principle at the core of synergetics which lead to a mathematical algorithm for the exact elimination of fast variables, whose first step is the adiabatic elimination procedure. The formalism yields nonlinear equations for the order parameters of the form of generalized Ginzburg-Landau-equations. In the meantime this line of investigation and this type of equations has found many applications in explaining the macroscopic space-time structure of physical, chemical and biological systems.

Let us now study the problem, whether the slaving principle and the elimination algorithm for fast variables can also be applied to systems of social science.

Haken's elimination procedure of fast variables presumes the existence of coupled equations of motion for fast (stable) and slow (becoming unstable) variables as well. In physics and chemistry a welldefined microlevel usually provides this set of equations. In social dynamics however, equations for the microlevel, namely for the motivations and activities of the individual steered by his braindynamics, are not available.

Nevertheless it is useful to speculate about the braindynamics and its outcome in order to see whether we are led to an alternative formulation of equations for the macrolevel of the society, which avoids the presumption of equations for the microlevel but still contains the essence of the slaving principle. Let us assume in this sense that the braindynamics in principle exists of a highly complex nonlinearly coupled dynamic system driven by endogenous and exogenous parameters. The manifold of "solutions" of this system, that means the possible trajectories of the relevant brain variables then can be classified into modes of motion (eventually converging to attractors), each of them of a certain regularity in the sense of a quasi-repetitive structure. Now we may assume that such modes correspond to quasi-stationary habits, actions, or quite generally "attitudes" with respect to different "aspects of life" of the individual. Endogenous fluctuations or changes of exogenous parameters however may induce transitions between different modes; they correspond to transitions between attitudes. Since motivations and intentions belong to the relevant parameters of braindynamics, such transitions between attitudes are perceived as "decisions between alternatives" by the individual.

Hence in a global sense the braindynamics amounts to ways of thinking of a certain persistency corresponding to quasi-stable attitudes, and to occasional transitions between these modes of thinking corresponding to decisions to change the attitude.

Without further information about the individual braindynamics it is only possible to model this global picture by introducing the space of possible attitudes of individuals and to describe transitions between attitudes probabilistically, that means in terms of transition probabilities. Such concepts will be introduced explicitly in section 2.

We are thus led in a natural way to a formulation of the dynamics alternative to equations of motion of the Langevin type, namely to a formulation in terms of a Fokker-Planck equation or master equation for the probability distribution of the relevant variables. Such equations also contain the transition probabilities introduced above.

It is at this point that the slaving principle comes in, but in a hidden and disguised form: In order to have a complete description on the microlevel, the probability distribution and the transition probabilities would have to depend on the (unknown!) variables of all individual brains as well as on the macrovariables of the society (in particular the socioconfiguration introduced in section 2). But it is highly plausible to assume that the braindynamics is fast in comparison to the slow macrodynamics of the society. The fast variables quickly adapt to the momentary values of the slowly varying macrovariables of the society. Hence, making use of the adiabatic elimination procedure, the brainvariables can in principle be eliminated and expressed as functions of the socio-macrovariables. As a consequence it is also possible to express the transition probabilities solely in terms of these macrovariables and to go over to a master equation for the socio-macrovariables only.

Our formulation of social dynamics anticipates this possibility of the adiabatic elimination of the brainvariables according to the slaving principle: Whereas the microlevel is only present in rudimentary form in terms of individual attitudes and individual transition probabilities, the dynamics starts from a master equation for the socio-macrovariables only, with transition probabilities solely depending on the latter macrovariables.

In section 2 we introduce the general principles of a quantification of social dynamics along these lines, and in section 3 we discuss the example of the migration of human populations, demonstrating that the characteristics of synergetics reappear in social systems, too.

2. GENERAL CONCEPTS OF QUANTIFICATION

Trying to achieve a quantification of social processes, we proceed from the microlevel of individuals with their attitudes and decisions to the macrolevel of quantitative collective variables and their dynamics. Finally, the interdependence of micro- and macrolevel must be discussed.

2.1 The Microlevel

Aspect Space. The space A comprises all (independent) aspects which the social system may provide for the individual. Examples are: politics, religion, education, habitation, economic supply and demand. Each aspect is considered as a dimension of the aspect space.

Attitude Vectors. The dimensions of $\mathcal{A}$ are further defined in such a manner, that the individual can have different attitudes with respect to each aspect. In the case of the above-mentioned aspects the attitudes are: political opinions, religious denominations, education levels, place of residence, production activity, consuming habits. If A aspects $a=1,2,\ldots A$ are considered and d_a attitudes $i_a=1,2,\ldots d_a$ belong to aspect a, the attitude vector $i \in \mathcal{A}$ of an individual is defined by $\mathbf{i}=\{i_1,i_2,\ldots i_A\}$. The space $\mathcal{A}$ can now be defined as the A dimensional lattice space with $L=d_1 \cdot d_2,\ldots d_A$ lattice points, whose coordinates are given by the attitude vectors.

Homogeneous Subpopulations. The society is composed of subpopulations P_α. A homogeneous subpopulation is defined by the (in principle verifyable) assumption that its members exhibit the same probabilistic decision behaviour with respect to the aspects. In general the classes of equal socio-economic background are candidates for homogeneous subpopulations. It should be clear, that a fine-grained description of the society requires its decomposition into more subpopulations than a coarse-grained description.

Individual Probability Transition Rates. The dynamics on the microlevel of a society is generated, if individuals decide to change their attitude. Such decisions can only be described in probabilistic manner. In this sense we introduce individual probability transition rates:

p_{ji}^{α} = probability transition rate per unit of time for a member of subpopulation P_α to change from attitude i to attitude j (2,1)

Transition Rates and Utilities. Let us now ask: What is the "driving force" for the transition of an individual between different attitudes? It is assumed that there exists a measure u_i^{α} and u_j^{α} of the subjective "utility" to a member of subpopulation P_α of the "old" attitude i and the "new" attitude j, respectively. The individual compares both utilities and for increasing difference $(u_j^{\alpha}-u_i^{\alpha})$ the transition rate in population P_α from i to j will more and more prevail over the reverse rate from j to j. Since p_{ji}^{α} must be a positive definite quantity by definition, its most plausible form in terms of the utilities of the old and new attitude is

$$p_{ji}^{\alpha} = \nu \exp\{u_j^{\alpha} - u_i^{\alpha}\}. \tag{2,2}$$

Arguing reversely, the formula (2,2) can be subject to empirical verification leading to the determination of the utilities, if the transition rates p_{ji}^{α} are known (see section 3.4).

Summarizing, we may say that the individual attitude vectors are the microvariables and that the individual probability transition rates describe the microdynamics of the system.

2.2 The Macrolevel

The microvariables and microdynamics now give rise in a natural way to macrovariables and macrodynamic concepts.

The Socioconfiguration. Let n_i^α members of subpopulation P_α take the attitude $\mathbf{i} = \{i_1, i_2, \ldots i_A\}$. Then the socioconfiguration

$$\mathbf{n} = \{n_1^1 \ldots n_1^P, n_2^1 \ldots n_2^P, \ldots n_i^\alpha \ldots, n_L^1 \ldots n_L^P\} \tag{2,3}$$

characterizes in a given moment the distribution of attitudes within the total population of the society. In general it can be assumed that $n_i^\alpha \gg 1$, and n_i^α can be treated as a quasi-continuous variable. The socioconfiguration is considered as a central (multicomponent) macrovariable of the society.

The Material Situation. Whereas the socioconfiguration quantifies the distribution of abstract attitudes in the society, it is also necessary (and is conventionally done in economics) to describe the material situation within the society. This is accomplished by introducing the situation vector

$$\mathbf{y} = (y_1, y_2 \ldots y_s) \tag{2,4}$$

consisting of quantified measures and indicators y_m of material variables. The y_m include prizes, capital, commodity stock, inventory, investment and production-rates, housing, infrastructure, etc.

The Configurational Probability Transition Rate. If one individual of P_α changes its attitude from i to j it induces a change of the socioconfiguration

$$\begin{aligned} \mathbf{n} &= \{n_1^1 \ldots n_i^\alpha \ldots n_j^\alpha \ldots n_L^P\} \\ \rightarrow \mathbf{n}_{ji}^\alpha &= \{n_1^1 \ldots (n_i^\alpha - 1) \ldots (n_j^\alpha + 1) \ldots n_L^P\}. \end{aligned} \tag{2,5}$$

Since each of the n_i^α members of P_α with attitude i can make this transition to j statistically independently, the configurational probability transition rate from $\mathbf{n}$ to $\mathbf{n}_{ji}^\alpha$ is given by

$$\begin{aligned} w(\mathbf{n}_{ji}^\alpha | \mathbf{n}) &\equiv w_{ji}^\alpha(\mathbf{n}) = n_i^\alpha p_{ji}^\alpha \\ &= n_i^\alpha \nu \exp\{u_j^\alpha - u_i^\alpha\}, \end{aligned} \tag{2,6}$$

whereas all transition rates from $\underset{\sim}{\mathbf{n}}$ to $\underset{\sim}{\mathbf{n}}' \neq n_{ji}^\alpha$ vanish:

$$w(\mathbf{n}'|\mathbf{n}) = 0 \quad \text{for} \quad \mathbf{n}' \neq \mathbf{n}_{ji}^\alpha . \tag{2,7}$$

2.3 Equations of Motion

We shall now see that the introduction of configurational probability transition rates leads in a natural way to equations of motion for the socioconfiguration.

Because of the probabilistic description of the microprocesses, we must expect that the exact description of the dynamics of the socioconfiguration can only be a probabilistic one, too. For this purpose we introduce the probability distribution over socioconfigurations:

$$P(\mathbf{n},t) = \text{probability, that the socioconfiguration } \mathbf{n} \text{ exists at time } t \, . \tag{2,8}$$

This probability is normalized according to

$$\sum_{\mathbf{n}} P(\mathbf{n},t) = 1 \, . \tag{2,9}$$

Master Equation. The probability distribution obeys a fundamental equation, the master equation which derives from a dynamic probability balance consideration:

$$\frac{dP(\mathbf{n},t)}{dt} = \sum_{\substack{\alpha,j,i \\ (j \neq i)}} \{w^{\alpha}_{ji}(\mathbf{n}^{\alpha}_{ij})P(\mathbf{n}^{\alpha}_{ij},t) - w^{\alpha}_{ji}(\mathbf{n})P(\mathbf{n},t)\} \, . \tag{2,10}$$

The meaning of (2,10) is intuitively clear: The change with time of the probability of configuration $\mathbf{n}$ is due to two counteractive terms on the r.h.s. of (2,10): The first term describes the probability flow per unit of time from all neighbouring configurations into the configuration $\mathbf{n}$, and the last term (with negative sign) subtracts the probability flow from $\mathbf{n}$ into all neighbouring configurations.

The solution of (2,10) not only yields the evolution of the most probable configurations but also the width and form of the probabilistic fluctuations around them. For most applications in social science it contains too much information, which cannot be compared with the poor empirical data. Therefore it is often sufficient to go over to equations for the meanvalues $\bar{n}^{\alpha}_{i}(t)$ of the variables n^{α}_{i}.

Meanvalue Equations. The meanvalues of n^{α}_{i} are defined by

$$\bar{n}^{\alpha}_{i}(t) = \sum_{\mathbf{n}} n^{\alpha}_{i} P(\mathbf{n},t) \, . \tag{2,11}$$

In a straightforward manner, exact equations of motion for the meanvalues can be derived from the master equation (2,19) (see /2/):

$$\frac{d\bar{n}^{\alpha}_{i}(t)}{dt} = \sum_{j(\neq i)} \{\overline{w^{\alpha}_{ij}(\mathbf{n})} - \overline{w^{\alpha}_{ji}(\mathbf{n})}\} \, . \tag{2,12}$$

These equations are not yet selfcontained, since the calculation of the right-hand side requires the knowledge of the probability distribution $P(\underset{\sim}{n},t)$. For sharply peaked unimodal distributions, however, it is justified to assume

$$\overline{w^{\alpha}_{ij}(\mathbf{n})} \approx w^{\alpha}_{ij}(\bar{\mathbf{n}}(t)) \tag{2,13}$$

so that the equations (2,12) take the approximate but selfcontained form:

$$\frac{d\bar{n}_i^\alpha(t)}{dt} = \sum_{j(\neq i)} \{w_{ij}^\alpha(\bar{\mathbf{n}}(t)) - w_{ji}^\alpha(\bar{\mathbf{n}}(t))\}. \tag{2,14}$$

Equations for Material Variables. In order to complete the dynamic description of the society on the macrolevel, it is necessary to set up equations of motion for the material situation vector too. We shall not go into details here, but make a remark which seems essential to the structure of the theory: Conventional macro-economic models start from phenomenological equations for the macrovariables. This procedure, however, disguises the relation between the micro- and macro-economic level.

An alternative way of finding equations for material variables is suggested by our conceptual framework: The evolution of all material variables - investment, production, prizes and consumption of commodities etc. - is without exception tied to the decisions and actions of subgroups of the population - managers, producers, merchants, consumers etc. - who may be partners or opponents. Therefore the dynamics of material variables can be traced back to the attitudes (including modes of action) and transitions between attitudes of members of subpopulations. Simultaneously this way of setting up macro-economic equations provides the link to micro-economics!

2.4 The Interdependence of Micro- and Macro-Level

At a first glance it might seem that the microlevel (individual attitudes, actions and decisions) determines the dynamics of the macrovariables, whereas no feedback takes place from macro- to microlevel. This is however not the case! Instead there exists a cyclic coupling between both levels: The individual activities merge into the collective state and dynamics of the society, and, vice versa, the latter leads to partial adjustment of individual behaviour, so that micro-behaviour and macro-dynamics are coupled selfconsistently.

Formally, the dependence on the macrostate of individual decisions is expressed by the fact, that the utility u_i^α of attitude i to a member of population P_α in general depends on the existing macrostate:

$$u_i^\alpha = u_i^\alpha(\mathbf{n},\mathbf{y}). \tag{2,15}$$

Hence, also the individual probability transition rates become a function of the macrostate

$$p_{ji}^\alpha(\mathbf{n},\mathbf{y}) = \nu\exp\{u_j^\alpha(\mathbf{n},\mathbf{y}) - u_i^\alpha(\mathbf{n},\mathbf{y})\}. \tag{2,16}$$

As a consequence the meanvalue equations (2,14) become nonlinear equations. The same holds in general for equations of the material situation.

The nonlinearity of the constitutive equations reflects the complexity of social processes. The "slaving principle" mentioned in the Introduction is implied by (2,16), since this equation directly ties the microdecisions arising from fast microvariables in the brain of the individual to the macrostate, that means to the order parameters. The nonlinear dynamics now leads - depending on initial conditions and on exogeneous control parameters - in general to a complex variety of possibilities for selforganizing dynamic patterns within a society.

3. AN EXAMPLE: THE MIGRATION OF HUMAN POPULATIONS

Migration of human populations is an appropriate example of dynamic social processes because of the following reasons:

a) Each individual takes a welldefined decision in a given interval of time either to stay or to change the area of its residence.

b) The interaction of populations on various psychological, social and economic levels leads to nontrivial migratory effects, for instance homogeneous intermixture of populations, ghetto formation, growth of metropoles etc.

c) The theory is amenable to comparison with the empiric situation, because migration data are available in many countries.

3.1 The General Migratory System

The system of P subpopulations $\alpha = 1,2,...P$ migrating between L areas $i = 1,2,...L$ is easily seen to be an application of the general concepts developed above, if the following identifications are made:

a) attitude i $\hat{=}$ "to live in region i"

b) utility u_i^α $\hat{=}$ "measure of attractivity of region i to a member of subpopulation P_α"

c) individual probability transition rate $p_{ji}^\alpha = \nu\exp\{u_j^\alpha - u_i^\alpha\}$
$\hat{=}$ probability per unit of time of a member of population P_α to move from i to j

d) socioconfiguration $\mathbf{n} = \{n_i^\alpha\}$
$\hat{=}$ regional population distribution, such that n_i^α members of subpopulation P_α live in region i.

A simple but nontrivial form of the utilities is found by considering them as functions of the socioconfiguration (see d), which may be expanded in a Taylor series up to first order:

$$u_i^\alpha(\mathbf{n}) = \delta_i^\alpha + \sum_{\beta=1}^{P} \sum_{j=1}^{L} \varkappa_{ij}^{\alpha\beta} n_j^\beta . \tag{3,1}$$

Here we consider the case, that the attractivity of region i only depends on the population in that region. That means, (3,1) is simplified to

$$u_i^\alpha(\mathbf{n}) = \delta_i^\alpha + \sum_{\beta=1}^{P} \varkappa^{\alpha\beta} n_i^\beta . \tag{3,2}$$

According to their meaning, the δ_i^α are denoted as preference parameters, and the $\varkappa^{\alpha\beta}$ as agglomeration parameters. After inserting (3,2) into (2,2) and (2,6) one obtains the explicit form of the master equation (2,10) and the meanvalue equations (2,14). In particular, the meanvalue equations now assume the form

$$\begin{aligned}\frac{d\bar{n}_i^\alpha}{dt} = & \sum_{j(\neq i)} \nu\bar{n}_i^\alpha \exp[(\delta_i^\alpha - \delta_j^\alpha) + \sum_\gamma \varkappa^{\alpha\gamma}(\bar{n}_i^\gamma - \bar{n}_j^\gamma)] \\ & - \sum_{j(\neq i)} \nu\bar{n}_i^\alpha \exp[(\delta_j^\alpha - \delta_i^\alpha) + \sum_\gamma \varkappa^{\alpha\gamma}(\bar{n}_j^\gamma - \bar{n}_i^\gamma)] .\end{aligned} \tag{3,3}$$

In the next sections we shall discuss the results obtained for special cases.

3.2 Two Populations Migrating Between Two Areas

One of the simplest cases is that of two interacting populations P_μ, P_ν migrating between two areas (for instance within a city). The socioconfiguration is $\mathbf{n} = \{n_1^\mu, n_2^\mu, n_1^\nu, n_2^\nu\}$. Since the total population numbers of P_μ and P_ν are constant,

$$n_1^\mu + n_2^\mu = 2\bar{m} \; ; \; n_1^\nu + n_2^\nu = 2\bar{n} \tag{3,4}$$

there exist only two dynamic variables

$$x = \frac{n_1^\mu - n_2^\mu}{2\bar{m}} \; ; \; -1 \leqq x \leqq +1$$

$$y = \frac{n_1^\nu - n_2^\nu}{2\bar{n}} \; ; \; -1 \leqq y \leqq +1 \tag{3,5}$$

and the transition rates (2,2) may be cast into the form

$$p_{12}^\mu = \nu \exp[\Delta u^\mu(x,y)]$$

$$p_{21}^\mu = \nu \exp[-\Delta u^\mu(x,y)]$$

$$p_{12}^\nu = \nu \exp[\Delta u^\nu(x,y)]$$

$$p_{21}^\nu = \nu \exp[-\Delta u^\nu(x,y)] \tag{3,6}$$

with

$$\Delta u^\mu(x,y) = \Pi^\mu + \varkappa^\mu x + \sigma^\mu y$$

$$\Delta u^\nu(x,y) = \Pi^\nu + \varkappa^\nu y + \sigma^\nu x \tag{3,7}$$

where $\varkappa^\mu$, $\varkappa^\nu$ describe the internal agglomeration trend within populations P_μ, P_ν respectively, and σ^μ, σ^ν the "sympathy" trends to live together with the other population.

Three qualitatively different classes of migratory dynamics can now be described within the model by different choices of the trend parameters $\varkappa^\mu$, $\varkappa^\nu$, σ^μ, σ^ν. (For simplicity we only consider cases with $\Pi^\mu = \Pi^\nu = 0$.) Characteristic cases of each class are depicted in figures 1a,b to 3a,b. The figures a) show the fluxlines and figures b) the stationary probability distributions belonging to the same trendparameters. For illustrative purposes small numbers $\bar{m} = \bar{n} = 20$ have been chosen in figures b), leading to a considerable width of the distribution.

<u>Case 1</u>, with moderate values of agglomeration trends $\varkappa^\mu$, $\varkappa^\nu$ and sympathy trends σ^μ, σ^ν leads to one stable fixed point, the homogeneous population mixture (see Fig. 1a,b).

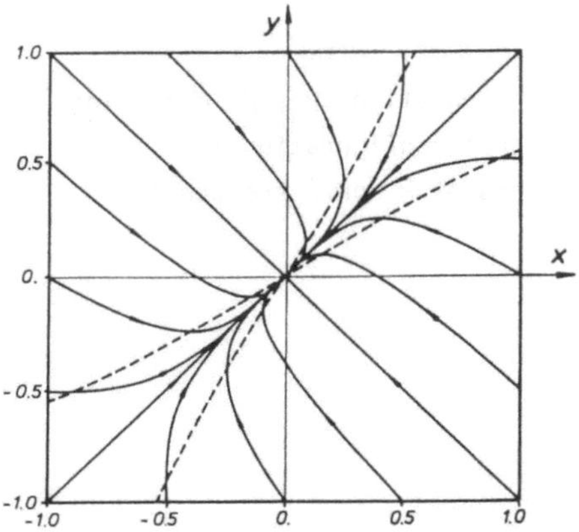

Figure 1a Fluxlines of meanvalues for agglomeration trends $\varkappa^{\mu} = \varkappa^{\nu} = 0.2$ and sympathy trends $\sigma^{\mu} = \sigma^{\nu} = 0.5$

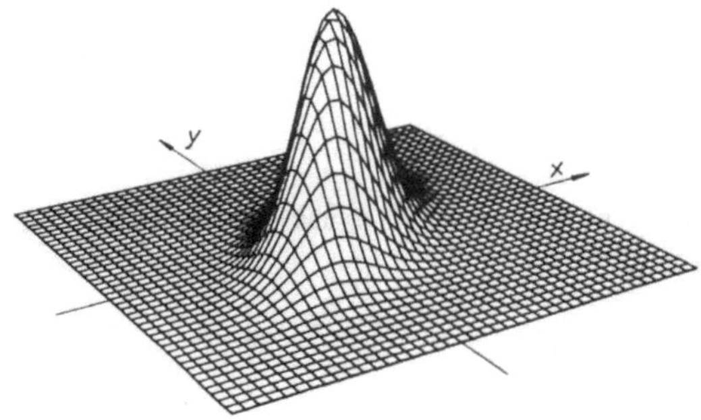

Figure 1b Stationary solution of the master equation for $\bar{m} = \bar{n} = 20$ and trend parameters as in Fig.1a

Case 2, with high values of agglomeration trends $\varkappa^{\mu}$, $\varkappa^{\nu}$ and positive sympathy trends leads to two stable fixed points in the first and third quadrant describing that both populations tend to agglomerate in the same region (see Fig. 2a,b). For negative sign of σ^{μ} and σ^{ν} that is for "mutual aversion", the fixed points would be in the second and fourth quadrant. This case describes ghetto formation, that means separate agglomeration of each of the populations in a separate area.

Case 3, with high values of agglomeration trends $\varkappa^{\nu}$, $\varkappa^{\mu}$ and asymmetric sympathy trends $\sigma^{\nu} = -\sigma^{\mu}$ leads to solutions of the meanvalue equations approaching a limit cycle and to a quadrumodal stationary probability distribution (see Fig. 3a,b).

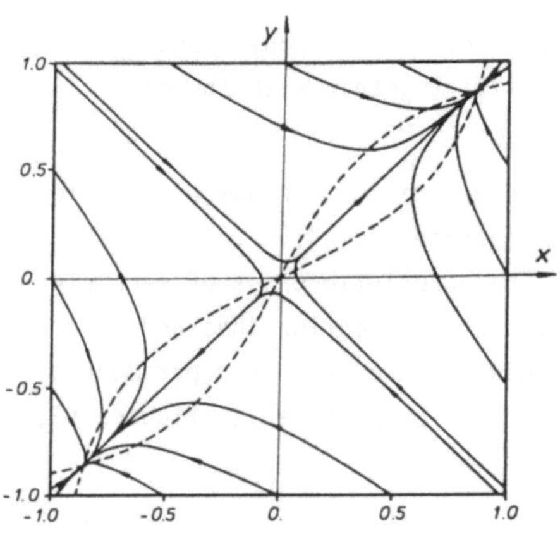

Figure 2a Fluxlines of meanvalues for agglomeration trends $\varkappa^{\mu} = \varkappa^{\nu} = 0.5$ and sympathy trends $\sigma^{\mu} = \sigma^{\nu} = 1.0$

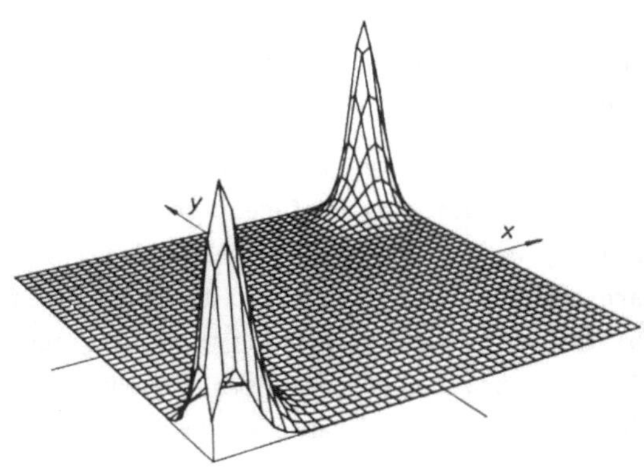

Figure 2b Stationary solution of the master equation for $\bar{m} = \bar{n} = 20$ and trend parameters as in Fig.2a

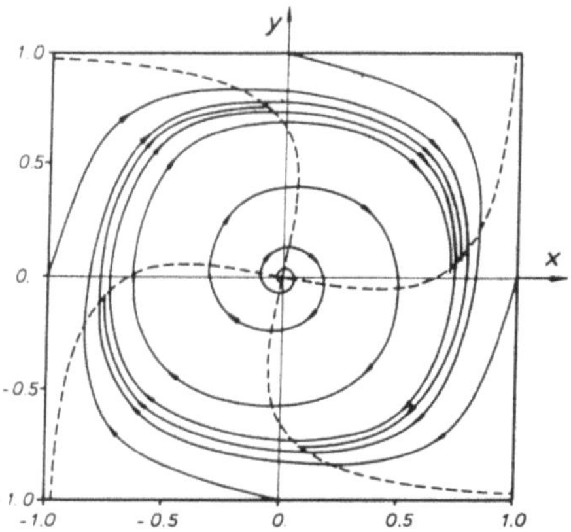

Figure 3a Fluxlines of meanvalues for agglomeration trends $\varkappa^{\mu} = \varkappa^{\nu} = 1.2$ and asymmetric sympathy trends $\sigma^{\mu} = -\sigma^{\nu} = 1.0$

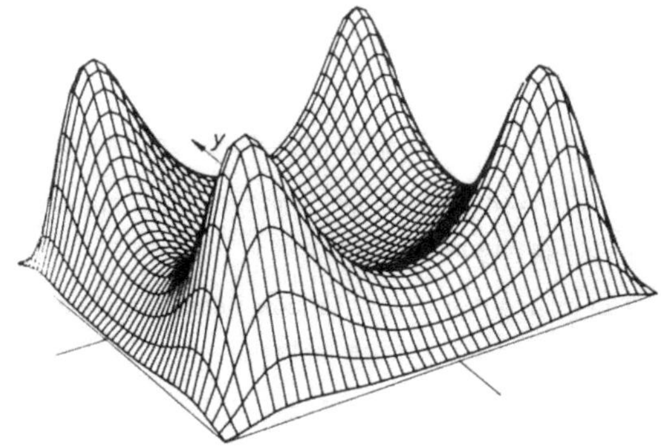

Figure 3b Stationary solution of the master equation for $\bar{m} = \bar{n} = 20$ and trend parameters as in Fig.3a

An interpretation of this restless migration process can be given as follows: Let us start from an initial state with both populations P_{μ} and P_{ν} living in region 1. Because of $\sigma^{\mu} = -\sigma^{\nu} = 1.0$, P_{μ} tends to live together with P_{ν}, whereas P_{ν} tends to evade P_{μ}. Therefore P_{ν} emigrates from region 1, settling in region 2. Thereupon P_{μ} follows P_{ν}, settling in region 2, too, and so on. In our simple model this process of P_{μ} chasing P_{ν} is continuing forever. More realistically the process describes the sequential erosion of suburbs of some big cities by migration of asymmetrically interacting populations of different social standards or different races.

3.3 Three Populations Migrating Between Three Areas

Everyday experience with politics teaches that chaos is a constitutive part of social processes. It must therefore be expected that the equations of motion for social dynamics also reflect this fact. We shall now demonstrate that already the migratory meanvalue equations with constant trend parameters comprise cases of deterministic chaos, as soon as the number of independent dynamic variables exceeds the minimal value 3 (for a more explicit discussion of the following results see /3/).
The analysis starts from (3,3) specialized to three populations P_{α}, $\alpha = 1,2,3$ migrating between three regions $i = 1,2,3$. Because of the constraints

$$\sum_{i=1}^{3} n_i^{\alpha} = N^{\alpha} = \text{const}, \tag{3,8}$$

six of the nine variables $n_i^{\alpha}(t)$ are independent, so that strange attractors could exist.

A numerical analysis with a set of representative values of the agglomeration matrix $\varkappa^{\alpha\beta}$ yields the following results: Chaotic trajectories appear, if

a) the intra-group agglomeration trends $\varkappa^{\alpha\alpha}$ are positive above a critical value,

b) the sign of inter-group sympathy trends $\varkappa^{\alpha\beta}$, $\alpha \neq \beta$ between at least two groups is asymmetric,

c) the interaction matrix $\varkappa^{\alpha\beta}$ is unsymmetric.

Only fixed points or limit cycles appear on the other hand for a fully antisymmetric agglomeration matrix $\varkappa^{\alpha\beta}$.

The route to chaos is now demonstrated in Fig. 4a,b until 8a,b , for which the concrete agglomeration matrix

$$(\varkappa^{\alpha\beta}) = \begin{pmatrix} 1.7 & 1.5 & -1.5 \\ -1.5 & 1.7 & 1.5 \\ \varkappa^{13} & -1.5 & 1.7 \end{pmatrix} \qquad (3,9)$$

has been chosen, where the parameter $\varkappa^{13}$ traverses the values 1.5, -0.5, -0.55, -1.5.

The Figs. 4a to 8a represent for different values of $\varkappa^{13}$ the projections of the asymptotic trajectories (attractors) into the n_1^1/n_1^2-plane, and Fig. 4b until 8b exhibit the corresponding Fourier spectra of $\ell n[n_1^1(f)]$.

Figure 4a,b and 5a,b show two limit cycles (with corresponding Fourier spectra) belonging to the same value $\varkappa^{13} = 1.5$. Both limit cycles merge into one for $\varkappa^{13} = -0.5$, as shown in Fig. 6a,b. Period doubling (or better period multiplying) of the limit cycle appears in Fig. 7a,b after the value of the trend parameter has slightly been shifted to $\varkappa^{13} = -0.55$. Finally the limit cycle has evolved into a strange attractor with continuous Fourier spectrum for the value $\varkappa^{13} = -1.5$, as shown in Fig. 8a,b.

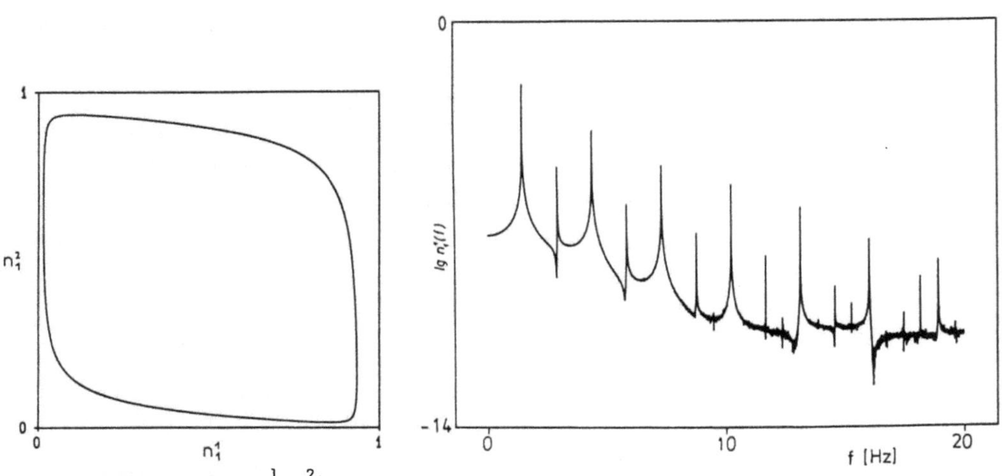

Figure 4a and b The (n_1^1, n_1^2) projection of a first limit cycle and the Fourier spectrum of $\ell n[n_1^1(f)]$ for $\varkappa^{13} = 1.5$

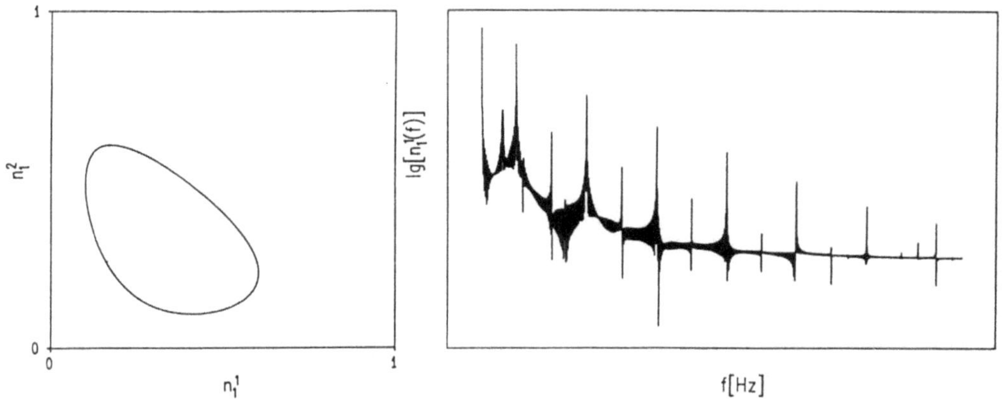

Figure 5a and b The (n_1^1, n_1^2) projection of a second limit cycle and the Fourier spectrum of $\ell n[n_1^1(f)]$ for $\varkappa^{13} = 1.5$

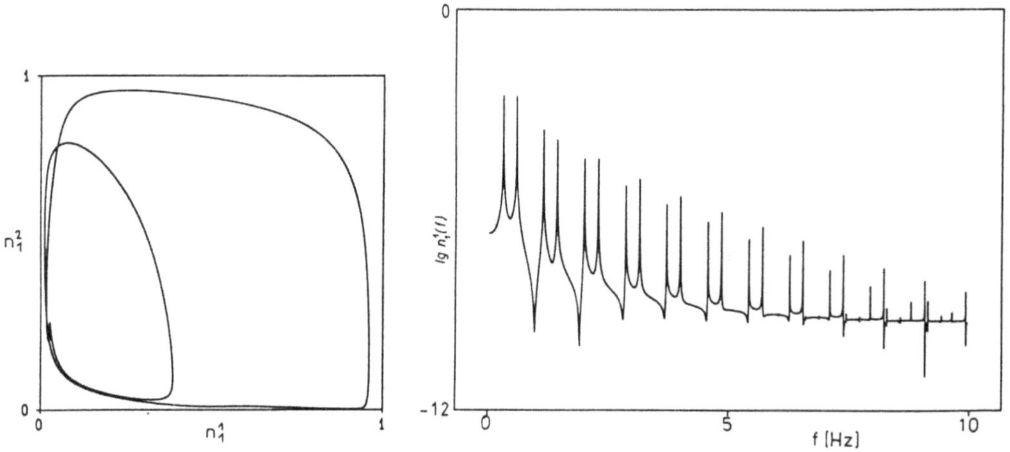

Figure 6a and b The (n_1^1, n_1^2) projection of one limit cycle with the Fourier spectrum of $\ell n[n_1^1(f)]$ for $\varkappa^{13} = -0.5$

The structure of the strange attractor can be further investigated by making a Poincaré map of the trajectory: The hyperplane $n_1^1 = 1/3$ is chosen and all points $\{n_1^1 = 1/3\ ,\ n_2^1, \ldots n_3^3\}$ where the strange attractor pierces this hyperplane are registered. The projections of these points into the planes

$$(n_2^1\ ,\ n_1^3)\ ,\ (n_2^2\ ,\ n_2^3)\ ,\ (n_1^3\ ,\ n_2^3)\ ,\ (n_2^1\ ,\ n_2^3)$$

are depicted in Fig. 9. They indicate that the trajectory of the strange attractor lies almost on a two-dimensional surface. Indeed, a detailed investigation shows that the correlation dimension of this strange attractor is D = 2.14, /3/.

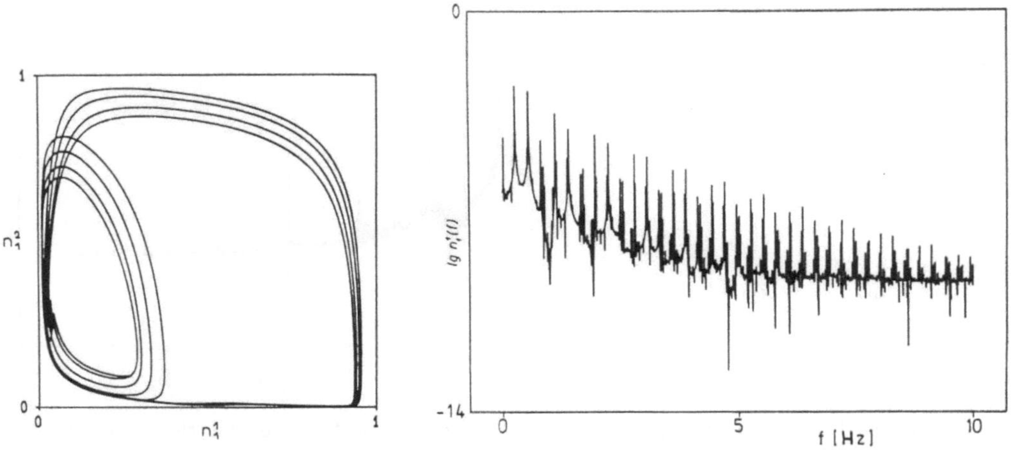

Figure 7a und b The (n_1^1, n_1^2) projection of a limit cycle with period multiplicity and the Fourier spectrum of $\ell n[n_1^1(f)]$ for $\varkappa^{13} = -0.55$

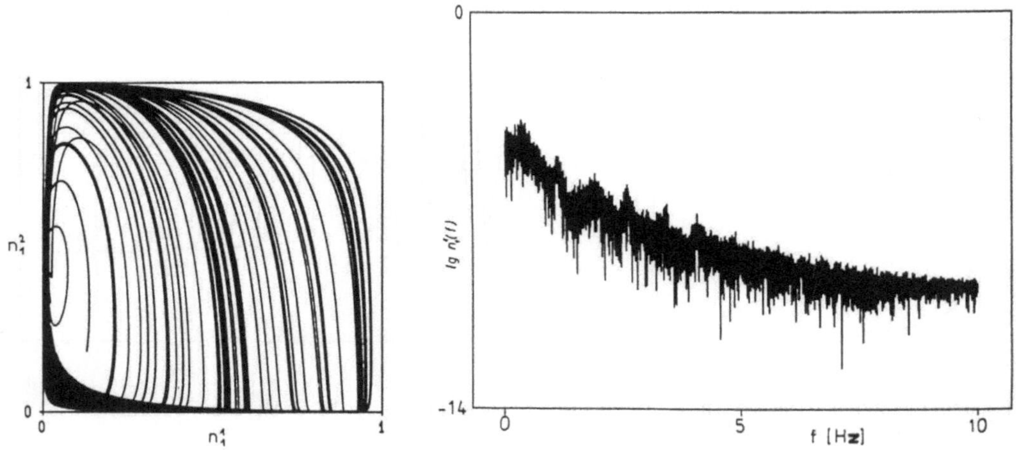

Figure 8a and b The (n_1^1, n_1^2) projection of a strange attractor with continuous Fourier spectrum of $\ell n[n_1^1(f)]$ for $\varkappa^{13} = -1.5$

3.4 One Population in L Areas: Empiric Evaluation

Whereas the preceding sections focussed on theoretical investigations demonstrating how the model description comprises various migratory phenomena, we shall now turn to concrete applications /4/.

In federally organized countries with, say, L states or regions, usually the following data for interregional migration are available year by year:

Region	Population	Number of transitions per year from i to j				
1	n_1	*	....	w^e_{j1}	....	w^e_{L1}
2	n_2	w^e_{12}	....	w^e_{j2}	....	w^e_{L2}
⋮	⋮	⋮		⋮		⋮
i	n_i	w^e_{1i}	....	w^e_{ji}	....	w^e_{Li}
⋮	⋮	⋮		⋮		⋮
L	n_L	w^e_{1L}	....	w^e_{jL}	....	*

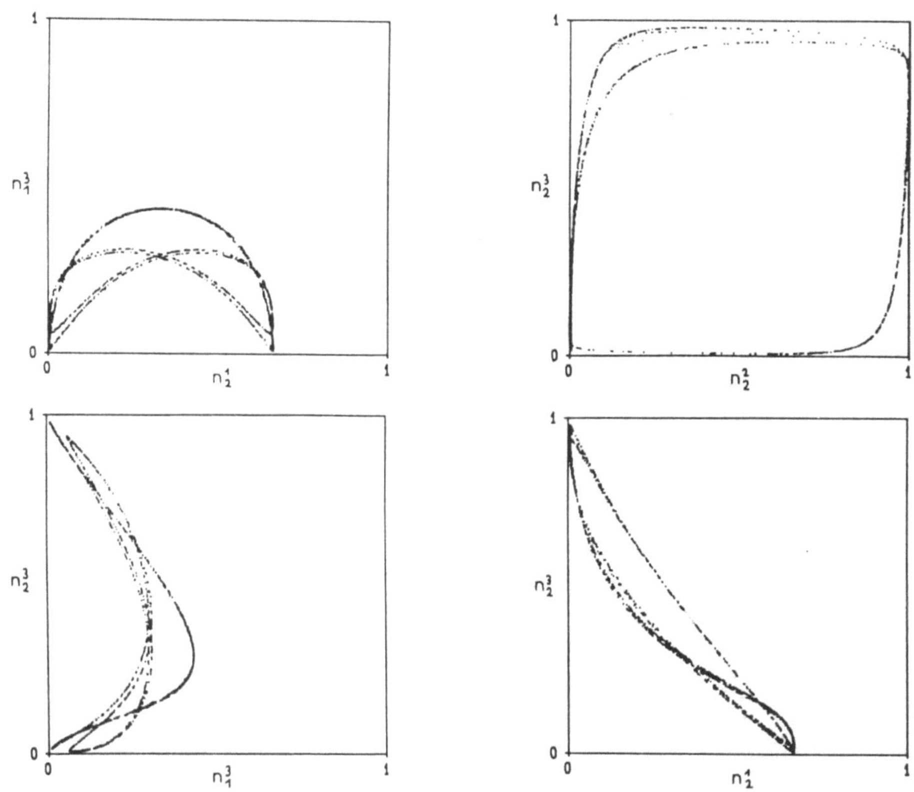

Figure 9 Projection of the traversing points of the phase trajectory of a strange attractor ($\varkappa^{13} = -1.5$) through the hyperplane $n^1_1 = 1/3$ onto different planes

In the migratory model the configurational transition rates

$$w_{ji}^{th}(t) = \nu_{ji}(t)\exp(u_j(t)-u_i(t))n_i(t) \qquad (3,10)$$

correspond to the empirical transition matrix elements $w_{ji}^{e}(t)$ listed above. Formula (3,10) is a generalization of (2,6) including only regional utilities $u\ (t)$ but also a mobility factor

$$\nu_{ji}(t) = \nu_{ij}(t) = \nu_o(t)e^{-D_{ij}} \qquad (3,11)$$

where $\nu_o(t)$ is a global timedependent mobility and $D_{ij} = D_{ji}$ an effective distance between regions i and j.

The utilities and mobilities in formula (3,10) now have to be chosen in a manner that leads to optimal agreement between $w_{ji}^{th}(t)$ and the empiric $w_{ji}^{e}(t)$. A standard procedure for optimal fitting is the least square method: The $u_i(t)$ and $\nu_{ji}(t)$ are determined by the requirement, that

$$\sum_{j,i=1}^{L} \sum_{t=1}^{T} [w_{ji}^{th}(t) - w_{ji}^{e}(t)]^2 \stackrel{!}{=} \text{Minimum} \qquad (3,12)$$

Figure 10 shows as a result of this regression analysis the timedependent "regional utilities" of the 11 states of the Federal Republic of Germany.

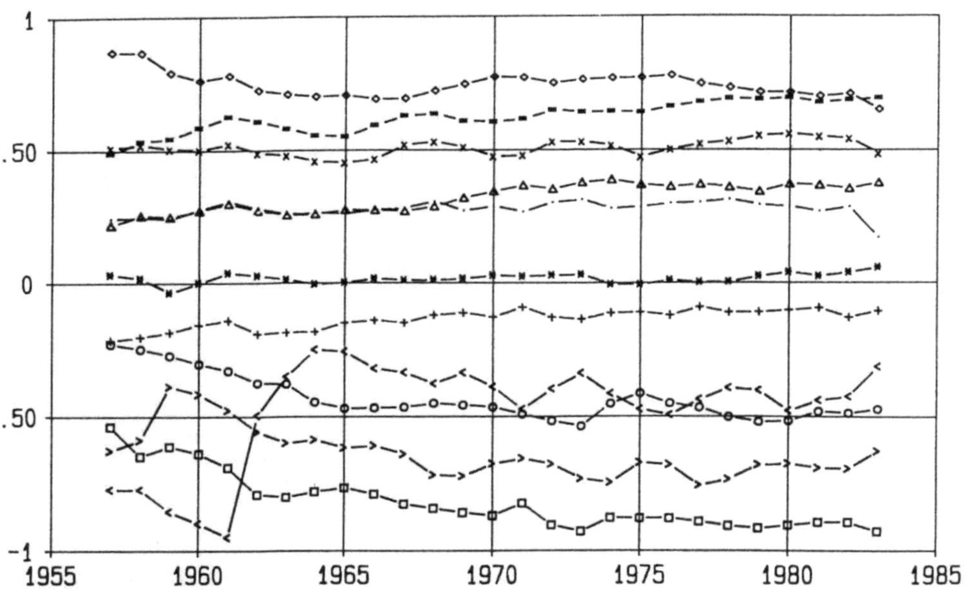

Figure 10 Regional utilities $u_i(t)$ for the 11 countries of the Federal Republic of Germany. Notations: + Schleswig-Holstein, o Hamburg, Δ Niedersachsen, □ Bremen, ◇ Nordrhein-Westfalen, • Hessen, * Rheinland-Pfalz, x Baden-Württemberg, = Bayern, > Saarland, < West Berlin

In the further analysis one tries to represent the regional utilities $u_i(t)$ in terms of the population numbers $n_i(t)$ and further socioeconomic key-factors. The following form can be substantiated:

$$u_i(t) = \varkappa n_i(t) - \sigma n_i^2(t) + \delta_i(t) . \qquad (3,13)$$

In this formula the dependence of $u_i(t)$ on the size of the region is taken into account by the agglomeration term $\varkappa n_i(t)$ and the saturation term $-\sigma n_i^2(t)$ whereas the "preference" $\delta_i(t)$ is a measure of the size-independent attractivity of the region. The preference can be fitted by appropriately selected socio-economic key-factors $\Omega_i^\alpha(t)$ as follows:

$$\delta_i(t) = \sum_\alpha a_\alpha \Omega_i^\alpha(t) . \qquad (3,14)$$

Figure 11 depicts the preferences that means the size-independent part of the regional utilities. The final figure 12 shows how all utilities found in Fig. 10 can simultaneously be fitted by using six regional socio-economic variables only: The population number $n_i(t)$ and its square $n_i^2(t)$ (see (3,13)), while $\delta_i(t)$ is fitted according to (3,14) by four key-factors: Number of overnight accommodations, export index, rate of unemployment and rate of employment in service sector.

The attempt to represent the regional utilities in terms of socio-economic key-factors is of course not made for its own sake but for deeper reasons: In complex

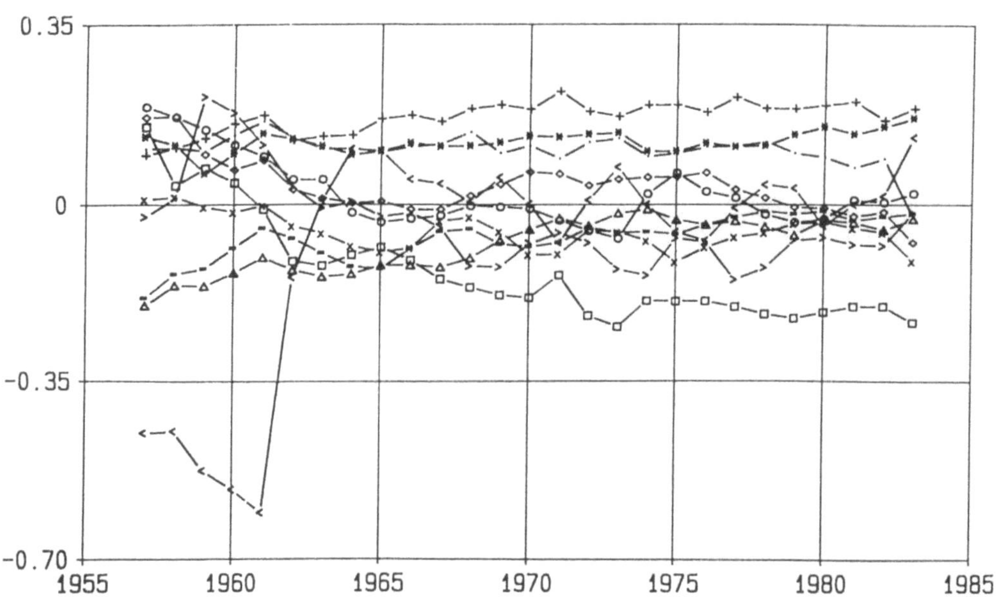

Figure 11 Regional preferences $\delta_i(t)$ for the 11 countries of the Federal Republic of Germany; notation as in Fig. 10

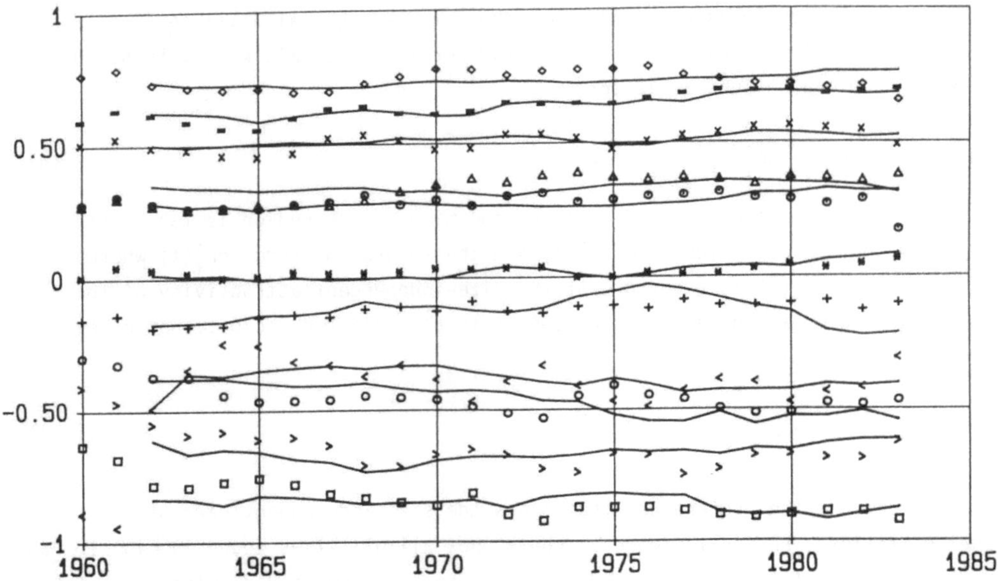

Figure 12 Representation of regional utilities by socio-economic variables. Symbols: regional utilities from migratory data as in Fig. 10 Straight lines: Representation of utilities according to (3,13) and (3,14) using $n_i(t)$, $n_i^2(t)$ and four key-factors

selforganizing systems like the society it is difficult to find cause-effect relations in a direct manner. Therefore the fact that the total space-time-dependence of utilities can be represented as linear combination of a few selected socio-economic factors only is taken as an indirect indicator of their causative nature for the migratory dynamics.

REFERENCES

1. H. Haken: Synergetics, an Introduction, 2nd ed. Springer Ser., Vol.1 (Springer, Berlin, Heidelberg 1977)
2. W. Weidlich, G. Haag: Concepts and Models of a Quantitative Sociology, ed. by H. Haken, Springer Ser. Syn., Vol. 14 (Springer, Berlin, Heidelberg 1983)
3. R. Reiner, W. Weidlich: Chaotic and Regular Dynamics in Migratory Systems, Proceedings of the 2nd European Simulation Congress, Antwerp, Belgium p.105(1986) SCS Europe, Ghent, Belgium
4. W. Weidlich, G. Haag: (ed.), Interregional Migration - Dynamic Theory and Comparative Evaluation, Springer Ser. Syn. (Springer, Berlin, Heidelberg 1987)

Part VIII

Hermann Haken's Publications

1. "Zum Identitätsproblem bei Gruppen", Math. Z. 56, 335 (1952)

2. (with H. Volz) "Zur Quantentheorie des Mehrelektronenproblems im Festkörper", Z. Phys./Chem. 198, 61 (1951)

3. "Eine modellmäßige Behandlung der Wechselwirkung zwischen einem Elektron und einem Gitteroszillator", Z. Phys. 135, 408 (1955)

4. "Über die Struktur der Lösung des Mehrelektronenproblems im Festkörper und ein Theorem von Bloch", Z. Naturforsch. 8a, 228 (1954)

5. "Zum Energieschema des Systems Elektron-schwingendes Gitter", Z. Phys. 139, 66 (1954)

6. "On the Problem of Radiationless Transitions", Physica XX, 11 (1954)

7. "Die Bewegung elektronischer Ladungsträger in polaren Kristallen", in Halbleiterprobleme II, ed. by W. Schottky (Vieweg, Braunschweig 1955)

8. "Die Unmöglichkeit der Selbstlokalisation von Elektronen im störstellenfreien Kristallgitter", Z. Naturforsch. 10a, 253 (1955)

9. "Zu Fröhlich's eindimensionalem Modell der Supraleitung", Z. Naturforsch. 11a, 96 (1956)

10. (with W. Schottky) "Allgemeine optische Auswahlregeln in periodischen Kristallgittern", Z. Phys. 144, 91 (1956)

11. "Theorie des excitons dans les cristaux polaires", J. Phys. et Radium 826, 17 (1957)

12. "Application of Feynman's New Variational Procedure to the Calculation of the Ground State Energy of Excitons", Il Nuovo Cimento 4, 1608 (1956)

13. "Die freie Weglänge des Excitons im polaren Kristall", Z. Naturforsch. 11a, 875 (1956)

14. "Kopplung nichtrelativistischer Teilchen mit einem quantisierten Feld", Nuovo Cimento 10, 1230 (1956)

15. "Der heutige Stand der Exzitonen-Forschung in Halbleitern", in Halbleiterprobleme IV, ed. by W. Schottky (Vieweg, Braunschweig 1957)

16. "Berechnung der Energie des Exzitonen-Grundzustandes im polaren Kristall nach einem neuen Variationsverfahren von Feynman I", Z. Phys. 147, 323 (1957)

17. "Die Theorie des Exzitons im festen Körper", Fortschr. Phys. 6, 271 (1958)

18. (with W. Schottky) "Die Behandlung des Exzitons nach der Vielelektronentheorie", Z. Phys./Chem. NF 16, 218 (1958)

19. "Über den Einfluß von Gitterschwingungen auf Energie und Lebensdauer des Exzitons", Z. Phys. 155, 223 (1959)

20. "On the Theory of Excitons in Solids", J.Phys. Chem. Solids 8, 166 (1959)

21. "Effective Particle Interaction via a Bose-Field", Ann. Phys., Fues Festheft

22. "Sur la theorie des excitons et leur role dans les transferts d'Energie a l'etat solide", J. Chim. Phys. 55, 613 (1958)

23. "Theory of Stark Effect of Excitons", in "Report of the International Conference on the Physics of Semiconductors", Exeter, p. 462 (The Institute of Physics and the Physical Soc., London 1962)

24. "Excitons", in Polarons and Excitons, ed. by C.H. Kuper and G.D. Whitfield (Oliver and Boyd, Edinburgh and London 1963)

25. (with H. Sauermann) "Nonlinear Interactions of Laser Modes", Z. Phys. 173, 261 (1963)

26. (with H. Sauermann) "Frequency Shifts of Laser Modes", Z. Phys. 176, 47 (1963)

27. (with H. Sauermann) "Theory of Laser Action in Solid State, Gaseous and Semi-Conductor Systems" , Proc. of the Summerschool "Enrico Fermi" (Soc. Italiana di Fisica, Bologna 1963)

28. (with E. Haken) "Zur Theorie des Halbleiterlasers", Z. Phys. 176, 421 (1963)

29. (with E. Abate) "Exakte Behandlung eines Laser-Modells", Z. Naturforsch. 19a, 857 (1964)

30. "Theory of Coherence of Laser Light", Phys. Rev. Lett. 13, 329 (1964)

31. "A Nonlinear Theory of Laser Noise and Coherence I", Z. Phys. 181, 96 (1964)

32. "A Nonlinear Theory of Laser Noise and Coherence II", Z. Phys. 182, 346 (1965)

33. (with Der Agobian and M. Pauthier) "Theory of Laser Cascades", Phys. Rev. 140, A 437 (1965)

34. (with W.Weidlich) "Quantum Noise Operators for the N-Level-System", Z. Phys. 189, 1 (1966)

35. "Theory of Intensity and Phase Fluctuations of a Homogeneously Broadened Laser", Z. Phys. 190, 327 (1966)

36. (with V. Arzt, H. Risken, H. Sauermann, Ch. Schmid and W. Weidlich) "Quantum Theory of Noise in Gas and Solid State Lasers with an Inhomogeneously Broadened Line I", Z. Phys. 197, 207 (1966)

37. (with W. Weidlich and H. Risken) "Quantum Mechanical Solutions of the Laser Masterequation I", Z. Phys. 201, 396 (1967)

38. "Dynamics of Nonlinear Interaction Between Radiation and Matter," in Dynamical Processes in Solid State Optics, ed. by R. Kubo and H. Kamimura (Syodabo, Tokyo, W.A. Benjamin, New York 1967)

39. (with W. Weidlich and H. Risken) "Quantum Mechanical Solutions of the Laser Masterequation II", Z. Phys. 204, 223 (1967)

40. (with H. Haug) "Theory of Noise in Semiconductor Laser Emission", Z. Phys. 204, 262 (1967)

41. (with H. Sauermann, Ch. Schmid and H.D. Vollmer) "Theory of Laser Noise in the Phase Locking Region", Z. Phys. 206, 269 (1967)

42. (with G. Strobl) "Exact Treatment of Coherent and Incoherent Triplet Exciton Migration", in The Triplet State (Cambridge University Press, Cambridge 1967)

43. (with R. Graham) "The Quantum Fluctuations of the Optical Parametric Oscillator I", Z. Phys. 210, 276 (1968)

44. (with R. Graham, F. Haake and W. Weidlich) "Quantum Mechanical Correlation Functions for the Eletromagnetic Field and Quasi-Probability Distribution Functions", Z. Phys. 213, 21 (1968)

45. (with R. Graham) "Quantum Theory of Light Propagation in a Fluctuating Laser-Active Medium", Z. Phys. 213, 420 (1968)

46. "Exact Stationary Solution of a Fokker-Planck Equation for Multimode Laser Action", Phys. Lett. 27A, 190 (1968)

47. (with M. Pauthier) "Nonlinear Theory of Multimode Action in Loss Modulated Lasers", IEEE J. QE-4, 454 (1968)

48. "Comments on the Interaction of Excitons and Phonons", in Excitons, Magnons and Phonons in Molecular Crystals, ed. by A.B.Zahlan (Cambridge University Press, Cambridge 1968) p. 185

49. "Exact Stationary Solution of a Fokker-Planck Equation for Multimode Laser Action including Phase Locking", Z. Phys. 219, 3 (1969)

50. "Exact Generalized Fokker-Planck Equation for Arbitrary Dissipative and Nondissipative Quantum Systems", Z. Phys. 219, 411 (1969)

51. "Laser Theory", in Encyclopedia of Physics, Vol. XXV/2C, ed. by S. Flügge (Springer, Berlin 1970)

52. (with A. Bivas, R. Levy, J.B. Grun, C. Comte, S. Nikitine) "Exciton Absorption and Emission of Cuprous Chloride During Excitation by a Powerful Light Source", Opt. Commun. 2, 227 (1970)

53. "Der heutige Stand der Theorie des Halbleiterlasers", in Halbleiterprobleme, ed. by H. Sauter (Vieweg, Braunschweig 1970)

54. "The Semiclassical and Quantum Theory of the Laser", in Quantum Optics, ed. by S.M. Kay and A. Maitland (Academic, London 1970)

55. (with R. Hübner and K. Zeile) "Coherence in Spontaneous Emission", in Quantum Optics, ed. by S.M. Kay and A. Maitland (Academic, London 1970)

56. "Nichtlineare Optik", in Physik (Teubner, Stuttgart 1970)

57. (with R. Graham) "Functional Fokker-Planck Treatment of Electromagnetic Field Propagation in a Thermal Medium", Z. Phys. 234, 193 (1970)

58. (with R.Graham) "Functional Quantum Statistics of Light Propagation in a Two-Level System", Z. Phys. 235, 166 (1970)

59. (with R. Graham and W. Weidlich) "Flux Equilibria in Quantum Systems Far away from Thermal Equilibrium", Phys. Lett. 32A, 2, 129 (1970)

60. (with R. Graham) "Laserlight - First Example of a Second-Order Phase Transition Far away from Thermal Equilibrium", Z. Phys. 237, 31 (1970)

61. (with R. Graham) "Microscopic Reversibility, Stability and Onsager Relations in Systems Far from Thermal Equilibrium", Phys. Lett. 33A, 335 (1970)

62. "Quantum Fluctuations in Nonlinear Optics", Opto-Electronics 2, 161 (1970)

63. "Laserlicht - ein neues Beispiel für einen Phasenübergang", in Advances in Solid State Physics, ed. by O. Madelung (Vieweg - Pergamon, Braunschweig 1970)

64. "Quantenoptik", Naturwissenschaften 58, 188 (1971)

65. (with R. Graham) "Synergetik - Die Lehre vom Zusammenwirken", Umschau in Wissenschaft und Technik Heft 6, 191 (1971)

66. (with R. Graham) "Generalized Thermodynamic Potential for Markoff Systems in Detailed Balance and Far from Thermal Equilibrium", Z. Phys. 243, 289 (1971)

67. (with R. Graham) "Fluctuations and Stability of Stationary Non-Equilibrium Systems in Detailed Balance", Z. Phys. 245, 141 (1971)

68. "The Theory of Coherence, Noise and Photon Statistics of Laser Light", in Laser Handbook, ed. by F.T. Arecchi, E.O. Schulz-Dubois (North-Holland, Amsterdam 1972)

69. (with R. Levy, J.B. Grun, and S. Nikitine) "Shift of the Exciton-Exciton Emission Line in Cadmium Sulfide", Solid State Commun. 10, 915 (1972)

70. (with S. Nikitine) "Spontaneous and Stimulated Emission from Excitons at High Concentration", in Luminescence of Crystals, Molecules and Solutions, ed. by F. Williams (Plenum, New York 1973) p. 4

71. (with S. Nikitine) "Spontaneous and Stimulated Emission from Excitons at High Concentration", Izv. Akad. Nauk SSSR, Ser. Fiz. 37, 220 (1973) (in Russian)

72. (with P. Reineker) "The Coupled Coherent and Incoherent Motion of Excitons and Its Influence on the Line Shape of Optical Absorption", Z. Phys. 249, 253 (1972)

73. (with P. Reineker) "The Influence of Exciton Motion on Spin-Resonance Absorption", Z. Phys. 250, 300 (1972)

74. (with A. Schenzle) "Giant Polaritons", Phys. Lett. 41A, 5, 405 (1972)

75. (with A. Schenzle) "Self-induced Transparency of Excitons", Optics Commun. 6, 2, 96 (1972)

76. (with E. Schwarzer) "The Moments of the Coupled Coherent and Incoherent Motion of Excitons", Phys. Lett. 42A, 317 (1972)

77. (with A. Schenzle) "Giant Polaritons and Selfinduced Transparency of Frenkel-Excitons", Z. Phys. 258, 231 (1973)

78. "Resonant Interaction Between Excitons and Intense Coherent Light", Z. Phys. 262, 119 (1973)

79. (with G. Strobl) "An Exactly Solvable Model for Coherent and Incoherent Exciton Motion", Z. Phys. 262, 135 (1973)

80. "Distribution Function for Systems Far from Thermal Equilibrium", Phys. Lett. 44A, 303 (1973)

81. (with H.G. Wöhrstein) "Atom-Field Correlation, Conservation Laws and the Phase Transition of the Laser", Optics Commun. 9, 2, 123 (1973)

82. "Distribution Function for Classical and Quantum Systems Far from Thermal Equilibrium", Z. Phys. 263, 267 (1973)

83. (with H.G. Wöhrstein) "Theory of Second-Order Mode Locking in Semiconductor Lasers", IEEE J. QE-9, 2 (1973)

84. (with H.G. Wöhrstein) "Stationary Solution of the Fokker-Planck Equations Near Detailed Balance", Phys. Lett. 45A, 231 (1973)

85. (with E. Schwarzer) "Theory of Line Shape of Optical Absorption and Emission Spectra in Molecular Crystals", Optics Commun. 9, 64 (1973)

86. "Synergetics - Towards a New Discipline", in Cooperative Phenomena, ed. by H. Haken and M. Wagner (Springer, Berlin 1973)

87. "Laser Light Emission from Interacting Chaotic Fields", Z. Phys. 265, 105 (1973)

88. "Correlations in Classical and Quantum Systems Far from Thermal Equilibrium", Z. Phys. 265, 503 (1973)

89. "Stability and Fluctuations of Multi-mode Configurations near the Convection Instabilitiy", Phys. Lett. 46A, 193 (1973)

90. "Quantenfeldtheorie des Festkörpers" (Teubner, Stuttgart 1973) ("Quantum Field Theory of Solids", North Holland, Amsterdam 1976); (Japanese, Russian, and Chechoslovakian editions)

91. "Laser Theory" in "Proceedings of the Winter College on Atoms, Molecules and Lasers, Trieste, 1973", ed. by A. Kastler (Int. Atom. Energy Agency, Vienna 1974)

92. "The Phase Transition Approach to Laser Statistics" in Photon Correlation and Light Beating Spectroscopy, ed. by H.Z. Cummings and E.R. Pike (Plenum, New York 1974)

93. "Solution of the Master Equation for Quantum Systems Weakly Coupled to Reservoirs and Far from Thermal Equilibrium", Z. Phys. 266, 265 (1974)

94. "Exact Stationary Solution of the Master Equation for Systems Far from Thermal Equilibrium in Detailed Balance", Phys. Lett. 46A, 443 (1974)

95. "Cooperative Phenomena in Systems Far from Thermal Equilibrium", in From Theoretical Physics to Biology, ed. by M. Marois (Karger, Basel 1973)

96. (with E. Schwarzer) "Theory of the Influence of the Coherent or Incoherent Motion of Triplet Excitons on NMR", Chem. Phys.Lett. 27, 1, 41, (1974)

97. (with J.C. Merle and S. Nikitine) "Temperature Dependence of the 1st Line of the Yellow Series of Cu_2O", Phys. Stat. Solidi. 61, 229 (1974)

98. (with A. Schenzle) "Giant Polaritons and Self-induced Transparency of Excitons", in "Proc. First Taormina Res. Conf. 1972", ed. by E. Burstein and F. de Martini (Oxford 1974) p. 371

99. "Synergetics, Basic Concepts and Mathematical Tools", in Cooperative Effects, ed. by H. Haken (North-Holland, Amsterdam 1974)

100. (with J. Goll and A. Schenzle) "Polaritons at High Light Intensities and in Bose Condensed Exciton Systems", in Excitons at High Density, ed. by H. Haken and S. Nikitine, Springer Tracts in Modern Physics, Vol. 73 (Springer, Berlin 1975)

101. (with S. Nikitine) "Theory of Stimulated Emission by Excitons", in Excitons at High Densitiy, ed. by H. Haken and S. Nikitine, Springer Tracts in Modern Physics, Vol. 73 (Springer, Berlin 1975) p. 192

102. "Exzitonen hoher Dichte", Phys. unserer Zeit 6, 78 (1975)

103. "Statistical Physics of a Chemical Reaction Model", Phys. Lett. 51A, 125 (1975)

104. "Statistical Physics of Bifurcation, Spatial Structures, and Fluctuations of Chemical Reactions", Z. Phys. B 20, 415 (1975)

105. "Generalized Ginzburg-Landau Equations for Phase Transition-like Phenomena in Lasers, Nonlinear Optics, Hydrodynamics and Chemical Reactions", Z. Phys. B 21, 105 (1975)

106. "Analogy Between Higher Instabilities in Fluids and Lasers", Phys. Lett. 53A, 77 (1975)

107. (with A. Wunderlin) "Scaling Theory for Nonequilibrium Systems", Z. Phys. B 21, 393 (1975)

108. "Higher Order Corrections to Generalized Ginzburg-Landau Equations of Non-Equilibrium Systems", Z. Phys. B 22, 69 (1975)

109. "Critical Fluctuations in Continuous Non-Equilibrium Systems", Z. Phys. B 22, 73 (1975)

110. "Cooperative Phenomena in Systems Far from Thermal Equilibrium and in Nonphysical Systems", Rev. Mod. Phys. 47, 67 (1975)

111. "Higher Order Corrections to Generalized Ginzburg-Landau Equations of Non-Equilibrium Systems", Z. Phys. B 23, 388 (1976)

112. "Generalized Onsager-Machlup Function as Solution of a Multi-dimensional Fokker-Planck Equation", Phys. Lett. 55A, 323 (1976)

113. (with H. Ohno) "Theory of Ultra-short Laser Pulses", Optics Commun. 16, 2, 205 (1976)

114. "Stimulated Emission at High Concentration of Excitons" in Molecular Spectroscopy of Dense Phases (Elsevier, Amsterdam 1976)

115. "Statistical Theory of Self-Organizing Structures", in Statistical Physics, ed. by I. Pal and P. Szepfalusy (Hung. Academy of Sciences 1976)

116. "Nonlinear Interaction between Excitons and Coherent Light", in "Proc. Int. School of Physics "Enrico Fermi" Course LXIV, ed. by N. Bloembergen (Soc. Italiana di Fisica, Bologna 1977) p. 350

117. "Laser and Synergetics", Phys. Bl., Heft 12 (1976)

118. (with P. Reineker) "The Coupled Coherent and Incoherent motion of Frenkel Excitons in Molecular Crystals", in Localization and Delocalization in Quantum Chemistry, Vol. II, ed. by O. Chalvets (Reidel, Dordrecht 1976)

119. "Generalized Onsager-Machlup Function and Classes of Path Integral Solutions of the Fokker-Planck Equation and the Master Equation", Z. Phys. B 24, 321 (1976)

120. "Synergetics", Europhys. News, Vol. 7 (1976)

121. (with H. Ohno) "Transient Ultrashort Laser Pulses", Phys. Lett. 59A, 261 (1976)

122. "The Laser - Trailblazer of Synergetics", in Coherence and Quantum Optics IV, ed. by L. Mandel and E. Wolf (Plenum, New York 1977)

123. "Die Kohärenzeigenschaften des Lichts", Acta Phys. Austriaca 47, 59 (1977)

124. (with A. Wunderlin) "New Interpretation and Size of Strange Attractor of the Lorenz Model of Turbulence", Phys. Lett. 62A, 133 (1977)

125. "Synergetics", Phys. Bull., London, Sept. 1977

126. (with W. Egler) "Theory of Damping and Fluctuations of the Polariton Due to Its Interaction with Lattice Vibrations", Z.Phys. B 28, 51 (1977)

127. (with E. Schwarzer) "Theory of the Influence of Exciton Motion on Proton Relaxation in Organic Solids", Phys.Stat.Solidi (b) 84, 253 (1977)

128. Synergetics. An Introduction (Springer, Berlin 1977); (3rd enlarged edition 1983; Russian, Japanese, Chinese, Hungarian editions)

129. "Synergetik", Nachrichten aus Chemie, Technik und Laboratorium, Bd.26, Heft 1 (1978)

130. "Nonequilibrium Quantum Statistics of Lasers and Nonlinear Optics", Ann. Israel Phys. Soc. 2, 99 (1978)

131. "Theory of Ultrashort Laser Pulses", Lectures at the Summerschool at Gowidlinko, Poland ed. by Held, 1977, to be published in the proceedings

132. "Coherence in Synergetics", in Coherence in Spectroscopy and Modern Physics ed. by F.T. Arecchi, R. Bonifacio, and M.O. Scully (Plenum, New York 1978)

133. "Synergetics: Some Recent Trends and Developments", Suppl.Progr. Theoret. Phys. Jpn. 64, 21 (1978)

134. "Nonequilibrium Phase Transitions and Bifurcation of Limit Cycles and Multiperiodic Flows", Z.Phys. B 29, 61 (1978)

135. (with J. Goll) "Exciton Self-induced Transparency and the Dispersion Law of Steady State Exciton-Photon Pulses", Optics Commun. 24, 1 (1978)

136. (with H. Ohno) "Onset of Ultrashort Laser Pulses: First or Second Order Phase Transitions?" Optics Commun. 26, 117 (1978)

137. "Nonequilibrium Phase Transitions and Bifurcation of Limit Cycles and Multiperiodic Flows in Continuous Media", Z.Phys. B 30, 423 (1978)

138. (with J. Goll) "Optical Resonance of Frenkel Excitons", Solid State Commun. 27, 371 (1978)

139. (with H. Olbrich) "Analytical Treatment of Pattern Formation in the Gierer-Meinhardt Model of Morphogenesis", J. Math. Biology 6, 317 (1978)

140. "Synergetics - A Field beyond Irreversible Thermodynamics", in Lecture Notes in Physics, Vol. 84 ed. by L. Garrido, P. Seglar, P.J. Shepard (Springer, Berlin 1978)

141. "Synergetics and a New Approach to Bifurcation Theory", in Structural Stability in Physics, ed. by W. Güttinger and H. Eikemeier (Springer, Berlin 1973)

142. "Nonequilibrium Phase Transitions and Instability Hierarchy of the Laser, an Example from Synergetics", in Synergetics Far from Equilibrium ed. by A. Pacault and C. Vidal (Springer, Berlin 1979)

143. "Theorie ultrakurzer Impulse und ihrer Wechselwirkung mit kondensierter Materie", Acta Physica Austriaca Suppl. XX, 107 (1979)

144. Licht und Materie, (2 volumes) (Bibliographisches Institut Mannheim 1979, 1981) (Light I, II (North Holland, Amsterdam 1981, 1985); (Russian edition))

145. "Synergetics and Bifurcation Theory", Ann. New York Acad. Sci. Vol. 316, 357 (1979)

146. "Die Synergetik. Ordnung aus dem Chaos", Bild d. Wiss. 3 (1980)

147. "Synergetics - Are Cooperative Phenomena Governed by Universal Principles?" Naturwissenschaften 67, 121 (1980)

148. (with H. Klenk) "Ordnung durch Instabilität - ein Beispiel aus der Plasmaphysik", Acta Physica Austriaca 52, 187 (1980)

149. (with J. Goll) "Theory of Optical Bistability of Excitons", Phys.Stat. Solidi (b) 101, 489 (1980)

150. (with H.C. Wolf) Atom und Quantenphysik (Springer, Berlin 1980); (English, Japanese, Italian editions)

151. (with G. Mayer-Kress) "Intermittent Behavior of the Logistic System", Phys. Lett. 82A, 4, 151 (1981)

152. "Synergetik: Nichtgleichgewichte, Phasenübergänge und Selbstorganisation", Naturwissenschaften 68, 293 (1981)

153. "Synergetik - Die Entstehung von Ordnung aus dem Chaos", Festvortrag zum 150. Jahrestag der Universität Stuttgart in Brennpunkte der Forschung ed. by W. Weidlich (Deutsche Verlagsanstalt, Stuttgart 1981)

154. "Selforganization Through Increase of Number of Components", Progr. Theoret. Phys., Suppl. 69 (1980)

155. (with G. Mayer-Kress) "Chapman-Kolmogorov Equation for Discrete Chaos", Phys. Lett. 84A, 159 (1981)

157. "Synergetics. Is Self-organization Governed by Universal Principles", AAAS Selected Symposium 61, 1981

158. (with G. Mayer-Kress) "Chapman-Kolmogorov Equation and Path Integrals for Discrete Chaos in Presence of Noise", Z. Phys. B 43, 185 (1981)

159. "Synergetics - a Study of Complex Systems at Points of Critical Behavior", in Continuum Models of Discrete Systems 4, ed. by O. Brulin and R.K.T. Hsieh (North-Holland, Amsterdam 1981)

160. (with A. Wunderlin) "Generalized Ginzburg-Landau Equations, Slaving Principle and Center Manifold Theorem", Z. Phys. B 44, 135 (1981)

161. Erfolgsgeheimnisse der Natur, Synergetik - Die Lehre vom Zusammenwirken (Deutsche Verlagsanstalt, Stuttgart 1981); (Italian, English, Spanish, Japanese, Chinese editions)

162. (with G. Mayer-Kress) "The Influence of Noise on the Logistic Model", J. Stat. Phys. 26, 1, 149 (1981)

163. (with C. Berding) "Pattern Formation in Morphogenesis", J. Math. Biology 14, 133 (1982)

164. (with A. Wunderlin) "Some Exact Results on Discrete Noisy Maps", Z. Phys. B 46, 2, 181 (1982)

165. (with A. Wunderlin) "Slaving Principle for Stochastic Differential Equations with Additive and Multiplicative Noise and for Discrete Noisy Maps", Z. Phys. B 47, 179 (1982)

166. (with H. Benk) "Theory of the Interaction of Frenkel Excitons with Local and Extended Perturbations in One-Dimensional Molecular Crystals. Application of the Green's Function Formalism to the Energy Tunnel and Cluster Problem", J. Chem. Phys. 77, 5730 (1982)

167. "Synergetics. Nonequilibrium Phase Transitions and Selforganization in Biological Systems", in Thermodynamics and Kinetics of Biological Processes ed. by I. Lamprecht and A.I. Zotin (de Gruyter, Berlin 1982)

168. "Synergetics - An Outline", in Mechanical and Thermal Behavior of Metallic Materials, (Soc. Italiana di Fisica, Bologna 1982)

169 Synergetik. Eine Einführung (Springer, Berlin 1982); (2nd enlarged edition 1983)

170. "Thermodinamica i Sinergetica", Enciclopedia del novecento, Enciclopedia Italiana (1983)

171. "Von der Laserphysik zur Synergetik", in Forschung in der Bundesrepublik Deutschland, (Verlag Chemie, Weinheim 1983)

172. "At Least One Lyapunov Exponent Vanishes if the Trajectory of an Attractor Does Not Contain a Fixed Point", Phys. Lett. 94A, 71 (1983)

173. (with T. Nagashima) "Chaotic Modulation of Correlation Functions", Phys. Lett. 96A, 385 (1983)

174. (with S. Nara) "An Approach to Pattern Formation in Crystal Growth", J. Cryst. Growth 63, 400 (1983)

175. (with H. Shimizu) "Cooperative Dynamics in Organelles", J. Theoret. Biology 104, 261 (1983)

176. (with M. Bestehorn) "A Calculation of Transient Solutions Describing Roll and Hexagon Formation in the Convection Instability", Phys. Lett. 99A, 265 (1983)

177. "Synergetik - Selbstorganisationsvorgänge in Physik, Chemie und Biologie" A.v.Humboldt Stiftung, Mitteilungen, Heft 43 (1983)

178. (with C. Berding, T. Harbich) "Pre-pattern Formation Mechanism for the Spiral-type Patterns of the Sunflower Head", J. Theor. Biol. 104, 53-70 (1983)

179. (with M. Renardy) "Bifurcation of Solutions of the Laser Equations", Physica 8 D, 57 (1983)

180. (with J. Goll) "Saturation of Interband Transitions in Semiconductors and the Effect of Optical Bistability", Phys. Rev. A28, 910 (1983)

181. Advanced Synergetics (Springer, Berlin 1983); (Russian,Japanese, Chinese editions)

182. "Synergetics - Theory of Nonequilibrium Phase Transitions and Formation of Spatio-Temporal Patterns" in Chemistry of the Future, ed. by H. Grünewald (Pergamon, New York 1984)

183. "Synergetik - Die spontane Entstehung von Strukturen in der belebten und unbelebten Natur", Helvetia Physica Acta 57, 140 (1984)

184. (with G. Mayer-Kress) "Attractors of Convex Maps with Positive Schwartzian Derivative in the Presence of Noise", Physica 10D, 329 (1984)

185. "Synergetics 83. Autowaves in Biology", Chemistry and Physics Meeting Report, Physica 11D, 265 (1984)

186. (with Z.C. Wang) "Theory of Two-Photon Lasers I", Z. Phys. B55, 361 (1984)

187. (with Z.C. Wang) "Theory of Two-Photon Lasers II", Z. Phys. B56, 77 (1984)

188. (with Z.C. Wang) "Theory of Two-Photon Lasers III", Z. Phys. B56, 85 (1984)

189. "Indeterminismus, Wahl und Freiheit - wie sind diese Begriffe im Bereich des Anorganischen zu verstehen?", in Civitas Resultate Band 5 (Hirzel, Stuttgart 1984)

190. (with M. Bestehorn) "Transient Patterns of the Convection Instability: A Model Calculation", Z. Phys. B 57, 329 (1984)

191. "Synergetics", Physica 127B, 26-26 (1984)

192. "Towards a Dynamic Information Theory", in Thermodynamics and Regulation of Biological Processes ed. by I. Lamprecht, A.I. Zotin (de Gruyter, Berlin 1984)

193. (with A. Wunderlin) "Some Applications of Basic Ideas and Models of Synergetics to Sociology", in Springer Series in Synergetics, Vol. 22, ed. by E. Frehland (Springer, Berlin 1984)

194. (with J.A.S. Kelso, H. Bunz) "A Theoretical Model of Phase Transitions in Human Hand Movements", Biolog. Cybernetics 51, 347 (1985)

195. "Pattern Formation and Chaos in Synergetic Systems", Physica Scripta, Vol. T9, 111 (1985)

196. (with Ji-yue Zhang) "Self-Pulsing Instabilities in the Generation of Higher Harmonics", Z. Phys. B 58, 337 (1985)

197. (with Ji-yue Zhang) "Self-Pulsing Instability in Inhomogeneously Broadened Traveling Wave Lasers", J. Opt. Soc. Am. B2, 141 (1985)

198. "Na puti k dinamicheskoi teorii informazii", in Termodinamika i regulazija biologitscheskich prozessov (Moskau 1984)

199. "Synergetics - Self-Organization in Physics", in Lecture Notes in Engineering, Vol. 13 (Springer, Berlin 1984)

200. "Synergetik - Selbstorganisationsvorgänge in Physik, Chemie und Biologie", Naturwiss. Rundschau 38. Jahrg., Heft 5 (1985)

201. (with B. Schnaufer) "A Theoretical Derivation of Cellular Structures of Flames", Z. Phys. B 59, 349 (1985)

202. "Some Basic Concepts of Synergetics with Respect to Pattern Formation. Morphogenesis of Behavior", Ber. Bunsenges. Phys. Chem. 89, 565 (1985)

203. (with G. Haubs) "Quantities Describing Local Properties of Chaotic Attractors", Z. Phys. B 59, 459 (1985)

204. "Synergetics - an Interdisciplinary Approach to Phenomena of Selforganization", Geoforum Vol.16, Nr.2, 205 (1985)

205. "Application of the Maximum Information Entropy Principle to Selforganizing Systems", Z. Phys. B 61, 335 (1985)

206. "Information, Information Gain, and Efficiency of Self-Organizing Systems Close to Instability Points", Z. Phys. B 61, 329 (1985)

207. "Order in Chaos", in Computer Methods in Applied Mechanics and Engineering,52, 635 (1985)

208. (with G. Schöner, J.A.S. Kelso) "A Stochastic Theory of Phase Transitions in Human Hand Movement", Biolog. Cybern. 53, 247 (1986)

209. "The Abiabatic Elimination Principle", in Dynamical Theories in Optical Instabilities, ed. by. R.W. Boyd, M.G. Raymer and L.M. Narducci (Cambridge University Press, Cambridge 1985)

210. "Towards a Quantum Synergetics: Pattern Formation in Quantum Systems Far from Thermal Equilibrium", Physica Scripta, 32, 274 (1985)

211. "Operational Approaches to Complex Systems" in Complex Systems - Operational Approaches in Springer Series in Synergetics, Vol. 31, ed. by H. Haken (Springer, Berlin 1985)

212. (with M. Bestehorn) "Pattern Formation and Transients in the Convection Instability", in Complex Systems - Operational Approaches in Springer Series in Synergetics, Vol. 31, ed. by H. Haken (Springer, Berlin 1985)

213. (with R. Friedrich) "Convection in Spherical Geometries", in Complex Systems - Operational Approaches, in Springer Series in Synergetics, Vol. 31, ed. by H. Haken (Springer, Berlin 1985)

214. "Information and Information Gain Close to Nonequilibrium Phase Transitions. Numerical Results", Z. Phys. B 62, 255 (1986)

215. "Excitons and the Electronic Polarization in Semiconductors", in Festkörperprobleme XXVI (1986), Advances in Solid State Physics 26 (Vieweg, Braunschweig 1986)

216. "The Maximum Entropy Principle for Non-Equilibrium Phase Transitions: Determination of Order Parameters, Slaved Modes, and Emerging Patterns", Z. Phys.B 63, 487-491 (1986)

217. (with G. Schöner) "The Slaving Principle for Stratonovich Stochastic Differential Equations", Z. Phys.B 63, 493-504 (1986)

218. "A New Access to Path Integrals and Fokker Planck Equations via the Maximum Calibre Principle", Z. Phys.B 63, 505-510 (1986)

219. (with R. Friedrich) "Static, Wavelike, and Chaotic Thermal Convection in Spherical Geometries", Phys. Rev. A34, (1986)

220. "Thermodynamics - Synergetics - Life", J. Non-Equilib. Thermodyn. 12, 1 (1987)

221. (with A. Wunderlin) "Synergetik: Prozesse der Selbstorganisation in der belebten und unbelebten Natur", in Selbstorganisation, ed. by A. Dress, H. Hendrichs, G. Käppers (Piper, München 1986)

222. (with A. Wunderlin) "A Macroscopic Approach to Synergetics", Midit-Workshop, Lyngdby (1986) (in press)

223. (with A. Wunderlin) "Processes of Self-Organization under the Influence of Gravitation" (German D1-Mission) (in press)

224. "Synergetic Information Versus Shannon Information in Self-Organizing Systems", Z. Phys.B 65, 503-504 (1987)

Edited books:

1. "Synergetics, Cooperative Phenomena in Multi-Component Systems" (Teubner, Stuttgart 1973)

2. "Cooperative Effects, Progress in Synergetics" (North Holland, Amsterdam 1974)

3. (with S. Nikitine) "Excitons at High Density", Springer Tracts in Mod. Physics Vol. 73 (Springer, Berlin 1975)

4. (with M. Wagner) "Cooperative Phenomena" (Springer, Berlin 1973)

5. "Synergetics - A Workshop 1977" (Springer, Berlin 1977)

6. "Pattern Formation by Dynamic Systems and Pattern Recognition", in Springer Series in Synergetics, Vol. 5 (Springer, Berlin 1979)

7. "Dynamics of Synergetic Systems", in "Springer Series in Synergetics" Vol. 6 (Springer, Berlin 1979)

8. "Chaos and Order in Nature", in Springer Series in Synergetics, Vol. 11 (Springer, Berlin 1981)

9. "Evolution of Order and Chaos", in Springer Series in Synergetics, (Springer, Berlin 1982)

10. (with E. Basar, H. Flohr, A.J. Mandell) "Synergetics of Brain", in Springer Series in Synergetics, Vol. 23 (Springer, Berlin 1983)

11. (with P. Reinecker and H.C. Wolf) "Organic Molecular Aggregates", Solid State Science, Vol. 49 (Springer, Berlin 1983)

12. "Complex Systems - Operational Approaches in Neurobiology, Physics and Computers", in Springer Series in Synergetics, Vol. 31 (Springer, Berlin 1985)

Edited book series: Springer Series in Synergetics

Index of Contributors

MIX
Papier aus verantwortungsvollen Quellen
Paper from responsible sources
FSC® C105338

If you have any concerns about our products,
you can contact us on
ProductSafety@springernature.com

In case Publisher is established outside the EU,
the EU authorized representative is:
Springer Nature Customer Service Center GmbH
Europaplatz 3, 69115 Heidelberg, Germany

Printed by Libri Plureos GmbH
in Hamburg, Germany